Free and Low-Cost Software for Scientists and Engineers

Other Books by Harley Bjelland

Writing Better Technical Articles
Using Online Scientific & Engineering Databases
Outrageous DR DOS® Batch File Programming
Create Your Own Desktop Publishing System
Configuring a Customized Engineering Workstation
Create Your Own Virtual Reality System

Free and Low-Cost Software for Scientists and Engineers

Harley Bjelland

McGraw-Hill, Inc.

New York San Francisco Washington, D.C. Auckland Bogotá
Caracas Lisbon London Madrid Mexico City Milan
Montreal New Delhi San Juan Singapore
Sydney Tokyo Toronto

Library of Congress Cataloging-in-Publication Data

Bjelland, Harley.
Free and low-cost software for scientists and engineers / by Harley Bjelland.
p. cm.
Includes bibliographical references and index.
ISBN 0-07-911896-8 (pbk.)
1. Science—Data processing. 2. Engineering—Data processing. 3. Science—Software. 4. Engineering—Software. I. Title.
Q183.9.B54 1994
502.85'5369—dc20 94-5085
CIP

1 2 3 4 5 6 7 8 9 0 DOH/DOH 9 9 8 7 6 5 4

ISBN 0-07-911896-8

The editors for this book were Roland S. Phelps and Sally Anne Glover, and the director of production was Katherine G. Brown. This book was set in ITC Century Light. It was composed in Blue Ridge Summit, Pa.

Printed and bound by R. R. Donnelley and Sons Company.

This book is dedicated to the software entrepreneurs who made this book possible. May your software continue to improve, your markets explode, and your bug-free programs find homes in millions of scientific and engineering computers all over the world.

Contents

Acknowledgments

My sincere thanks to Roland Phelps, Electronics Acquisition Editor at TAB/McGraw-Hill, for his encouragement and for working with me on this book. Also, many thanks to Sally Glover, who accomplished an outstanding editing task on some very difficult material.

Introduction

This book has been a labor of love and toil in equal proportions. I have been, and continue to be, impressed by the quality, quantity, and versatility of the software created by the authors featured in this book. I sincerely hope that my book contributes a small part in bringing their work to the attention of the many scientists and engineers who can benefit from these programs.

I've always felt that most software compendiums are too broadly based, aiming at what is called a "horizontal market," or "something for everyone," directed to the entire world of computer users. This makes it difficult to find specialty technical and scientific programs. As an engineer and scientist with many years of experience (more than I care to admit), I felt that a book directed to the "vertical market" of engineers and scientists was needed. Thus, this book.

I've also felt that software descriptions given in catalog and compendium listings do not provide adequate information to help potential customers decide whether they should procure and evaluate specific programs, so many of the descriptions in this book are rather long. However, there is no correlation between the length of the description and the worth of the program.

The descriptions contained in this book have been largely "abstracted" from the manuals and text that accompanied the software. The personal pronouns used in the descriptions are those of the author of the software, not me. I have proofed, edited, abbreviated and, I hope, improved some descriptions.

No Version Numbers

Version numbers are not included with the software descriptions because the length of the writing and publication cycle (six months to a year or more) quickly outdates this information. So some of the descriptions might also be outdated. However, with the ingenuity of the software authors at work, newer versions of these programs are undoubtedly available and much improved. Regrettably, however, some software still might not be available.

I have not attempted to rate these programs competitively because one person's diamond is another person's pebble. Also, because several hundred programs are covered, I have not had the time for the enormous task of personally evaluating all of them in my computer. However, the 50 or so I did evaluate provided me with an excellent sampling of, and an appreciation for, the many gems available.

One of my best software sources for this book was the PC SIG 11 CD-ROM created by the outstanding software vendor PC SIG. This is an excellent collection. For those of you who have a CD-ROM drive and want to evaluate thousands of software programs, I strongly recommend this disk, or a later version if it is available.

Summary of Book

Chapter 1 introduces the various types of free and low-cost software available: public domain, freeware, and shareware. Also covered are file extensions, archiving, the use of wildcards, plus the basic copy and format commands available.

One unavoidable fact about working as an engineer or a scientist is: we have to keep learning during our entire professional life to attempt to keep pace with the continual and often revolutionary developments in our fields of expertise. (Personally, I don't think I will ever want to quit learning; it's too much fun.) To assist in this vital endeavor, chapter 2 discusses a variety of typical "computer tutor" software programs that help continue this learning process.

Chapter 3 covers computer utilities and describes dozens of ways that you can make your computer work better when you perform specific applications. Topics covered range from archiving to file and disk managers to diagnostics.

Next, chapter 4 describes several software programs that can aid you in your personal and professional advancement. These programs range from showing you how to write an excellent resume to searching for a job to managing many of your home activities. (This doesn't include the management of kids, however. Even after raising five of my own, I haven't found any software for that.)

Word processing is the single most-often-used software application on a personal computer. Chapter 5 covers word-processing programs that range in size from simple ASCII text editors to complex, full-featured word processors. The chapter also discusses presentations and that ultimate in word processing—desktop publishing.

Then, chapter 6 covers a number of software programs that are specifically directed to the engineering and scientific professions. These range from structural analysis to statistics, electronics, computers, physics, math, astronomy, and chemistry.

Much of engineers' and scientists' lives are concerned with data of all types: recording and analyzing data from lab experiments, design studies, test results, analyses, etc. To help organize, assemble, and record this data for later retrieval, research, and analysis, computerized databases are the ideal medium. Chapter 7 covers the various types of databases that can be used in engineering and science.

More Vital Tasks

Other important tasks for scientists and engineers are estimating the cost of, and creating a schedule for the completion of, a project or program. Then these cost and schedule estimates must be periodically updated to compare actual dollar and milestone accomplishments with those originally estimated in order to assess the progress of the project. (Estimated and actual are often worlds apart. The estimated gets you started; a credible actual keeps you employed.) Chapter 8 covers cost-estimating software, the use of scheduling to manage programs, spreadsheets for estimating, and methods to track finances.

Some of the most significant advances in engineering and science designs, illustrations, drawings, and documentation have been made possible by the advent of the PC and accompanying software in the areas of drafting, design, artwork, graphics, etc. Chapter 9 covers some of the sophisticated CAD, graphics tools, drafting, and paint-and-draw programs. These have turned many who cannot draw a straight line with a steel ruler (I count myself in this category) into "artists" (I'm working on that) who can create, store, recall, and revise 2-D and 3-D graphics of considerable complexity.

As more and more reports, manuals, books, periodicals, ads, descriptions, and memos are being created, we are being inundated by published (electronically and on paper) engineering and scientific documentation. There is so much material being generated that it has become impossible to review all of the pertinent material relating to one's specialties. To provide rapid access to the specific information you need in this mountain of information, telelibraries were created. Chapter 10 shows you how to use telecommunications software to tap into these gigantic electronic archives. These libraries of the present and the future store up-to-date (as well as archival) data and help you to find all the specific information you need in a minimum of time.

For those of you who are concerned with or interested in computer programming (whether you are an amateur or a pro), chapter 11 covers a number of software language learning, tutorial, help, and utility programs that are available on diskette. These range from writing simple batch files (which even I can do) to the pros (whom I envy) who can write complex programs for creating and rotating a complex three-dimensional, multicolored object.

Time for a Break

Finally, it's time to take a break for a little R & R (relaxation and recuperation), something we all need but often neglect. Chapter 12 covers how to have fun with your PC. (You can even learn while you're having fun.) Topics include using your PC to help you make your vacation plans, compose music, enjoy sports, and play board, card, arcade, and adventure games. These excellent relaxers help banish your problems and let you escape into a world of fantasy (even if only for an hour or two); these games can let your subconscious mind mull over and perhaps solve some of your most perplexing problems.

A bibliography lists several reference texts for you to delve into further to learn more about your specialties. (Remember, I said I don't know everything.) The glossary defines most of the uncommon (and some of the common) terms you'll encounter in this book.

Vendors who supply free and low-cost software are listed in appendix A. Appendix B lists some of the major telelibraries that maintain large archives of public-domain software, freeware, and shareware. Appendix C is dedicated to the heroes and heroines of this book, the entrepreneurs who have created this software. In appendix C I've given some suggestions on how they, and all others who write of technical matters, can better create, document, and market their innovative ideas.

Chapter

1

How to Get Started

"Man is a tool-using animal." THOMAS CARLYLE

There are more than 20,000 free and low-cost software programs available for the IBM PC and its compatibles. These programs range in size and capability from a simple utility like DCAT3 (which catalogs floppy disks) to a full-fledged, major-league spreadsheet like AsEasyAs (which accomplishes much of what Lotus 1-2-3 does). Free and low-cost software also goes all the way up to a dynamic, full-featured, outstanding, major-league word-processing program called Super WP (which emulates many WordPerfect functions).

However, the basic dilemma with having such a plethora of free and low-cost programs available is the enormous task of locating, procuring, researching, testing, and evaluating them to determine which software will be of the greatest utility to engineers and scientists. That's what this book is about. As an author (8 published books so far) and an engineer/scientist (with more than 25 years of experience, much of it in the computer profession), I have endeavored to research, evaluate, and narrow this huge databank of available software down to those free and low-cost programs that will be of the most benefit to all engineers and scientists for use in their professional and personal lives.

Commercial Software

One basic dilemma with commercial software is that you have to invest a significant amount of money to purchase it, even though you're not certain what you are receiving. You must make your purchases based on the recommendations of others. Or the software might have been recommended by a friend or co-worker, or you might have received a flashy, difficult-to-resist, multicolor brochure and exaggerated sales pitch in the mail (I always fall for those). Or the software might have been enthusiastically reviewed in a periodical, or perhaps an aggressive salesperson might have talked you into buying a program.

But you're always taking a chance when you invest several hundred dollars in commercial or "retail-only" software. It simply might not accomplish everything you intend it to do, and only you know what you want. You really have no way to determine if it will suit you ahead of time. Some of your huge investments in software might thus end up stored in your closet as "shelfware," never to be used again (as for me, my closet runneth over).

Another important consideration is not only the high initial cost of commercial software, but also the recurring high costs of having to periodically update the software to its latest version every few months whenever the software company needs more revenue. During a few years' time, with all these costly upgrades, you'll discover that you have expended three to five or more times the initial cost of the software.

Still another serious deficiency in the commercial software market is that much useful scientific and engineering software is so inexpensive and has such a small market share that commercial companies decide it's not worth their effort to develop and market it. Examples of this are the many thousands of useful free and low-cost utility programs needed for specialty applications, such as simple but useful formula and unit conversion programs, structural analysis, chemical and physical reference data, etc.

Yes, commercial software can be costly because you have to help pay the salaries of the company's employees, the dividends to the stockholders, and the company's operating costs. And you have to part with a bundle of your hard-earned dollars to pay for a lot of fancy packaging and costly periodical and mail advertising.

There are, however, some very attractive alternatives to buying costly commercial software, options which this book covers. These other avenues of purchasing software are:

- Public domain.
- Freeware.
- Shareware.

Public-Domain Software

Public-domain (PD) software is a type of software to which the author has renounced all rights. The author has, in effect, donated the software program to the general public to use, free of charge. You are free to do anything you want to do with public-domain software. You can:

- Give copies to your friends and co-workers.
- Use it in your home or office.
- Change it around any way you wish.

Public-domain software is available from numerous bulletin boards, PD/shareware vendors, and telelibraries such as those listed in appendices A and B. Although this might seem to be an ideal source, only a limited amount of software programs have been declared to be in the public domain. So, let's consider the next step up in quality and quantity: freeware.

Freeware

Freeware is a class of software in which the author permits anyone to use the program at no cost. However, the copyright for the software program is still maintained by the author, principally to retain control over it. You can use the software at no cost, but you can't alter the software. There is a large volume of freeware available, much more than is available in the public domain. A Shareware Hall of Fame has been created to honor shareware authors (see chapter 11). From freeware, the next step up in quality and quantity is shareware.

Shareware

Shareware might be best described as "loaned" software. The shareware authors "loan" their software, at no cost to you, so you can test and evaluate it for your applications. You can obtain shareware at no cost from bulletin boards, or certain telelibrary vendors (listed in appendix B) or by purchasing diskettes from PD/shareware vendors (listed in appendix A) for the nominal handling cost of from one dollar to three or four dollars per diskette; this fee covers the cost of the diskette, duplication, administrative costs, mailing, etc.

If you determine that the shareware program meets your requirements and you decide to continue to use the program after your evaluation period has expired (typically one to two months), you are then requested to send a nominal "registration fee" to the shareware author. This registration fee can range from a few dollars to more than a hundred dollars for the more complex programs, but that's still a terrific bargain when compared to commercial programs.

This "try-before-you-buy" concept gives you an excellent opportunity to thoroughly evaluate the shareware program in your own computer and test it by performing the specific applications you require it to accomplish. Once proven by your evaluation, your registered shareware programs should not end up as shelfware.

In addition to the software program(s), shareware diskette(s) usually have an ASCII manual stored on the disk, a manual you can print out on your printer. And when you decide to keep using the software and register it with the author, you usually receive a printed (and often fully illustrated) manual, as well as periodic updates as the software is improved; you also receive additional benefits such as the opportunity to consult with the author. And, very importantly, you'll know that you will be encouraging other software authors to continue to make their creative endeavors available as shareware. Finally, you'll have a clear conscience that you are not pirating software that an author has devoted many, many hours to creating.

Crippleware

Crippleware, another source of software, is covered here merely for completeness, and is not too desirable a source. *Crippleware* is a type of shareware in which the author offers a limited or "crippled" version of the software. They are often called "demos." Crippleware might have a limited set of commands, be able to handle only a smaller amount of data, only be used for a limited amount of time, or it might not have a manual stored on the diskette(s), etc. Thus, you end up with only a limited demo program.

As inducements for sending the author a registration fee after evaluating a crippleware demo, you might be offered a full "bells-and-whistles" version of the program, a printed manual, free updates, and/or commissions for inducing others to try out and register the program after it has been recommended by you. Fortunately, a group called the Association of Shareware Producers (ASP) is prominent in discouraging this crippling practice.

Shareware and the ASP

The Association of Shareware Producers (ASP) is a trade group of professional programmers who have joined together to observe certain standards of programming and support. The primary goals of the ASP are to:

- Promote the shareware idea.
- Establish standards of programming and support for programmers.
- Set standards for advertising and copyright compliance.
- Provide a means for shareware authors to communicate with each other.
- Help settle any disputes that might arise between author, shareware vendors, and customers.

ASP requires that its members conform to the following standards:

- Acknowledge the license fees paid by users with a letter or other means.
- Provide guaranteed support to registered users.
- Do not market trivial software programs.
- Do not market crippled, time-limited, or usage-limited software as shareware.
- Treat the customer in a professional manner.

ASP authors are generally designated in the shareware vendors' catalog listings by the letters ASP after the author's or company's name.

How to Obtain Software

The major sources for obtaining PD, freeware, and shareware are:

- Friends and acquaintances.
- Bulletin boards.
- Telelibraries.

- Commercial PD/shareware vendors.
- CD-ROMs.

Friends and acquaintances (including computer clubs) are permitted to make copies of PD, freeware, and shareware and give them away. For freeware and shareware, the rules are slightly different. They can give freeware and shareware away as long as they copy the programs in their entirety and do not modify the programs in any way.

Telelibraries (listed in appendix B) and bulletin boards can also give away copies of programs under the same restrictions as mentioned in the previous paragraph. Telelibraries and bulletin boards are the best sources if you only need a program or two, or if you are in a hurry to obtain a specific program. Bulletin boards generally have a limited variety of programs, however.

Telelibraries, such as CompuServe and America Online, have very large libraries of PD, freeware, and shareware. You can obtain this software by downloading it using your modem communications program via the telephone lines. Even though your copy of the software is ostensibly "free," you still might have to pay to access the bulletin board or telelibrary and also pay for the telephone-line time required to download the program. If a program takes 30 minutes or so to unload, and it costs 15 cents a minute, the phone time alone is only $4.50; it's still quite a bargain if you're in a hurry.

The best sources of PD, freeware, and shareware, however, are the commercial PD/shareware vendors. Some are listed in appendix A. They have huge libraries of PD, freeware, and shareware; some of the largest have several thousand programs listed. The cost for their disks is nominal, ranging in price from one to three or four dollars. And many of these vendors often archive the programs so they can pack more software on a single disk, thus reducing the cost to you.

Bulletin Boards

As previously mentioned, one of the key sources for obtaining PD, freeware, and shareware is bulletin boards (BBs). Bulletin boards are usually not-for-profit enterprises set up to serve two major functions for computer users, as:

- A storage library for PD, freeware, and shareware for distribution to its subscribers.
- A means for computer users to communicate with each other to ask questions, offer advice, counsel, and receive the latest information on happenings in their fields of interest.

To obtain software from these sources, you need an internally or externally mounted modem for your computer, plus telecommunication software that allows you to automatically dial up a BB and download software directly into your computer. Several shareware telecommunication programs, and a brief discussion of modem hardware, are covered in chapter 10.

Thousands of bulletin boards are in operation; they are located all around the world. Check with your library or current periodicals for the name, location, telephone number, and the contents of their software libraries. One excellent source that lists many of the bulletin boards in the United States is the monthly periodical, *Computer Shopper*. Another source (although not as up to date) is the diskette, National BBS List, which is available from the Public Software Library and which lists more than 7,000 BBs across the United States.

Online Databases

Another excellent source of obtaining PD, freeware, and shareware programs is online databases, which I prefer to designate by a more appropriate name, telelibraries (that is, libraries located at a distance). You also need a modem and telecommunications software to download PD, freeware, and shareware from telelibraries. Telelibraries are commercially managed enterprises that operate at a profit, so you'll be charged a nominal fee for your online time (and telephone-line time) when you query and download software from these vendors. However, telelibraries have enormous and diverse libraries of PD, freeware, and shareware programs. If you contact them and download these programs during their off-hours (usually during the night and on weekends), connect charges are minimized.

Among the most popular and large-volume telelibraries that provide copies of PD, freeware, and shareware are:

- CompuServe.
- Genie.
- Prodigy.
- America Online.

These telelibraries, and how to contact them, are covered in more detail in chapter 10. (First commercial: see Bjelland, Harley L. *Using Online Scientific & Engineering Databases*, Blue Ridge Summit, PA: Windcrest/McGraw-Hill, 1992.)

CD-ROMs

Another excellent source of public domain, freeware, and shareware is CD-ROMs. This "library-of-the-future" media has an enormous storage capacity, about 680MB, and is still growing. Several outstanding collections of this free and low-cost software have been stored on CD-ROMs and are offered for a very low cost ($20 to $50). Considering the fact that several thousand programs can be stored on a single CD-ROM, the cost per program is less than one cent.

The outstanding software vendor PC-SIG, for example, has an excellent collection on their CD-ROM, PC-SIG Library. The subject matter is divided into specific classes of programs, so it's easy to research to find just what you are looking for. PC-SIG also has a short description of each program to aid in your evaluation.

CD-ROM software is stored in an archived format (more about archives later in this chapter). The disks usually have the unarchiving program also stored on the disk, so you can unarchive the program after you download it to your floppy or hard disk for evaluation.

How to Download from a Bulletin Board

To make sure that the software you download from a BB or Telelibrary is 100 percent correct, and to compensate for the fact that your telephone line might be noisy and garble the data, all software downloads should be transmitted to you utilizing what is called a *file transfer protocol*. This protocol performs an error-checking procedure, which ensures that what you receive at your computer is identical to that which was transmitted by the BB or telelibrary. Some of the most commonly used protocols are: XModem, YModem, and ZModem.

Most telecommunications software programs can accomplish downloads using any of these protocols. Before you download, you have to select a file protocol transfer that you and the vendor have in common. These are often specified by the BB or telelibrary. My personal preference, if I have a choice, is ZModem.

The specific downloading process might vary from one vendor to the another, but they usually utilize the same basic procedure. The general process (assuming you're logged on to the vendor) is:

1. Select the special interest group (SIG) forum, or whatever library in which the vendor stores the specific software you're interested in (e.g., IBM SIG).
2. Select the protocol you have in common with the vendor. While error-checking protocols are not essential when downloading straight text, they must be used when you're downloading software programs because the loss or garbling of a single character could make the program inoperable (e.g., Zmodem).
3. Specify the directory in which you want to store the downloaded program in your own computer. You can use the vendor's file name or assign your own filename and directory (e.g., C:\temp\Newfilename). (Note: Every computer should have a C:\temp, or similarly named directory for purposes such as this.) Alternatively, if it's a short program you can download it directly on your floppy drive.
4. Select the PD, freeware, or shareware program you want to download, then sit back and relax. The vendor informs you of the progress of the download, then lets you know when it's complete. Then you can log off.

Some telelibraries have simplified and largely automated their downloading procedures. CompuServe, for example, presents a series of easy-to-use menu choices for accomplishing this download. America OnLine has one of the best menus and easy-to-use procedures for downloading software, although their library isn't as extensive as some of the other telelibraries.

Back Up Your Software

After you download, back up your new software, whether it was obtained on a disk, or you downloaded it into one of your directories (such as C:\TEMP). Before you back up, however, you might have to format some disks. The next section summarizes the various formats required.

Format

You can format your diskettes using one of these commands:

FORMAT A: This command formats only one side of a 360K diskette in a standard 360K double-sided drive.

FORMAT A: Formats a high-capacity diskette in a high-capacity drive.

FORMAT A:/4 Formats a 360K diskette in a high-capacity, 5.25-inch drive.

FORMAT A:/F:<SIZE> (Don't type the brackets.) Formats a diskette for the capacity specified by <SIZE> and can be used for 160K to 1.44MB diskettes. e.g., FORMAT A: /F:1.2 to format a 1.2MB diskette.

FORMAT A:/S Makes your diskette a system diskette. This formats the diskette and copies the files that must be on a system diskette, including COMMAND.COM and a few hidden files that are needed for DOS to operate and which must be stored at specific locations on the diskette. You can boot up your computer with a system diskette. It's a good idea to always have a system diskette handy in case your computer is reluctant to boot up some time.

The /S added to the last command is called a *switch*. Switches modify and augment the basic commands.

Basic DOS Copy Programs

Before you load and use the PD, freeware, or shareware programs, make backup copies and store the disks away in a safe place. The basic DOS copy programs are:

- COPY
- DISKCOPY
- XCOPY

Use COPY when you copy a file from one location (source) to another (target) A:. To copy FILENAME.EXT from drive C to drive A, use:

```
COPY C:FILENAME.EXT A:
```

However, COPY is a slow command to use for multiple files because it reads and copies only one file at a time, even if you specify a group of files by using wildcards. COPY reads a single file into RAM, copies it to its destination, then moves on to the next file and repeats the procedure. Slow, slower, slowest.

You can also use COPY to print an ASCII text file (e.g., MANUAL.TXT) stored on A, where PRN is the designation for the printer:

```
COPY A:MANUAL.TXT PRN
```

DISKCOPY

To copy the precise contents of one diskette in drive A to another diskette of the same size and same capacity in drive B, use the DISKCOPY command and this format:

```
DISKCOPY A: B:
```

This DOS command copies the entire contents of diskette A onto diskette B. The target diskette need not be previously formatted before you use this command. DISKCOPY automatically formats the diskette, if needed. But remember that this command only works when copying from one diskette to another diskette of the same size and capacity.

You can still use DISKCOPY, even if you have only a single disk drive, drive A. Just input:

```
DISKCOPY A: A:
```

The computer prompts you when to remove the source diskette from drive A and when to insert the target diskette as needed to complete the copy process. You might have to interchange the source and target diskettes a few times, depending on how much RAM you have.

XCOPY

To copy all the files from one diskette type (e.g., 5-inch disk) in drive A to a different diskette type (e.g., 3½-inch disk) in drive B, use the XCOPY command and this format:

```
XCOPY A: B:
```

The advantage of this command is that it's much faster because it copies as many of the files as your RAM can hold, then transfers all of these files stored in RAM to the target disk. Often XCOPY completes a copy on a single pass if you have enough RAM.

However, to make sure you copy all the files from A to B, including the subdirectories, use this format:

```
XCOPY A: B: /S /E
```

The /S switch makes sure that all of the files in the source directory are copied to the target directory, along with all of the subdirectories. The /E switch does the same thing, but it also creates subdirectories at the destination, even if the subdirectories are empty. The /E switch works only if the /S switch is also specified.

If the disk in drive A contains hidden and system files that you want to copy, you have to first strip off the hidden and system attributes. To accomplish this, input this command:

```
ATTRIB -H -S A:\*.* /S
```

These files can then be copied via XCOPY.

Every time DOS creates or alters a file, it sets a bit, called the *archive bit*, in the attribute byte that is stored with the file's name. XCOPY has another switch, /M, that directs it to copy a file only if its archive bit is set. It then clears that bit after the file has been copied. So if you utilize XCOPY for backup purposes, use this command:

```
XCOPY C:\DIRNAME C:\BKUP\DIRNAME /S/M
```

This copies only the files in DIRNAME that are new or modified. Then it clears the files' archive bits so the files won't be copied the next time you use XCOPY, unless the files' contents have been changed.

Several PD, freeware, and shareware utility programs are available that considerably enhance the above DOS commands. They're covered in chapter 3.

Before You Run a Program

If you've received your software on disks and made backup copies of your programs on floppies, store the original disk(s) in a safe place and use the copies to run the program. If you obtained the software on a download and stored it on your hard disk, back up the software on floppy disks and store your floppies in a safe place.

To take a look at the filename extensions to see what types of programs you have on your disk, input DIR/p for this listing. The /p switch causes the listing to pause for every full page; DIR/w gives you a four-column-wide listing of the directory.

Filename Extensions

Three filename extensions designate programs that are directly executable. They are specified by the following letters positioned to the right of the period in the filename:

- .EXE
- .COM
- .BAT

EXE designates an "executable file." This file extension is most commonly used to execute larger programs. EXE programs are divided into separate modules. Because of this, they are slower to load than COM files. However, they can accommodate extremely large files. To run the HANS.EXE program, for example, simply input HANS at the prompt and hit the Enter key. It's not necessary to input .EXE.

COM designates a "command file" and is the oldest of the executable file formats. However, COM files can be no larger than 64K in size, so they are used only for the smaller programs. To run a DAVID.COM program, for example, simply input DAVID at the prompt and hit the Enter key.

BAT designates a *batch file*. A batch file is a series of DOS commands stored in a text file. Batch files are easy to write and are very useful for accomplishing many of the mundane and repetitive tasks you require of your computer. (Second commercial: See *Outrageous DR DOS Batch Files* in the Bibliography; the book is valid for both DR DOS and DOS.) Batch files are also used in shareware, often for starting the main EXE program, or for printing out manuals, etc. To run a SANDRA.BAT program, for example, simply input SANDRA at the prompt and hit the Enter key.

You will also encounter a wide variety of other file extensions. Some of the more common nongraphic file extensions are listed in Table 1.1.

TABLE 1.1 Nongraphic File Name Extensions

This ext.	Usually defines
$$$	A temporary file, or one in which an error occurred when the file was created
ARC	Archive file (multiple files, data compression)
ASM	Assembly language source file, program code
BAK/BK!	Backup file
BAS	A program that has to be run with BASIC
BAT	A DOS batch file, series of commands executed by DOS
BIN	Binary file
C	A C-language source file, program code
CAL	SuperCALC spreadsheet data file
CBL	COBOL source code (also COB)
CFG	Configuration file
COM	Program file, less than 64 Kbytes in size
DAT	Data file
DBF	dBASE data file
DEF	Definition file
DIF	Data Interchange Format in ASCII for sharing spreadsheets
DOC	Text file containing documentation or a Word document
DRV	Driver, a file that defines hardware specifics
EXE	Executable program
BAS	A program that has to be run from BASIC
FON	Font file
FOR	FORTRAN source code (text)
FOX	A FoxBase compiled file
HLP	Help file
INI	Windows INIT or initialization file
LZH	File compressed by LHA
LIB	Library file
MNU	Microsoft Mouse Menu file
OVL	Overlay file
OVR	Overlay file

TABLE 1.1 Continued

This ext.	Usually defines
PAS	PASCAL source code (text)
PIF	Program Information File for Windows
PRS	Printer definition file for WordPerfect
TXT	Text file
TMP	Temporary file
WKS	Spreadsheet (also WK1 and WKE)
ZIP	Archive/file compression formatted file

One very positive attribute of PD, Freeware, and Shareware authors is that they delight in continuing to improve their programs. Their updates are indicated by increasing numeric version numbers. For example, 1.0 is usually the initial version. Updates might range from 1.01 (a minor one) to 1.1 (a bigger one) to 2.0 (a quantum jump). When you obtain PD, freeware, or shareware, double-check to make sure you are receiving the latest version.

Archived Software

When you review the filenames of your PD, freeware, or shareware programs, you might note several unusual filename extensions such as .ZIP, .LJH, or .ARC. These extensions designate that the files have been *archived*; that is, several files have been combined and compressed into a single file to save space, often requiring less than half of their original disk storage space. Also, because the files are compressed, they can be transmitted to you in a much shorter time, thus reducing the cost of your online connect time.

Software obtained from PD, freeware, and shareware vendors on diskettes is often archived using the same techniques. This allows the vendor to pack more software on each floppy diskette, reducing the cost to you.

Here are some of the archiving programs currently in use: ARC, ARJ, USQ, LIB, LU, SQ, PKZIP, LHA, and ZOO. The most frequently used archiving programs are: PKZIP (designated by FILENAME.ZIP), ARC (designated by FILENAME.ARC), and LHA (formerly LHARC) (designated by FILENAME.LZH). PKZIP is the most common one at the present time, but LHA is gaining fast because it not only offers the speed of PKZIP, but it also has the best compression rate of the group. Another reason that LHA is becoming used more and more frequently is because there is no charge to utilize it for noncommercial, personal use, whereas PKZIP users must pay a registration fee.

Help files are usually provided with the unarchiving programs, and several special utilities (see chapter 3) are available that allow you to examine and determine the names, contents, and sizes of all the files that have been archived, without having to unarchive them.

Unarchiving

Before archived files can be used in your computer, they must be unarchived, that is, expanded and returned to their original form. Each of the above archive programs also has an unarchiving program that is usually included with the PD, freeware, and shareware diskette(s). However, even before unarchiving, some programs, called *archive shells*, let you look at the files in an unarchived mode so you can ascertain if they contain the information you are seeking. SHEZ and ArcMaster, for example, give you this capability.

The basic unarchiving procedures are essentially the same. I'll use the popular PKZIP to illustrate how to unarchive. Let's take the file CAROL.ZIP as the file to be unarchived to its original files. The program was initially archived by the program PKZIP.EXE. To "unzip" the archive, use the program, PKUNZIP.EXE.

Assume the shareware diskette is in drive A and has CAROL.ZIP and PKUNZIP.EXE stored on it. Let's store the unarchived program in C:\TEMP.

First make sure you are at the A> prompt by inputting A: at the C> prompt and hitting Enter. To restore the files to their original state and store the files in C:\TEMP, input at the A:\> prompt:

```
A:\> PKUNZIP CAROL C:\TEMP
```

That's all there is to it. Sit back and relax while CAROL is being unarchived and stored in your C:\TEMP directory. To check your results, return to the C:\TEMP directory and input DIR/P to list the newly unarchived files. The process for archiving files is covered in more detail in chapter 3.

Wildcards * and ?

An extremely useful (and dangerous if you're not careful) technique in manipulating files is the application of "wildcards." Two wildcards are used. They are: * and ?.

The asterisk (*) is truly a wildcard because it can be substituted for anywhere from zero to eight letters or numbers in a filename or extension. For example:

```
A:\> COPY *.* C:\TEMP
```

copies every file on the diskette in drive A to the C:\TEMP directory.

```
A:\> DELETE *.BK!
```

deletes all files on the diskette in drive A with the BK! extension.

It's not common knowledge, but *.* has a simple abbreviation: a single period. For example:

```
DIR *.*
```

and

```
DIR .
```

both display all of the files in the current directory.

The wildcard procedure could have been used in the preceding unarchiving example to unarchive all of the archived files in the specified directory and store them in the C:\TEMP directory with the command:

```
A:\> PKZIP *.* C:\TEMP
```

The tamest of the pair of wildcards is the question mark (?). It's used to represent a single character or number, or it can substitute for no character. For example, DELETE MAN?.TXT deletes all files having the first three letters MAN, plus none or one more letter or number. DELETE ???.??? (this one is really dangerous, almost as disastrous as *.*) deletes all files having from none to three letters or numbers before the period, to any files having none to three after the period, a situation that can cover a multitude of files.

The asterisk and question mark could also be used in combinations. For example:

```
DIR READ??.*
```

displays a directory containing all filenames that begin with READ and that have zero to two more letters and that can have any extension.

To View/Read Documentation Files

One other command you'll find very useful is the one needed to view a text file. Use this format:

```
TYPE FILENAME.EXT/P
```

The TYPE command displays the text on the screen; /P pauses the display one page at a time so the information doesn't scroll off the screen and out of sight. (Why this dumb command needs the /P command to stop the scrolling is beyond me. Stopping the scrolling should have been built into the TYPE command.)

PD, freeware, and shareware diskettes usually have various types of README files also stored on the diskette. These ASCII text files might have the file name format of READ.ME or README.1ST, or some similar designation. Whatever filename is used, you can display the files on your screen with this command:

```
TYPE READ*.*|MORE
```

(where READ is the partial name of the file).
Or you can print these files out on your printer with this command:

```
COPY READ*.* PRN
```

Be sure to read and/or print out these documentation files before you run the program. These text files usually have O.B.T.W.s (oh, by the way, or afterthoughts), more up-to-date information on the software. They might also have information on how to install/start/use the programs.

You can also view text files using the outstanding LIST program (see chapter 3) to browse through the file forwards and backwards, and to print selected screens on your printer.

Another file you should print out after you have evaluated and decided to continue to use the shareware is the REGISTRATION form. Fill out the printed form and mail it in with your check to help support the shareware concept.

Loading Unarchived Software

The earlier sections showed how to transfer archived software from a diskette in one of your floppy drives to your hard disk. If your programs have not been archived, you can install them by one of these methods:

- Self-loaders.
- Manual loaders.

Self-loading programs usually have an INSTALL, SETUP, or similar program on the diskette to load the program automatically, with little needed intervention on your part. These self-loaders usually specify a directory in which to store the program, give you an opportunity to change it if you desire, then automatically load the software into the specified directory. You might be required to answer a couple of questions while it loads, such as the type of printer you are using, your monitor's type and resolution, the port your modem uses, etc. Or you might have to answer these questions the first time you run the program. When the program is completely loaded, it informs you that it's ready to be used.

One installation problem you might encounter is that occasionally an install (or setup) program insists on only being installed from drive A. (Some dumb install programs still work this way.) If you have your install program in drive B, you have to trick the computer into believing drive B is drive A by inputting:

```
ASSIGN A=B
```

Then you can continue to install from drive B, referring to it as drive A. To cancel this assignment when you're finished, input:

```
ASSIGN
```

with no drive letters. Alternatively, the assignment is automatically cancelled the next time you boot up. But a better solution is to use the SUBST command as follows:

```
SUBST A: B:\
```

When you've finished the installation, bring things back to normal with:

```
SUBST A: /D
```

The D switch cancels the drive redirection.

For a manually loaded program, you have to first make (create) the directory (MD) (we'll use BOOK) in which you want to load the program:

```
C:\>MD BOOK
```

Assuming your diskette is in drive A, switch to the A drive and input:

```
A:\>XCOPY *.* /S/E C:\BOOK
```

Hit Enter, then sit back and relax while your copy is processed.

Testing Your Programs

If you can evaluate your programs by running them from one of your floppy drives, I recommend it. This way you won't mess up your hard disk with files you might later need to erase. Or you might end up with some unwanted and mysterious files that you don't know the origin of and you might be afraid to erase them. When you load and unload files often, you fragment your disk, which slows down your disk access. Then you'll have to run your de-fragmenting program (see chapter 3) to reorganize your disk files.

However, if the programs you want to evaluate are so large that they can't be run from a floppy, run them from your C:/TEMP directory. (See? I told you this is a handy directory.) Many of the programs that have an INSTALL capability let you select the directory in which to install the program, so just change the default directory to C:/TEMP and run it from there.

Of course you have to unarchive the files to run them; that's why you should always keep copies of your favorite unarchiving files in the C:/TEMP directory. That way you can unarchive and store the unarchived files, then run them from this directory.

Wherever you run your programs from, make sure you have this command or something similar in your AUTOEXEC.BAT file:

```
PROMPT $P$G
```

Sometimes you can get lost among the branches (and leaves) of your directory tree, so this simple but very useful prompt tells you what directory branch you are in so you won't be left out on a limb.

How to Use This Book

Consider this book as a workbook and a reference tool. I planned it so that everyone would read this entire first chapter to get properly oriented, even though some of the material might seem elementary. That's why I placed this "how-to-use" section here, near the end of this chapter.

Each chapter following this one is pretty much self-sufficient and can be used on a stand-alone basis. You can read and refer to the chapters in any order that suits you. For example, if you need a good graphics program, just go to chapter 9.

I urge you to be an active reader. Underscore or highlight any information you want to remember. This helps you in two ways. First, it highlights, reinforces, and stores the pertinent information in your personal memory bank. Secondly, when you return to locate this information, it's much easier to spot. Thirdly, you're forced to buy your own personal copy of this book so my publisher and I . . . (Oops! Forget I said that.)

Feel free to make notes in the margins, to expand on, question, or clarify any of the ideas and information. Far be it from me to claim to know everything about anything. Add your own ideas, your creative gems, to my thoughts to reap the synergism of two great minds.

Checklist

When you obtain your public-domain, freeware, and software programs, it's a good idea to keep track of them with one of the databases listed in chapter 7. Actually it's a good idea to use one of those databases to keep track of all of your software, not only to help you locate it, but also to have on record what version you have, who you purchased it from or registered it with, its cost, etc. Here's a suggested list of information to include in your software database.

Name of program
Version
Type of program
Where obtained
Date obtained
Registration (purchased) cost
Software author (company) name and address
RAM required
Operating system required
Monitor type required
Is manual included on diskette?
Is context-sensitive help available?

Pull-down menus?

Mouse operation?

What do you get for registering?

Ready, Set, Acquire

With this background, you should now be ready to procure and evaluate several PD, freeware, and shareware programs for your professional or personal applications. So why not invest some of your time in learning/reviewing some of the basic material that will help you? Chapter 2 introduces you to the vital topic of using computer tutors to learn/review new topics.

Chapter

2

Computer Tutor

A large variety of computer tutor programs are available via free and low-cost software channels. Some of the computer tutor topics covered in this chapter include: creativity, reading better and faster, typing, computer basics, and learning foreign languages.

Why Cover These Topics?

Whenever you study a new topic, everything you learn must be based on what you understand and remember from related topics. When each new idea or fact is added to your personal memory databank, you understand each new concept based on what you earlier learned about a related topic by analogy, similarity, contrast, etc. For example, you must first understand and remember your basic principles of algebra and trigonometry before you can learn calculus. Thus, mastering any new topic means to understand how the new knowledge relates to what you learned earlier, then to remember and store the new information in your personal databank. The section on memory enhancement later in this chapter shows you how to manage the second half: improving your memory's ability to remember what you read.

Our personal memory serves as an active database for inspiring creative activities. Creativity requires that you use the knowledge stored in your personal database to solve any new problem. For example, if you can't remember the basic rules of grammar and composition, how can you expect to write readable memos, reports, and books? You must have grammar and composition principles so deeply ingrained in your mental database that you shouldn't ever have to consciously think about them when you write. Similarly, when you want to create a new design for a bridge, you must first understand the basic principles of bridge construction and how similar bridges were designed so that you can build upon this knowledge and design a better and perhaps more unique bridge.

Robert H. Nett, a man who trained people to improve their memories, said, "In all my years of teaching, I have met few individuals who possess a naturally good memory." So, unless you're one of the very few so blessed, you have to train your memory to obtain the maximum benefit from it. Programs covered later in this chapter show you how.

Improve Reading Habits, Too

It's also important to improve your reading habits. Engineers and scientists often have to devote several hours a day to reading not only company memos and reports, but also the many articles that cover new developments in their areas of expertise. So, reading is a skill that can and should be improved as much as possible. Better reading skills result in improved learning and retention, better jobs, faster advancement, and more leisure time (the reward). For these reasons, several programs are included in this chapter to help improve your reading speed. However, increased reading speed without comprehension is useless, so these exercises also help you to increase both your reading speed and comprehension. The synergism of enhancing your memory and improving your reading speed and comprehension helps you master any new topic.

Typing Too?

The principal method of inputting data into a personal computer is, and will continue to be for the foreseeable future, via a keyboard. The faster and more accurately you can type, the more you can accomplish, and the more efficient and mistake-free your work will be. The old "hunt-and-peck" system is no longer adequate in our modern computer age. You have to learn "touch-typing"; that is, you have to be able to input your information into your computer without looking (except for an occasional peek) at the keyboard. So I have included some programs that tutor you not only to learn touch-typing, but to also help you increase the speed and accuracy of your keyboard inputs. These very important techniques will help you improve that most basic and oft-used of all your computer activities: keyboard inputs.

Computer Basics

Before getting too deep into using the software discussed later in this book, consider pausing a little to learn and/or review some computer basics. If you're already highly skilled in this topic, just skim through this material. Please don't skip it entirely; I guarantee you'll learn some things that will make the effort worthwhile.

The topic of computer basics is followed by a number of tutorials covering basic DOS and Windows. Again, even if you consider yourself an expert in this area, I encourage you to set aside a few hours and invest some time in studying and/or reviewing these tutorials. You'll be surprised at all the new information you'll learn and be able to incorporate into your daily activities.

Although this book concentrates on software, and even though you don't necessarily need to know how the electronic and electromechanical components in your computer function, you should be curious enough to want to see what some of the components look like when you encounter hardware terms. For this reason, I highly recommend the excellent tutorial/graphical program, What's In That Box? This program takes the cover off a typical PC, illustrates the contents of the inside of a computer chassis, and shows you what its major components and many of the smaller components look like, where they fit, and how they function.

It's a Small World

Finally, as our world grows smaller and smaller, and as we begin to market and communicate among foreign countries, we need to learn the elements of some of the world's cultures and understand some of their languages, not only to converse, but also to be able to read scientific and engineering publications in our areas of expertise. To help with this development, several language tutorials, ranging from conversational to business, are included here.

How to Use These Programs

I've tried to include the best of the breed for the various categories of public domain, freeware, and shareware. There are several sources for obtaining this software; most programs are available on 3½- and 5¼-inch disks. Some come in archived form; some are ready-to-use. For the archival disks, follow the directions in chapter 1 to unarchive them.

I usually dearchive my software on one of my floppies and run them from there. By using a floppy drive, you can thoroughly evaluate the software program's merit for your specific applications without cluttering up your hard drive with programs you'll have to later erase, or you might have to run a disk-optimization program later to reorganize your scattered clusters.

However, if the programs are too large, I unarchive my programs into my C:\TEMP directory. In this way I can evaluate them using this temporary directory and determine if I only need to use the program a few times, or if I want to register the software and install it on my hard disk in another directory for permanent use.

Creativity and Memory Enhancement

Let's begin the tutorial sections of this book with two of the most basic and vital topics of all: creativity and memory enhancement. These topics are closely tied together, since creativity is useless if you can't remember what you hear and read. The creativity and memory enhancement software listed here provide techniques to help you brainstorm, organize, and store information in your brain's memory cells for permanent storage and ready recall.

For each program listed, I've given a short summary of what the program does. The programs are listed in alphabetical order.

MaX

MaX is a program to enhance your vocabulary. It runs only once a day, gives you a short quiz on a certain word, and tells you the definition of the word. The quiz helps reinforce the definition in your memory. MaX keeps a record of your score and displays it, if you wish. It checks each time it is run to determine if it has been run for the day. If it has, it tells you and exits. If not, it proceeds with the quiz. MaX uses the date that your computer keeps, so if your machine doesn't set the date automatically, you have to set the date using the DOS "date" command before MaX is run. MaX is set up in this way so you can trigger it from your AUTOEXEC.BAT and not have to remember to run it every day. Of course, this works best if you turn on your computer every morning. If you leave your computer on all the time, then you have to remind yourself to run MaX daily.

This software is also designed to accommodate more than one user, keeping separate records for each. If you run MaX without any command line parameters, it asks for your name. You can also tell MaX your name on the command line. MaX comes with a vocabulary file, MAX.VOC, which is used by default.

Memory Builder

Memory Builder helps you memorize any type of information that can be expressed as a question and answer. Some examples of suitable topics are foreign language vocabulary, engineering and scientific facts, and math formulas. Memory Builder is essentially a group of computerized flash cards. It allows you to make up a learn file, which is a Windows-compatible file containing questions and answers. Because of the built-in editor, learn files can be created with an ease and speed not found in any other memorization program.

Memory Builder has many advantages over ordinary flash cards:

- Learn files can be created and learned quickly.
- Learn files can be stored for any length of time, so you can come back and review them periodically.
- Learn files can be made into icons that can be started from the Program Manager.
- Input errors are easy to find because you can look at, and correct, many questions and answers on the screen at the same time.
- Memory Builder has a consistent foreign accent interface, so that once you become familiar with the procedure, you can easily enter any type of accent. Accents can be used in all areas of Memory Builder, except when entering file names.

Memory Improvement System

Graphic flash cards, used in the Memory Improvement System, help you memorize a variety of information. This is an outstanding memory tool for all purposes, including vocabulary improvement, remembering facts, class work, business information—anything you want to commit to memory. This system maximizes the amount of information you can learn, while minimizing the effort it takes.

With these techniques, you can learn a new language or memorize formulas in record time. There are literally thousands of applications for this versatile learning tool. A small Spanish vocabulary is included on the disk to help you practice with and evaluate this memory improvement program.

Memory Master

Memory Master helps you create study files for learning vocabulary, topics, or basically anything you want to master by memorizing it. You can operate this program in one of two ways: Flash Cards or Written Study. Both modes use study files that you can create using your favorite editor or word processor, or by using the built-in frill-free editor.

You can build in your own "verbal pats on the back," encouragement when you provide the right answers. You can bring up the flash cards in regular or reverse order. The program helps you remember the correct answer by requiring you to type in the correct answer if you miss a question. The learning pairs (study/answers) you answer incorrectly are displayed at the end of the program for your further review. It uses a keyboard or a mouse.

Nimonik

Nimonik is a complete, computerized learning environment that helps adults learn and memorize almost any subject. Nimonik is not a game or children's educational program. Fast, efficient, and easy to use, everything is menu-driven; it really works. Here's how: You enter into Nimonik everything you need to learn or memorize. It then drills, reviews, and tests you on the material quickly and effortlessly. Nimonik automatically keeps track of where your weaknesses are and gives you additional practice in those areas, thus saving you hours of time and work.

Nimonik helps you organize and manage your learning, too: you can print, edit, add, copy, merge, etc. Multiple users can study multiple subjects. Files can be as large as your disk space allows.

The most important difference between Nimonik and other learning programs is that Nimonik keeps track of where your weaknesses are. This feature alone saves you untold numbers of hours of work and frustration when learning or memorizing your subjects. The way you learn is, first of all, by being forced to condense and organize your material during the ADD option, which also forces you to understand the material. TEST lets you know how well you know the material.

DRILL, the single most important activity in Nimonik, drills you on those terms that you need the most help with. You should spend most of your time in the DRILL option. Finally, REVIEW makes sure that you have covered all the terms you need to know. All these activities (REVIEW, TEST, and DRILL) keep track of which terms you know and don't know, but only DRILL actually uses that information. TEST chooses terms at random; REVIEW chooses terms sequentially as you entered them.

Omni-Cognitive Self Test

Omni-Cognitive Self Test helps you determine which side of the brain you use most often and gives a description of what the numbers mean. Once you know which side of your brain you use predominantly, you can figure what you will be good at.

Speed Reading

It seems as though the volume of published information accumulating in all science and engineering specialties doubles every few years. There is so much to read and so little time. So why not learn how to read faster, skim technical topics skillfully, increase your concentration, and retain more of what you read? These programs show you how.

Readfast

Readfast increases reading speed/efficiency. Reading is a vital activity for almost everyone in today's "civilized" world. We read in formal schooling, on the job, for relaxation, and for entertainment. We read to keep up with the news and to learn things that were not taught in school, or that we missed. Most of us would like to be able to read faster and to retain more information from our reading. Even though we might want to read in a leisurely manner sometimes, it is a definite advantage to be able to read rapidly when we need to read a lengthy report or a book to get information in a hurry.

This software explains some of the causes of slow reading and shows how to overcome those obstacles. You are paced as you read to force you to increase your speed. Your reading speed is measured to show you the progress you make.

One of the major reasons for slow reading habits is the way we were taught to read—word by word. Of course, that is necessary when learning to read for the first time, but reading word by word is at about the same rate as your speaking speed.

Another obstacle to efficient reading is regression—looking back at something you have already read because you didn't understand and remember it the first time. This often happens because your attention wanders.

Some people move their lips as they read—again, because that's how we learned to read initially. If we didn't read aloud, the teacher couldn't tell if we were reading it right. Forcing yourself to read faster by using the pacing provided by this program helps you overcome many of these habits that slow you down.

Typing Tutors

If you still look at the individual keys when you input data from your keyboard, you're not functioning very efficiently. One of the most compelling reasons for learning touch-typing (that is, typing without looking at the keys) is that you not only can increase your typing speed, you also increase the accuracy of your input. When you touch-type, you can watch the words and data appear on your screen and correct them instantaneously if you err.

Nontouch typists have to continually search for the keys; their eyes have to jump back and forth between the keyboard and the screen to see if they're hitting the proper keys. They spend a lot of time using the Backspace key to erase and correct their frequent errors. A few hours invested in these software programs will save you light-years of inputting effort in the future.

PC-Fastype

PC-Fastype is available in mono, CGA/EGA, and VGA versions. The mono version works on any monitor; the CGA and above versions require a color monitor. This program functions with the three most popular styles of keyboards.

This excellent, interactive tutorial for beginning, intermediate, and advanced typists utilizes excellent graphics to help you learn. A picture of the keyboard is graphically displayed on the bottom half of the screen. The displayed timer starts ticking when you enter the first keystrokes. You type each letter as you are prompted for it; the program selects the letters randomly.

If you hit the wrong key, an X appears on the proper key on the screen to illustrate your error. The number of characters you have to achieve is displayed before you start. When you finish the drill, your WPM and other statistics are displayed in the text panel. You can record your history files and recall them later so you can resume the tutorial from where you left off.

The program is designed so you have to keep increasing your speed and graduate to faster typing levels. For example, you have to exceed 10 WPM to advance from the beginner stage. PC-Fastype can also be used on networks. The VGA version provides a 3-D view of the keyboard.

Quictype

Quictype is the perfect game for polishing your touch-typing skills. This program works in CGA, EGA, VGA, or Hercules graphics systems. The user can specify the graphics mode and the practice text file for typing.

When run, Quictype defaults to the highest resolution of your system. If you have a dual display system, both monochrome and color, you can specify QUICTYPE 3 to practice typing on the Hercules display. If you wish to specify the practice typing filename, the mode must also be specified on the command line.

Quictype presents a menu with four options. Options 1 through 3 select increasing levels of typing speed, and option 4 quits to the DOS prompt. If you have a very fast AT system, you might wish to select a slower game play mode, even if you are a fast typist. Regardless of what game-playing speed you choose, your word per minute (WPM) typing speed ratings are always accurate.

Once a typing speed has been selected, the game play screen is presented as a field of twinkling stars. The first sentence is selected and immediately begins smooth scrolling down the display. There is a dividing line, below which your typing cursor appears, at the bottom of the display. You must successfully type the sentence before it reaches this dividing line. If it reaches bottom, every character not typed is counted as a typing error.

As you type the sentence, your input appears at the bottom of the screen. Typing errors are displayed, followed by a brief chirping sound. The error is then immediately erased, at which point you can retype the key correctly. This encourages correct typing the first time, disallowing the use of the backspace key for corrections. (A good typist is measured by both speed and accuracy.)

Quictype is a great program for touch-typists who need to improve their typing speed or zero in on typing accuracy. From the error reports, you can see the problem keys. At this point you can create different practice text files that concentrate heavily on those problem areas.

Type Trek

Type Trek is a typing tutor in a game format. As a game, it follows (loosely) the classic "Star Trek" games that originated on mainframe computers, where the goal was to seek out and destroy Klingon ships. (That was long before "Star Trek: the Next Generation" appeared, when Klingons were the enemy.) As a typing tutor, Type Trek introduces the keys on the keyboard, starting with the home-row keys, and adding keys until the keyboard is mastered. Controlled drills are provided through the use of "code words" that fire phasers and photon torpedoes, move the Enterprise, dock and refuel, etc.

Type Trek can be enjoyed by experienced typists as well as novices. For example, level 14 provides drills for the number keys (usually the top row of keys). Because it isn't essential, typing classes might not cover this row, so here is a chance to finally learn it. Level 15 provides a general review of all the keys.

Touch-Type Tutor for Windows

The Speed mode is for typists and companies to measure typing speeds. The Tutorial Mode has eight graduated lessons and adapts automatically to the user's skill level. Bar graphs showing progress are automatically available after each set of lessons. The Game mode presents a random series of characters in the form of a game, forcing the user to type faster and faster to win. The program can be operated with or without a mouse.

Each menu item has an underlined character for selection without a mouse. Certain menu items also have "accelerator" keys for quick selection without a mouse. Speed test is an easy way to measure the speed of a typist using selectable styles of text in various categories.

The game mode helps familiarize you with finger placement and provides some lighthearted relief. This simple game teaches both familiarity with the positions of keys and encourages consistency of typing speed. The key to winning this game is to type consistently while gradually increasing your speed. The object of the game is to move the person who seems to be trapped on a moving conveyor belt heading toward a rather fearsome-looking spear. You notice a white letter appear above the person, and typing the letter allows the person to jump away from the spear. As more letters appear, you can type them until the person makes it all the way to the left end of the belt and thus to safety. You might also notice, however, that the belt seems to get faster too, and the spear gets closer and closer.

Typing World

Starting with an excellent tutorial covering typing in general, you quickly move on to specific topics such as speed typing. Two great exercises help you develop that all-important "typing rhythm" and increase your speed. In FALLING LETTERS, you must press the key corresponding to a falling letter. TRIVIA TYPE challenges you to beat the clock in duplicating a displayed sentence. Variable speed and ability settings make this graphics-packed program perfect for beginners and experts alike.

Hardware

If you've never looked inside your computer chassis and wondered what RAM, ROM, CPUs, Motherboards, and other computer components look like, then it's about time you took a good look. The following software program lets you peer inside your computer chassis from the comfort and convenience of your monitor. This interesting program doesn't require any screwdrivers, pliers, or mechanical skill. All you need is the shareware disk listed. Although this disk might not display the precise part numbers that your PC has installed, it does illustrate what typical computer components look like so you'll be able to recognize a RAM and a CPU the next time these hard-working components stare you in the face.

What's In That Box?

This excellent tutorial conducts you on an educational tour of your PC. It is also available in a VGA version. The easy-to-use menu system moves you quickly ahead, one page at a time, or you can use the fast search function to locate any keyword.

First, the cover of the computer case is taken off and you're given a look at the major components mounted inside your PC. This first illustration shows the relative size, outline, and mounting positions of the motherboard, power supply, hard and floppy disk drives, RAM chips, and all the major internal components. The next screen display, accessed by hitting the NEXT menu selection, gives you a quick summary of what the major components are and what they do.

Continue by hitting NEXT, and the detailed layout of a typical motherboard is illustrated, along with the real estate it occupies and the relative size and outline of the major components mounted on the motherboard, including the CPU, the plug-in cards, RAM chips, math co-processor, etc.

Hit NEXT again and you get more and more details about how the computer operates, what bits and bytes are, and a short course on the functional operation of the components, from boot-up with BIOS to running programs.

The hardware tour concludes with an internal drawing of the monitor and its major components and an animation sequence that illustrates how characters are painted on your monitor. This tutorial is highly recommended, requires CGA, EGA, or VGA to function, and can be run from one of your floppy drives or your hard drive.

Computer/DOS Basics

If you're like most computer users, you probably use only about one or two dozen of DOS's 100 or so basic commands. You might not even be using those limited few in the most effective manner, and there are probably other commands that you have not used, commands that can simplify to expedite your work. In any event, a little review will help you. So peruse the following reviews and try out a few of the programs. I guarantee that you will be amazed and pleased at the new information you'll learn and begin to use. Some of these programs might not have been updated to the latest version of DOS; however, they do cover the basic commands that still exist in DOS.

DOS Helper

This interactive DOS tutor with an easy interface provides online help while you use DOS. The program gives the user a choice of two modes of operation: a menu-aided "assist" mode and a direct "command line" mode. The assist mode uses drop-down menus. In either mode, help screens can be displayed with just the press of a key. Users can also browse through a command list. Help screens describe the DOS command, the command syntax, all of the switches (options), and provide several examples of usage.

Included in the package are six very useful utilities described in the following.

- LF—Locate File: Locates and displays the complete path for each occurrence of specified files. Wildcards can be used.
- LCD—Locate\Change Directory: Locates a specified directory, then changes default to that directory.
- MOVE—Moves files from one directory to another. Wildcards can be used.
- HIDE—Hides files (sets hidden attribute). Wildcards can be used.
- UNHIDE—Unhides files. Displays hidden files. Clears hidden attribute of any specified files. Wildcards can be used.
- SIZE—Displays "true" size of a directory, including slack. Size is given in sectors and bytes. The /s switch gives sizes of all subdirectories.

HyperDOS

HyperDOS is a comprehensive reference work on DOS. It's a very large and extensive program. To have all the information presented by HyperDOS available in printed form, you'd have to buy three to five reference books.

HyperDOS is not aimed at the novice who doesn't even know what a keyboard is, although even that is explained. It's aimed at the person (new or experienced) who wants to know more about how to control a computer. The methods are explained in a number of commands you have to key in, and, hopefully, they will dispel the anxiety lying behind that little, blinking DOS command-line prompt.

Professor PC Laptop Tutorial

Professor P.C. Laptop is a comprehensive tutorial system that provides laptop, notebook, and portable computer owners (or those interested in buying a laptop or notebook) hundreds of powerful tips, tricks, and reference articles. Tutorials include:

- Travelling with laptops and notebooks.
- Clever equipment purchasing strategies.
- Battery care and maintenance.

- Telephone and manufacturer support numbers.
- Software selection.
- DOS lessons.
- History of computers.
- Hard drive secrets.
- Laptop batch files.
- Modem use.
- Little-known CompuServe resources for laptop owners.
- Custom reading list.
- Computer virus protection methods, and much more.

An amazing collection of laptop and notebook computer reference material is provided. This program contains a high-speed search system, the capability to turn sound on/off, tutorial printing, help screens, and color or monochrome light-bar menus.

Tutor Com

This shareware minicourse in microcomputing covers the basics of a variety of topics, including:

- Computer history.
- Introduction to DOS.
- Computer terminology.
- Subdirectories.
- Hardware.
- Batch files.
- The keyboard.
- Introduction to programming.

Material is presented in a series of screens; the user can move forward or backward through the tutorials, or jump to selected locations using a menu. Short quizzes test you on your knowledge, but you can bypass these when reviewing.

Language Tutors

Cantones

The language tutor Cantones teaches spoken Cantonese, but the program can readily adapt to Mandarin or to any other language that the IBM keyboard supports, sometimes with no modifications. Cantones serves mainly to teach vocabulary and pronunciation. The program accommodates the tone system used by Parker Po-Fei Huang and Gerard P. Kok in Speak Cantonese (New Asia—Yale-in-China Chinese Language Center, 1981), who distinguish the seven tones of Cantonese by using Western script in combination with three superscript markers.

Cantones keeps track of the user's success with four categories that the user must master: word recognition, tone recognition, measure recognition, and sentence translation. It allows the user to review in different ways: prompted by Chinese or by English; prompted with words or with sentences; completing whole lessons or special categories of lessons, testing only those items that the user missed last time or testing the entire lesson again. Cantones prints lessons, including lists of those words the user still does not know from a lesson. With a standard word processor, the user can also MergePrint the lessons into other formats such as flash cards, vocabulary tests, etc. The program is based on Crew's earlier version, MailMerge Cantonese (*Hong Kong Computer Journal*, February 1985, awarded "Best Article of 1985" by the Hong Kong Computer Society).

Language Teacher

Language Teacher is an outstanding series by Cindy and Andrew Bartorillo. It is a series of menu-driven language tutorial programs designed to run on the IBM PC or any compatible clone. Programs are available for French, Spanish, German, and Italian. Each program in the series contains hundreds of word combinations and verb conjugation forms. A line printer can be used with Language Teacher to take advantage of its versatility.

Also included is the ability to: select language-to-English or English-to-language combinations, choose multiple-choice answers, be retested on misses, print a multiple-choice test, and run a full quiz diagnostic routine with line printer output. Of particular interest to non-English students is the capability to select the direction of the combinations. All words, phrases, and verb conjugations are selected randomly by the program; the user is unable to predict what the next selection will be.

The program package is not designed to teach conversational skills or pronunciation, but instead provides practice drills to increase your working vocabulary and to perfect your use of the conjugation forms. In most cases, only a high-quality color monitor or green screen provides the resolution needed to display the special characters used with Language Teacher. All other program functions are suitable, though, if a low-quality color composite monitor is used.

Gakusei

Gakusei is a serious Japanese language tutorial. It is designed to help serious Japanese language students master the grammar and usage typically taught in a first-year, college-level language course. In contrast to other Japanese computer-based learning systems that present lessons as "situations," Gakusei presents grammatical patterns and makes them easy to find through extensive cross referencing. You should be able to quickly locate them in the tutorial, learn to use them, then get on with the business of learning the subtleties and nuances of Japanese language and culture, which cannot so easily be reduced to such simple "formulas."

A student who uses this product will become familiar with reading Hiragana and Katakana display and using the Hiragana and Katakana keyboards. Rather than select a numbered choice to a question in the exercises, the student can respond to a question by typing Japanese characters.

The ability to type, view, and print Japanese characters need not cost hundreds of dollars. You do not need a keyboard specially labeled with Japanese characters, or a CD-ROM drive, or an adapter board, or a color VGA monitor, or even a laser printer. Gakusei is designed to run on relatively inexpensive PC hardware.

There are many things that Gakusei cannot do. It cannot replace your Japanese language teacher or class. It is not a Japanese word processor, and it does not provide automatic sentence translation. It does not use dazzling audio-visual sound effects. It does not teach stroke order for characters or teach Kanji; to do so would mean to add a graphics capability at considerable time and expense. The print quality of the fonts is not intended to match that of professional desktop-publishing systems.

Japanese for Business and Travel

Japanese for Business and Travel is an introduction to Japanese. This tutorial performs as a teacher when no teacher is available. It keeps track of your errors and emphasizes those areas. It should only be used by one individual at a time, so as to not confuse the built-in teacher.

This tutorial is composed of questions about vocabulary, grammar, and culture. The questions are of the fill-in type, multiple choice, and matching. The course assumes no prior knowledge of Japanese. Be advised that the material is challenging, but anything worth knowing has its price. Don't get discouraged. It is claimed that Japanese is one of the ten most difficult languages on earth to learn.

Russian Tutor, Verbs, Numbers, and Letters

This is only a beginner program. It teaches the alphabet, numbers, some verbs, some greetings, and some information about how to address someone by their name. The individual files are:

- Learn the Cyrillic alphabet.
- Learn some numbers.
- Say hi and bye to someone in Russian.
- Learn about Russian objects.
- Pronounce a list of male and female names.
- Learn sixty verbs in the past, present, and future tenses.

Finally, there is a text file that contains the verbs for the verb program, as well as a backup copy of the verbs list. To make the best of this program, you should by all means study the alphabet program until it becomes second nature. You should use pencil and paper and copy every letter you see, the sound it makes, and its name. The rest of the aims depend on you being able to identify characters of the Cyrillic alphabet, so this is an important step.

Spell and Pronounce Spanish Words

This software helps you learn how to spell and pronounce Spanish words in a short period of time. It's a great brush-up tool for refresher training and helps students learn the Spanish language quickly and easily. The program is mouse-supported.

This software is also a great program for a refresher course, or it can be used by anyone who plans to travel to a Spanish-speaking country. Not only are the study words alphabetized; they are also listed by categories. If you'd like to learn about words associated with items found around the house, simply select that category. A Spanish dictionary can be easily printed by letter (A to Z), or by category. A list of all misspelled words can also be printed to be used for review.

Chapter

3

Computer Utilities

The designation "utilities" covers a very broad category of software. The definition I prefer states that utilities operate only on the computer or operating system, such as the disk drive, RAM, DOS, etc. Other categories of software manipulate items in the outside world: a word processor operates on words, a spreadsheet on numbers, etc.

The utilities in this chapter are your dedicated assistants. Their purpose is to make your work easier, more thorough, and faster to accomplish. Some are small programs, some medium; some are very, very large. You're sure to find many gems among these listings.

General

9 Utility Programs

This software includes a program for accomplishing any DOS function on an entire directory tree at once. It helps find lost files and can delete all .BAK files, etc., across many directory boundaries. One of the nine, CPY, can be used as a direct replacement for DOS copy, with the added benefit that it does not quit when a floppy fills up with a partial file. Instead, it stops, erases the partial file, prompts you for a new floppy, and continues on its way. This way you won't end up with partial files split up across a pair of floppies.

DIRS gives the total disk space occupied by a set of files. MOVE moves a file from one directory to another without an intermediate copy. MEM displays the amount of free memory, and BEEP generates a beep. PUSHDIR saves the current directory from a batch file. POPDIR restores the current directory from the previous PUSHDIR. MTCONV converts a MaxThink outline into the structured file format, which can then be loaded into PC-Outline.

Batmouse

Batmouse allows you to write DOS batch files with mouse support. Batch files used with small enhancement programs, such as Batmouse, can make batch files extremely powerful. Batmouse is simple to use, and an example batch file and ANSI screen are included to show how simple it is to utilize. Batmouse has been tested with MicroSoft and Logitech mouse drivers with no problems. This program should work with most mice.

Binary Toolshed

This software includes several programs to help you work with DOS, programs that equally apply to novices and power users. They not only make using traditional DOS features easier, but they also provide features and facilities not available through DOS. ToolBOX provides numerous features:

- File directory displays.
- Point-and-shoot operations.
- Single key activation of sets of commands.
- Pass through DOS command-line operation.
- Reuse of command-line text.
- Quick access to recently entered directories.
- Simultaneous directory and DOS window views, and much more.

This shell program provides a platform for initiating other program operations, whether from the Binary Toolshed, or other independent parties. TBOXCFG is the configuration capability for the ToolBOX program, including colors, directory display or sort formats, key bindings, title banner, and more. Configuration files are usable by the VIEW and TREE utilities, and BTSPATCH sets the default command-line arguments for all Binary Toolshed programs.

TOASCII converts text files that use the extended IBM PC character set to use the standard 95 printable ASCII characters. TREE graphically displays the directory structure. The user can cursor through this structure to change directories or examine files. VIEW displays the contents of text files for viewing or browsing. All of the programs are independent and can be used without the others, if desired.

Colorado Utilities Disk Manager (CUDM)

Colorado Utilities Disk Manager (CUDM) is a full-featured disk catalog system. With it you can manage thousands of disk files from your hard disk, floppy disks, and CD-ROM. It creates a master catalog of files, libraries, and volumes and allows you to add comments of up to 410 characters per entry.

With CUDM you can locate any file on any diskette or hard disk in seconds. You can find and eliminate duplicate files across all types of disk storage and manage your free space to use fewer diskettes. Other features of CUDM include:

- Updating or deleting previously cataloged entries.
- Sorting entries based on different characteristics.

- Searching for specific entries or classes of entries based on "strings," wildcards, or categories.
- Flexibility to lay out your own list screens and choose your screen colors.
- Support for more than 20 printers, including lasers, ink jets, dot matrix, and impact.

Colorado Duplicate File Manager

This program identifies and manages duplicate file creep on multiple hard or floppy disk drives. Colorado Duplicate File Manager views text files, deletes unwanted duplicates on any drive, and uses the most current compression archiving program to manage, view, and delete files contained in archives. This software manages from one to four drives with duplicates containing names, sizes, dates, and times, and it uses any screen configuration (25/43/50 lines) and any DOS-supported disk type.

Control 3

This is a utility like Norton Commander, but it is faster. Control 3 is a short and fast program that performs all commonly used file management operations with a minimum of bother, and from the convenience of a multimenu environment. For the first time, you can display and control three drives or directories on the screen simultaneously (hence the name "Control 3"). Features:

- Color-coded directory lists (optional).
- Immediately update directories after changes on screen.
- Type any normal DOS command you wish.
- Treat files with certain extensions in specified ways (for example, load .DOC files into your favorite editor or text listing program).
- Execute, copy, delete, rename, sort, make, and remove directories.
- View and print text files.
- Shell to DOS.
- Touch-up file dates and times and change file attributes.
- Change disk volume labels.
- Delete and copy whole directory structures complete with any contained files.
- Mini-editor, ideal for composing or modifying batch files or small text files.
- File encryption.
- Find and display strings of characters within both text files and executable files.
- Move effortlessly through directories and between drives.
- Proportional disk space indicator.
- Fast file and directory find.

DAYOBrain

DAYOBrain is a Job Scheduler and/or menu system for use on PCs. You can use the Brain to run any program or batch file during the day or night, manually, or per a schedule. The Brain can also act as a menu system, allowing you to select, then run a job or program. This software was developed to facilitate the maintenance, reporting, and archiving that many businesses experience.

DAYOBrain schedules and runs programs and routines at preselected dates and/or times. It gives you the freedom to run tedious and/or time-consuming programs (i.e., reports, backups, archives, and indexes) at what would normally be nonproductive times (i.e., at night and weekends). This makes you more productive because your work continues when you aren't even around. A mouse is highly recommended, but not required.

Directory/File Management and DOS Shell

This program presents a tree diagram of the directory structure of the current disk/drive on the left side of the display screen. A listing of the files in the current directory is displayed on the right side of the display screen. The mouse or function keys allow movement up and down through the directory tree to "highlight" and enter directories and display the appropriate files of the directory on the right side of the display. A pop-up mouse menu or option keys are available for copying, deleting, moving, renaming, tagging, untagging, finding, executing, and doing many other functions on files.

It handles up to 2000 directories per drive and up to 2500 files per directory. The program uses EMS memory to store the directory and file data. Online context-sensitive help is available. A Microsoft mouse is supported.

File Commander

File Commander is an extender for the Win File Manager. It allows you to add four menu items to the File Manager menu bar. From each of these top-level menu items, File Commander allows you to make up to 99 menu items, spread up to 5 levels deep.

File Commander comes with four sample menu files, each of which are associated with a new File Manager menu item. Although these files are intended as code samples, you can peruse to make your own add-on menu items; many users, especially those averse to plunging in and learning the batch language, find that the sample menu items do all sorts of utilities, and these users can be perfectly content to live with the sample menus. This version also allows you to add the same menu items to the Norton Desktop for Windows File Manager menu. File Commander attempts to determine if there are any compatibility problems. If so, it displays a message with further instructions, and then exits.

Gepkat

Gepkat is an easy, comprehensive, practical, and very effective system for cataloging a large collection of diskettes. Many features set this system apart from all others, both in terms of what it does today, and also in terms of what it can do for you. Unlike other systems, this system can handle your diskette collection, even if it becomes large.

This system has the following major points:

- Handles very large collections, of up to 9999 diskettes.
- Uses hard disk space reasonably efficiently.
- Understands .ZIP, .ARC, .LZH, .ARJ, and other popular types of compressed archive files, and also catalogs the individual members of those archives so that you can easily find them again.
- Gepkat's database structure and programs using it are designed to be very robust, so the database is unlikely to be damaged by accident.
- Its lookup program is user-friendly, powerful, and fast. The program has many more features.

List

List lets you scroll up, down, left, and right through documentation or data files. You can search for specific data. List uses wildcards to let you browse a series of related files with only one command. It also has hex display, screen wrap, strip high bits, split screen, rulers, and much more. It supports 43-line and 132-column modes on EGA, lists prints files, and has a phone dialer. Other features include:

- If you have PKZIP, LHARC, and a hard drive, you can view contents of .ZIP, .ARC, and .LZH files (including self-extracting files).
- Also available is a small file management utility to copy, delete, rename, etc.
- Can call your favorite text editor from this DIR listing.
- Bookmark, DOS shell are callable from within the program.
- Supports mice and LANs.
- Can paste data from the file being viewed to another file.
- Can execute programs from within List itself.
- List also can search a group of files for hex data or text strings, with one command.

Other niceties are the split-screen display, extensive documentation, and the 3-program setup. Small, medium, and large versions are included. Pick the one with just enough features and a small enough .COM file size (from 9K to 20K) for you.

PC File MasDir

This software is a combination of two programs: a file cataloging program referred to as D\LIB, and DOS utility functions, referred to as DOS\U. PC File MasDir features pull-down menus, pop-up windows, and context-sensitive on-screen help. DOS\U is a file utility program providing DOS functions for directories, making, renaming, and deleting. Features:

- File functions for copying, deleting, moving, renaming, attribute checking, and date/time changing.

- Hard-disk file searching and printing by name and extension, date and time, between dates, and for hidden files.
- Printing by directory and tree printing of all files on the hard disk.
- Built-in file viewer for both ASCII and hex so you can verify the files before you make changes.

D\LIB is not a simple disk-labeling program; it is a very powerful floppy cataloging program that builds libraries of your diskettes. It allows you to:

- Update and delete diskettes.
- Search and print the libraries by name and extension, date and time, hidden files, as well as printing by diskette number.
- Browse and search the entire library, as well as individual diskette numbers.

Very powerful searches quickly locate files once the library is built by simply numbering your diskettes as you swap them in and out of your drive. You can search using wildcards in both file name and extension, time of creation between two dates, and by file types. You can generate printed copies to keep by your PC for those times when it is inconvenient to accomplish an electronic search.

WnScroll

This text file viewer/browser for Windows allows viewing of files of any size in either text or hex format, in a moveable window with full horizontal and vertical scrolling. WnScroll has very fast file loading and scrolling, even with very large files. All standard keyboard and mouse conventions are available for scrolling, resizing and viewing files. WnScroll overcomes Notepad's 52K file size limitation. Find, Find Next, Print, and Clipboard Copy are implemented.

Diagnostics

InfoPlus

InfoPlus generates 18 screens of information about your computer. Some of the techniques InfoPlus uses are not documented or officially supported by either IBM or Microsoft. Using the program is easy; the Page Up and Page Down keys move from page to page. Home and End move to the first and last pages, respectively.

For a full record, start InfoPlus using the AP option. This sends a copy of every screen to a file or the printer. You can add InfoPlus to the Windows Program Manager. InfoPlus runs safely under Desqview and is "Desqview aware." If DV is detected, InfoPlus writes to the screen using BIOS calls, unless you use the D command-line switch. This allows windowed screens, even on non-386/486s.

Intest Pattern Generator

Intest Pattern Generator is released into the public domain. This test pattern generator was originally developed for testing computer interfaces, video decoders, distribution amplifiers, video, and matrix switchers. This test program needs no setup or special installation. When the program is started, you see a menu of test pattern choices. Press the letter corresponding to the pattern you would like to use. Pressing the letter a second time inverts the pattern.

This program is useful for testing the output of any computer-driven display device, including CGA, HGC, MCGA, EGA, VGA, and SVGA cards. It also works with portable displays, including LCDs. It is especially useful when setting up a display system consisting of a computer driving a projection television, or other external display (such as an LCD panel) using a computer interface device.

Formatting

BGFormat

BGFormat is a background diskette formatter. With this program, you can accomplish other tasks while formatting a diskette in the background. The program is menu-driven. When first loaded, the program checks for the type of diskette drive connected to the system.

BGFormat only formats 360K and 1.2M for 5.25-inch diskettes and 720K and 1.44M for 3.5-inch diskettes. It can format 11 diskettes at a time. At the end of each completed format, it displays CHANGE and beeps to alert you to change diskettes. Once it senses change, it continues formatting until it reaches the total format count. Then it displays FINISH and beeps.

Copy Programs

Although DOS has a few basic copy programs, many entrepreneurs have considerably improved on these programs. Here are some of the better ones.

CopyX

This program copies a group of files from your hard disk to a group of floppies in the most efficient manner. CopyX produces a group of floppies in such a manner as to store the largest number of files on the smallest number of floppies. In the process of software development for many businesses, hard drives keep filling up. In looking through the contents of a hard drive, if you notice that a particular project has been completed and you feel the need to keep a copy around to support a client in the future, then floppies are one of the least expensive, easiest, and most reliable storage methods available.

In looking through the hundreds of files of different sizes associated with a project, you might be frustrated as to how to completely fill the smallest number of diskettes with the maximum number of files. CopyX was invented for just that purpose. This product is quite intuitive in itself, and most users who are fairly familiar with the normal DOS interface will find the user interface is quite familiar because it is patterned after the normal DOS command conventions.

DCopy

DCopy is a replacement for the DOS DISKCOPY command. It overcomes some of DISKCOPY's more annoying shortcomings. DCopy meets a situation that is becoming a typical system—only one drive of a particular media size. In these situations, particularly with larger disk capacities such as 1.44MB, DISKCOPY involves a very frustrating series of diskette swaps. DCopy overcomes this limitation and allows you to copy a diskette with only one swap. DCopy also allows you the option of making multiple copies of a diskette after reading the source diskette only once. Here is a summary of DCopy's features:

- Copies diskettes with a single swap.
- Copies any size and format to 1.44MB.
- Recognizes drives identified through an external device driver such as DRIVER.SYS.
- Makes multiple copies with a single read of the source diskette.
- Automatic allocation of system resources to create "swap pool."
- Manual control over allocation of system resources.
- Detects format of target diskette and reformats if necessary.
- Error checking ensures correct copies.
- Automatically eliminates bad track mapping on target diskette.
- Optionally duplicates bad track map on target diskette.
- Optional sound cues available for diskette swapping and errors.
- Can be forced to display monochrome for LCD displays and monochrome monitors.

Disk Copy Fast

This one-pass diskette duplication utility makes copies of 1.44MB, 720KB, 1.2MB, and 360KB diskettes. Disk Copy Fast is incredibly fast. For example, assuming diskette shuffling doesn't take any time, using the DOS command DISKCOPY to duplicate a 360KB, 1.2MB, 720KB, or 1.44MB diskette takes 43%, 50%, 42% or 57% longer than using Disk Copy Fast. Add up the shuffling time, and the saving becomes unbelievable. More importantly, this program achieves its high performance by precise timing, not by sacrificing compatibility.

Every single read, write, or format is accomplished following strict industry standards. It performs the DOS commands FORMAT, DISKCOPY, and DISKCOMP in one single pass and has a friendly and informative user interface. No need to go through a lengthy DOC file and still be unable to remember all the options and switches. Hot keys make your backup much easier.

Extended memory (XMS) is utilized for faster speed and to save your hard disk from unnecessary wear and tear. The software creates an exact image file of the diskette. Fast mode reads and writes only the portion of the diskette containing data and does not waste time reading and writing the empty portion. When copying a 1.2MB diskette in a system with 590KB free memory, Disk Copy Fast easily stored 34 tracks, while another diskette copy program could only store 17 tracks of data in conventional memory.

EQuicopy

EQuicopy allows a user with a 1.2MB floppy drive and a 1.44MB floppy drive to copy disks between the two drives, as if they were equivalent. The resultant disk is a copy of the original, down to the sector level. In short, if you copy a 1.44MB 3.5-inch diskette to your 5.25-inch drive, you end up with a 1.44MB 5.25-inch diskette.

EZ-DiskCopy

This powerful diskette duplicator allows you to read a diskette into memory and make as many copies as you desire, without rereading the source. It also verifies the copy during each operation. The program maps out a diskette usage, finds out information about a diskette (sides, sectors, etc.) and accomplishes it all without cryptic commands. Other features include format only, compare, directory information, and more.

OnePass

OnePass is a replacement for the DOS DISKCOPY command. Unlike DISKCOPY, OnePass is not limited to available memory and does not require any disk swaps. Here are some of the features of this program:

- Single-pass copying of any standard floppy disk, including 720K.
- 1.2MB, 1.44MB, and 2.88MB diskettes.
- Multiple copies of the same disk can be made without having to reread the original.
- The swap file can be saved, archived, or transferred over a modem, then it can build an exact copy of the original diskette.
- It can swap to a hard disk or a RAM disk.

Besides making quick disk copies without playing the floppy shuffle, OnePass can also duplicate software disks. Diskettes created by OnePass are identical to the original; boot disks actually boot, etc.

PCopy

PCopy is an advanced replacement for the DOS COPY command, one of the most used of the DOS commands. PCopy features:

- Copies files to and from any disk or directory.
- Allows you to save older versions of files with new names before replacing them.
- Moves files by copying them or renaming them to the new location.
- Deletes the original file if /X is specified.
- Creates target directories if needed.

- Makes sure the DOS archive flag is set correctly for each file as it is moved.
- Preserves the DOS date and time for each file moved.
- Allows you to pause processing at any time by pressing any keyboard key.

VGA-Copy

Some of VGACopy's features are:

- High-speed disk copies, verifies, and formats in one pass.
- Many different formats, including 1.44MB on 5.25-inch and 1.72MB on 3.5-inch disks and Atari ST formats.
- New boot-sector for your disks. If booting from a VGA-copied or formatted disk, you'll be asked for booting from HD, from A:, or from B:.
- A: and B: can be swapped.
- Creates optimal sector interleave on target disk.
- Disk read/writes are up to 100% faster.
- Soundblaster support for VGA-COPY messages.
- Supports McAffee's Virus scanner.
- Nice graphics under VGA (VGA Copy also works under Hercules, EGA, CGA).
- Mouse support.

Backup

DRS

DRS reads bad disks! If you have a disk that returns GENERAL FAILURE READING DRIVE X, DRS will read it anyway and will even read a disk that you poke a hole in. Of course you cannot read the data in the bad area, but you can recover the rest. Mark the areas you want to recover, then save those marked areas in files you name. Instruct DRS where to put these files, on another floppy or your hard drive. Also instruct DRS to simply print out your marked areas.

How safe is DRS? If it reads bad disks, then it can also write on them. DRS is safe. It does not write to logical sectors anywhere; it only reads. The files you declare and save are normal DOS files with your own names. If it is possible to write on a bad disk, why did this program not give you this ability? Because DRS is an entry-level recovery package. Had casual users accidentally told DRS to save data at a logical sector on the hard drive, they could wipe out their FAT table, etc.

Backup/Restore Help

This program is a great help in backing up your directories to diskette and restoring directories from diskettes to your hard drive. You can use the keyboard, the mouse, or both to select functions, directories, or to page through the directories. This program deals with whole directories only; it doesn't go down to the file level.

Simple Backup

This program is a file backup and restore system made up of three separate EXE files:

- File backup.
- Restore.
- Script file creation.

Simple Backup has a complete help system; just press F1 for help. The backup utility works quite nicely for backing up specific work drives or directories. Features:

- Supports floppy disks, hard disks, and network backups.
- Compresses files 40–80% or more.
- Automatically formats floppy disks.
- Splits files across multiple floppy disks.
- Uses all disk space.
- Backs up multiple drive\directory in a single pass.
- Has a built-in help system.
- Has script files support for automating backup operations for users.
- Has message customization for international support and special needs.

Flexibak Plus

This is a flexible, easy-to-use system with a logical, simple, and unique approach to hard disk backups. With a conventional backup system, you have to make a full backup about once per week, plus daily incremental backups onto separate disks. Should you need to restore from the backup, you must first restore the full backup and then every incremental backup. With Flexibak Plus, all this is different as the concept of full backups and incremental backups is removed. Instead, you only take a full backup once. All subsequent backups are placed on the same backup disks, giving long-term speed increases, simpler file restoration, and improved backup management capability.

Here are some outstanding features of Flexibak Plus:

- Selective backups allow you to specify, globally and by directory, which file groups should be included in, or excluded from, the backup, potentially dramatically reducing the number of required backup disks.
- Incremental backups are placed on the same disks as the full backup, so you must take a full backup only once.
- The file management system means that Flexibak Plus keeps track of what has been backed up and informs you if any files are missing from your hard disk since the last backup to allow you to recover accidentally deleted files.

- File compression and the capability to handle up to 20 hard disk drives are parts of the system.
- An alarm system can be set to warn you each day that you have not taken a backup for a specified number of days of your choice.

BakupWiz

Is your vital data stored on your hard disk safe? What would happen to your business or your work if your hard disk "crashed" and you lost most or all of your data? You say you don't back up your hard disk very often because you're too busy and it takes too long? BakupWiz solves these problems.

The benefits of owning BakupWiz are many:

- Speed—Your hard disk is backed up to diskette(s) quickly and reliably.
- Flexibility—You can back up all files on your hard disk, only selected files, or any files that have changed since a specified date and time.
- Reliability—BakupWiz has been thoroughly tested on many different machine configurations and operating conditions. For additional assurance, you might instruct BakupWiz to verify all the data that it has backed up.
- Network Compatibility—Novell Netware and any NetBIOS compatible networks are supported. The network's fileservers can be backed up without requiring the users to logout from the network. Under Netware, BakupWiz backs up and restores both file and network security information. Netware 286 and Netware 386, including Apple File Protocol, are supported.
- Security—The security of your data is maximized by BakupWiz's unique password system. This system is so secure that we strongly recommend you devise a fail-safe method of retaining your password.
- Ease of use—BakupWiz is operated via a pull-down menu system. Help lines and message windows provide additional information.
- Compatibility—It is 100% compatible with MS-DOS and most compatibles.
- Backup device support—The program can use 1.2MB and 360K 5.25-inch formats, as well as 1.44MB and 720K 3.5-inch formats. BakupWiz can also back up to any arbitrary DOS path.
- Automatic formatting—When you insert a new diskette at BakupWiz's request during a backup operation, it automatically formats the disks.

Point and Shoot Backup/Restore

With this program, you can back up as much as 1MB per minute. It provides safe, quick backups of hard disk files to floppy disks and efficient restoration of backup disks to hard disk. Features include:

- Automatic disk formatting.
- Extensive error checking.

- Up to 10 user-defined backup configurations.
- Multiple subdirectory definitions, with or without trailing directories.
- Restoration to different drives and/or directories.
- Modified file.
- Pattern-matched file selection.
- File compression.

DOS Enhancers/Replacements

4DOS

4DOS was developed when entrepreneurs realized that computers could be much more powerful and helpful than they were. Whether you are a computer novice or an experienced power user, 4DOS helps you get the most out of your IBM PC or compatible computer.

Technically, 4DOS is a command interpreter or "DOS shell," which means it reacts to the commands you type at the C> prompt. 4DOS was designed so you don't have to change your computing habits or unlearn anything to use it. If you know how to display a directory, copy a file, or start an application program, you already know how to use 4DOS. This software understands all of the commands you know and adds to them. Its purpose is to make DOS friendlier, easier to use, and much more powerful and versatile, without requiring you to use or learn a new program, a new set of commands, or a new style of work.

Bang!

Bang! provides a variety of powerful and flexible hard-disk management functions. These functions are provided in one utility with a consistent command syntax and option. Bang! is an information utility designed especially for hard-disk users. It provides enhanced versions of the following DOS commands: CHKDSK, DIR, and TREE. Enhancements include:

- Ability to search across several or all disk drives.
- Optional file selection criteria such as date or size.
- Optional sorting by any field.
- Optional arbitrary formatting of the output lines.
- Optional execution of the formatted output lines.
- CHKDSK of network drives.
- /P (pause) option supports any number of screen lines.
- /W (wide) option supports any number of screen columns.

There are also several additional commands:

- DUP—List only filenames that occur more than once.
- DUPCMD—List only commands occurring multiple times on the PATH.
- FREE—An abbreviated CHKDSK showing total and free space.
- LABEL—List volume label(s).
- LS—Similar to DIR.
- Alphabetical list of filenames only.
- UNIQUE—List only filenames that occur once.
- WHERE—Similar to DIR but also searches subdirectories.

Cat.EXE

Cat.EXE is a replacement for the limited DOS directory (DIR) command. A very powerful utility, it is loaded with features. Display Format Options are: /1 gives a 1 column, full display of all file information, including the path to the file. /1V gives a 1 column, full display. This is the same as the /1 option, but also includes the volume label on each line. This can be used with the /V option and DOS command line redirection to send output to a catalog file of your floppy disks. The volume label identifies the disk each file came from. /2 provides a 2-column display option, with all information but the file attributes. The path is shown in headings but not on the individual file lines. This is the DEFAULT option and does not need to be specified. /2A provides a 2-column display, with attributes shown instead of time.

/3D gives a 3-column display with filename and date. Subdirectories are displayed in lowercase in this display to distinguish them from files that are shown in uppercase. /3S provides a 3-column display with filename and size. Sub-directories will have "<DIR>" shown in the file size field. /3A provides a 3-column display with filename and attributes.

Subdirectories can be identified by the attributes. /W or /6 gives a 6-column display (names/extensions only). Subdirectories are displayed in lowercase in this display to distinguish them from files that are shown in uppercase.

DOSMax and DOSM86

DOSMax is a device driver that maximizes the use of low memory by moving DOS system data to upper memory blocks provided by your upper memory manager. It is not necessary to use any other programs to create upper memory data types like FILES, BUFFERS, FCBS, or LASTDRIV from Quarterdeck's QEMM386. In addition, if you are loading DOS high, DOSMAX prevents DOS from loading into the HMA and moves the DOS kernel to an upper memory block. This improves performance and allows programs to use the HMA more efficiently.

Such programs include Microsoft Windows and Quarterdeck's DESQVIEW. DOSMAX can also move a portion of COMMAND.COM into an upper memory block, instead of the HMA. This feature recovers space in low memory, which is available when DOS is in the HMA.

Support is automatically provided for Microsoft Windows so that all files in your CONFIG.SYS can be loaded high. Normally, this prevents Windows from running, but the problem is circumvented. DOSMax automatically moves all the data that can be moved, depending only on the availability of upper memory. It automatically stops moving data when it determines that memory would be fragmented if the move was completed. Finally, a command-line mode is provided to move the Master Environment, to control Windows support, and to check and report on the status of the high memory area reserved for COMMAND.COM.

Top

Top is not just another DOS shell. It combines many popular DOS utility functions into one fast, small package, but adds several unique and productive new functions to enhance anyone's PC usage. Featuring movable, sizable pop-up windows for most of its commands, Top allows aliasing, command-line history and enhanced editing, file viewing (both ASCII and hex), screen color and mode control, and enhanced functionality for many DOS commands. Any or all Top windows can be constantly displayed concurrently with typing at the command line, including the 4000+ line context-sensitive hypertext help tool.

You can use EMS, XMS, RAM disk, or hard disk for swapping. A complete online setup utility called TOPSET is provided and can be used to edit/maintain Top's extensive configuration capabilities. Complete online help is available. Extended screen mode support for Super VGA text modes is provided. Many Top functions accept multiple parameters and "enhanced" wildcards.

Archivers

Archivers are software programs that combine and compress several files into a single file. Compressing files has three benefits:

- They use less disk space than normal files, often less than half of the original storage space required.
- Several individual files can be compressed into a single compressed file. This keeps files in a given set together and makes file group identification, copying, and transporting faster and easier.
- Compressed files can be transmitted and received in less time via a modem, which reduces telecommunications charges.

After one or more files are archived, the original files still remain on the disk, but they are in a compressed form and must be uncompressed before being used. After uncompressing files from an archive file, the compressed file still remains in the archive. What you are doing when you unarchive a file is extracting a copy, or in the case of creating an archive, you are archiving a copy of the original file.

You can recognize archived files by their filename extension. For example, files with an extension of .ZIP are archives created by PKWARE or a compatible program. Those with extensions of .LZH were created by LHA or a compatible.

Archivers were popularized by BBSes, which use them to save storage space on BBSes and to reduce the time it takes to transfer files. Archivers are also used to conserve space on hard disks and on backup floppies.

ARJ

This relatively new archiver has a few extra features such as allowing extraction to alternate file names, multiple volume creation, merging of archives, file extraction lists and exclusion options, and string searching. The archives are slightly smaller than the others.

LHA

This archiving program makes the smallest archive files of any general-purpose archiver, and the beauty of it is that it's free, unlike ARC and PKZIP, which are shareware and require payment to the author for continued use. LHA is also convenient in that it creates archive files that can self-extract without the end user needing any other extracting programs, and you can update the time stamps of all the files in an archive.

PKWare

This archiver reduces most files by about half; it also creates self-extracting, compressed .EXE files. PKZIP.EXE is the program that quickly compresses files and creates files with a .ZIP extension. PKZIP also handles file maintenance, including adding and deleting files and reporting on technical information from within the compressed file.

PKUNZIP is the program's other half; it uncompresses or extracts compressed files. In addition to extracting a complete ZIP file, it can selectively release individual files, show files on the screen for fast viewing, or print them out on a printer. If you don't want to extract all of the files, or if you are limited in storage space, you can selectively extract files from a ZIP file. Instead of extracting everything at once, you can view file size information, then select specific files to be extracted.

This program has many options; for example, you can specify a date, then only the files with dates equal to or later than that specified will be included in the PKZIP operation. If no date is specified, the current date is used by default. You can selectively release individual files, show them on the screen for fast viewing, or print them out on your printer. Password encryption is included so sensitive data files can be scrambled with password protection.

Files contained in a ZIP file can be viewed in several different ways. For example, they can be sorted by: compression ratio, size, date, name, or extension. Individual file comments (up to 60 characters in length) can be programmed for each file in the ZIP file. PKZIP compresses up to 3,900 files into a single ZIP file.

SHEZ

SHEZ is a shell that uses PKZIP, PKPAK, LHA, and List to browse and execute programs from the file list. It unarchives files that you need. The majority of people who use a computer are not, and probably have no desire to ever become, intimately acquainted with the inner workings of that expensive metal and plastic box, which some call a time saver. As a result, a great deal of time and money has been spent to write software that allows the user to accomplish the most with the least effort. After all, a computer is meant to decrease your frustration level and lighten your work load. To achieve these high aims, programmers attempt to write software that can solve problems without users ever realizing that they might have experienced any difficulty.

SHEZ is an exceptional piece of software for both the novice and experienced computer user. It presents the novice with an easy means of accessing rather cryptic DOS utilities, but at the same time provides the experienced user full access to the more advanced (and sometimes arcane) features of a program. SHEZ is a program whose utility and ease of use will appeal to everyone. In simple terms it saves you precious space on your computer disks; or rather, SHEZ uses other programs to expertly accomplish this for you.

SHEZ is a shell program. To those fluent in computerese, shell describes the program's purpose and function. Don't be intimidated if you are not sure what a shell is. A shell is simply a program designed to make using other programs easier.

SHEZ's main purpose as a shell is to simplify the compressing and decompressing of files. It cannot do this by itself, but it uses several other programs. At this point it would be best to list the other programs that SHEZ requires. You will need at least one, and preferably more, of the following to make SHEZ complete:

- PKZIP, PKUNZIP by Phil Katz for *.ZIP files. This is probably the most widely used compression program around. You should definitely get this program!
- ARJ by Robert Jung for *.ARJ files. This is a newer compression program used widely. Recommended.
- PKPAK, PKUNPAK by Phil Katz for *.ARC files. An older program but still used by many. Recommended.
- LHARC, LHA by Haruyasu Yoshizaki for *.LZH files. Distributed free through many sources and recommended.
- PAK by NoGate Consulting for *.PAK files. Not as common as other programs. Optional.
- ZOO by Rahul Dhesi for *.ZOO files. Not as common as other programs. Optional.

A list program is recommended for SHEZ. This lets you view files before decompressing them—a very useful feature.

There are many list programs in existence, but LIST.COM by Vernon Berg is strongly recommended. It is available through most shareware outlets and electronic bulletin board systems (BBS). The ability to scan compressed files for computer viruses is another strong feature of SHEZ. It tests compressed files for viruses before they can damage your system. This program has been extensively tested for compatibility with SHEZ and is available from nearly every shareware outlet and BBS. SHEZ also edits both compressed and regular files.

Archive Utilities

2all-Ultimate File Compression Converter

Anyone who has telecommunicated for any length of time knows about the archiver wars. You have your favorite archiver, and maybe your favorite isn't big with your local SysOp. So you do like everyone; you toast your SysOp's hard drives, downloading 8 gigabytes of files (and you send one, of course), then you curse under your breath and type until your fingers bleed from converting each one to the format you like. Or, you run 2all.

This program takes all the archived files in a directory and converts them to any archive format you want, stamps each one with the original archive's date and, if you like, deletes the original file.

ACZAR

In very easy ways, ACZAR helps manage files that are in archived (compacted) format. ACZAR offers a variety of helpful additions such as general file and disk management functions. All archiving and unarchiving programs currently in use are difficult to utilize because they are command-line driven.

ACZAR manages these archiving and unarchiving programs. Also, you can use ACZAR as a DOS shell to manipulate files and directories. The following programs do the actual archiving or unarchiving. All are available on electronic bulletin board systems throughout the United States. These files are: PKUNZIP, PKZIP, LHA, ARJ, ARC, PAK, ZOO, and ARCA/ARCE. The following are additional files: SCAN; DIET; LZEXE; and PKLITE. Features include:

- Ease of use.
- Fully mouse driven.
- All major archives.
- Built-in viewer.
- EMS memory support.
- Help system.
- 43/50-line display.
- Scans for viruses.
- Extensive switches.
- Automatic. Unlike other archive shells, ACZAR locates archives and searches all drives and directories automatically.
- SAA interface.
- Complete file and disk management systems built in.

ArcMaster

ArcMaster is short for the term Archive Master, where the word archive means a place where documents and/or records are kept. ArcMaster (AM) is an archive program interface control system. It provides a clean interface that removes the burden of remembering all the various command-line switches and options that might be used with archiving programs. In addition, AM is a powerful and flexible DOS shell that provides numerous ways in which the user can maintain and manipulate a DOS directory and file structures.

AM can be used as a stand-alone program in its useful role as a DOS shell; however, to manipulate archive files, users must have present in their system at least one of the supported archive programs. You will need at least one of the following archive programs to utilize the full capabilities of AM: ARJ, LHA, PAK, ZIP. AM supports several utilities designed for the specific purpose of compressing executable (EXE) files. These utilities serve to compress an EXE file but allow the compressed file to be executed normally from DOS. Their trick is that they insert code into the start of the target EXE file that decompresses it as it loads into memory for execution.

The recommended program is a state-of-the-art browser from Vernon Buerg named LIST.COM. AM can be configured to use any file browser that accepts a file path name on its command line.

CMPvrt

CMPvrt makes it possible to convert a file from one archiving format to another easily. In the spirit of promoting shareware, you need to acquire the appropriate file-unarchiving utility for the type of file that you want to convert from/to. Most, if not all, are available on many BBs. Note that because of the different conversion possibilities, the unarchiving of subdirectories is not supported. But, CMPvrt converts compressed files within compressed files (providing they use the same type of archiving as the main file being converted).

Conversion is possible to a maximum of two levels. CMPvrt makes it possible to convert a file from one compression format to another easily. You will need to acquire the appropriate file (de)compression utility for the type of file that you want to convert from/to.

RView

RView is a free program. It is a directory lister and utility for several archive types. Currently supported are ARC, ARJ, LZH, LZS, PAK, ZIP and ZOO. RView displays either ZIP- or LZH-style compression ratios and has various sort options. RView converts self-extracting archives to normal archives. It provides a single, stand-alone program capable of listing directories of several archive formats. Filename extensions are ignored in determining archive types.

This utility also contains various features not conveniently provided elsewhere at the time of their inclusion. The type of archive is displayed before listing each archive. For self-extracting archives, the letters "SFX" follow the name of the base archive type. The program further distinguishes between small- and large-model LZH SFX archives and shows which ZIP SFXes are of the "mini" (junior) type.

RVShell

RVS is a free program for listing and viewing the contents of several archive types. It employs external extractors and an external viewer (all user-selectable) to extract and view compressed files from within the program. Currently supported are ARC, ARJ, LZH, LZS, PAK, ZIP and ZOO. It displays either ZIP- or LZH-style compression ratios and has various sort options.

RVS was designed to complement use of RV. As with RV, filename extensions are ignored in determining archive types. The program displays either in color or black-and-white, based on the video mode at program start-up. However, if the user has a color graphics card, but no color monitor, the display can be improved by using the DOS command "MODE BW80."

TranScan

TranScan provides an easy way to scan compressed files for viruses using ViruScan by McAfee and Associates. TranScan supports the nine most popular compression programs, including ZIP, ARJ, LZH, and ARC. It converts files to your default compression method, deletes unwanted files, adds selected files, and inserts ZIP and ARJ file comments. It supports embedded compressed files and also retains directory structures saved in ZIP, ARJ, and LZH files. TranScan tests TeleDisk files, converts GIF files to JPG, and converts JPG files to GIF, GIFLite, and GifTest GIF files. Basic features:

- Uncompresses ARC, ARJ, DWC, HYP, LZH, PAK, SDN, ZIP and ZOO files.
- Scans compressed files using ViruScan by McAfee and Associates. Virus name as reported by SCAN.EXE is written to screen and log.
- Checks TD0 files using TeleDisk File Checker.
- Converts files to preferred compressed file format.
- Deletes unwanted files inside a compressed file.
- Adds optional BBS.ADD or other files to compressed file.
- Adds ZIP or ARJ comments.
- Compresses noncompressed files like TXT and DOC.
- Tests and compresses GIF files using GIFTEST, PKLITE, and GIF2JPG.
- Converts JPG files using JPG2GIF.
- Processes a single file, directories, subdirectories, or entire disks.
- Accepts DOS wildcards in filenames like "*" and "?."
- Rejects files based on age or date.
- Menu- or command-line driven.
- Easy internal SetUp.
- SetUp screen scrolls both directions using PgUp, PgDn, arrows, tab keys, or mouse to move around the menu.

- Context-sensitive help available at every prompt in Menu and SetUp.
- File-pick lists and editors included in SetUp.

Windows UnArchive

Windows UnArchive extracts and decompresses files from ZIP and ARJ format archives. It takes the place of both command-line DOS unarchivers and Windows-based front ends for those utilities. Windows UnArchive is the successor to Windows Unzip, which is now discontinued. Features and improvements over Windows Unzip include:

- 60%–90% faster exploding than Windows Unzip.
- Support for the ARJ archive format.
- Ability to run a member file using its association.
- A new "Try Out" feature, perfect for trying downloads.
- Cooperative multitasking while unarchiving. To use Windows UnArchive, you must have Windows and you must run Windows in standard or enhanced mode. Windows UnArchive does not operate in real mode.

File Management

Baker's Dozen

Baker's Dozen has more than 13 handy utility programs. It includes a one-page spreadsheet, a programmable date calendar, a disk utility program, and a program to compare two ASCII files. GKEY provides all keyboard information-key number pressed or released, a hexadecimal and decimal table of the ASCII character set, scan codes, and valid parameters. A file-find locates in which file(s) a given string occurs; another program prints files at 90 degrees. Another program sorts files with up to four fields, where each file can be in ascending or descending order, case sensitive or case insensitive.

Another program redirects all information sent to a printer to a disk file of your choice. Remove Subdirectory removes a subdirectory, including all files and all children directories. Screen border, color, and default screen colors can be set. A snapshot can be taken of any 80-column screen (not graphics). COM1 and COM2 ports can be switched, and printer ports LP1 and LP2 can be switched. Baker's Dozen has a calendar, file finder, a sideways printer, minispreadsheet, trig, business functions, hexadecimal/decimal conversion, and more.

File Find

Fast File Find

Fast File Find does fuzzy searches and more. It can be run on any DOS-based PC having a CGA, MDA, EGA, VGA, or HERC monitor. Token ring networks or Novell networks are also supported. DOS memory requirements range from a 160K to 400K, depending on how the FFF program is configured.

You can also perform a fuzzy search. This is useful if you try to locate a file but don't know the exact filename. The fuzzy search routine attempts to locate files that sound like the filename entered on the command line. To indicate that you want a fuzzy search performed, place the ~ character in front of the filename. Multiple file-specs can be entered along with multiple drives.

Find

Find is an ultra-high-speed file-finding program. It searches the disk drive(s) for the specified files using the standard DOS drive, file name, and wildcard structure. Standard features:

- Install program.
- Speed.
- Automatic search of multiple drives.
- Automatic wildcard names.
- Colors.
- Elapsed time.
- Display uses DOS directory format.
- Intelligent I/O.
- Long screen support.
- Dynamic memory allocation.
- EMS memory support.
- Extended disk partitions.
- SNAPSHOT debugging aid.
- DesqView aware.

WhereIs

WhereIs has a multitude of search options, some of which are:

- The ability to search inside ARC, ZIP, PKA, PAK, ZOO, and LZH archived files.
- The ability to search for files that have been compressed with PK-Ware's PKLite program.
- True regular expression searching on filenames.
- The ability to display duplicate entries across all of your disk drives (even those duplicates that are in the archive files previously mentioned).
- Up to 25 filenames can be excluded from the search.
- Built-in ZIP archive error detection. Every time you do a search that includes ZIP files, every ZIP file on your hard drive is checked for integrity.

- The ability to do restrictive searches on such criteria as file dates and file sizes.
- The ability to do a verified (or unverified) deletion of any files found in the search is great for periodically deleting those .BAK files or files with a size of zero.
- Support for Novell's "volume" configurations.
- Default switches can be set in an environment variable.

WhereIs helps hard drive owners retain (some of) their sanity by locating files on their hard drives. We've all had situations where we know that a file is somewhere on the hard drive; we just can't remember where. Another added feature of WhereIs is that it can also be told to look inside of ARC files. This is a very important feature with the common acceptance of archived files these days including .ARC, .ZIP, .LZH, .ZOO, .PAK, and .PKA. Other advanced features of WhereIs include:

- Grep-type searching on filenames (i.e., find filenames with a 4 in the fifth character).
- Date and time range specifiers.
- File size range specifiers.

Switching

Back and Forth Professional

Back and Forth Professional is a versatile program management tool that gives you the power to load up to 20 programs at once. Now you can switch between your favorite word processor, database, graphics program, and TSR utilities at will, without having to exit one to open and use another. Features include:

- Delivers a flexible task management environment where up to 20 programs can be open in memory, with each provided with as much memory as needed.
- Occupies only 23–40K of memory, less than 1K if loaded high.
- Offers an easy-to-use, colorful, and customizable environment.
- Supports CGA, EGA, Hercules, VGA, SVGA, and 1024×768 video modes.
- Enables Microsoft-compatible mouse.
- Uses expanded memory (EMS), extended memory, RAM disks, hard disks, and conventional memory to store swapped programs.
- Lets you select the hot keys used to pop up Back and Forth, other B&F commands, and all defined programs.
- Accepts the definition of 50 programs for use within Back and Forth.
- Provides an alternative menu list for defining menus and programs for use within Back and Forth.
- Runs a selected list of programs automatically when first loaded.

- Provides a set of desktop accessories including an address book, time and event scheduler, graphics screen capture, editor, tape, and RPN calculator.
- Signals when specific appointments occur.
- Tasks can also be set to run at scheduled times using the Overtime! appointment handler.
- Lets you cut text from a running program and either place it in a folder for later pasting, print it, or write it to a separate file.
- Allows you to edit the contents of the cut buffer and save text.
- Dials a number from the screen if a modem is available.
- Assigns a sequence of keystrokes to a single macro key to simplify command execution within a program or DOS.
- Replaces special abbreviations with commonly used words, names, and phrases to save typing time.
- Permits color customization where you can choose the colors that suit your tastes rather than someone else's.
- Supports all video modes provided by UltraVision.
- Switches the [Ctrl] and [Caps Lock] keys.

Screen Savers

Blank-It

This memory resident screen-saver program has several special features. *Screen savers* are programs that can be loaded into memory and then forgotten. After a certain period of keyboard inactivity, the program blanks the computer screen. Pressing the space bar makes the display reappear, and the users are right where they left off before the screen blanked out.

The entire purpose of a screen blanker is to stop screen "burn in," which can result from having a computer screen displaying the same image for long periods of time. When an image becomes burned in, it can be read even when the computer is turned off. This burn-in ruins the monitor, necessitating its replacement, a very expensive proposition. For those who have never heard of screen burn-in before, it is not uncommon. Burn-in risks are especially high where a computer is often waiting at a data entry screen, or where one particular program is used often and the same information stays on the screen for long periods of time. Companies with many computers that run for long periods of time are extremely susceptible to problems with screen burn-in, so Blank-It is strongly recommended.

Blank-It defaults to a time-out of 10 minutes; however, this time-out can be changed by entering a new time-out on the command line. Valid time-outs are specified in minutes, and can range between 0 and 59. As long as keys are being pressed, the screen will not blank, and there is no evidence that Blank-It is even running. However, once the time-out period is exceeded since the last key press, the screen blanks. When the space bar is pressed, the screen reappears and the users are right where they left off.

Hard Drive

ActaeOn hard-drive manager

While hard disks have several advantages over floppies, their very size can cause problems. Simple tasks such as finding a particular file, or moving a group of files, take on a new dimension as the volume of data increases. It was this (coupled with the fact that MS-DOS is not the most friendly of operating systems) that prompted the design of ActaeOn. It is a hard-disk management program designed to make life easier by providing a menu interface to most of the DOS commands and allowing users to see their directory structure at a glance. In addition, this program provides a number of facilities that are either not possible or are extremely difficult to do using native DOS.

ActaeOn is a hard-disk manager with a point-and-shoot interface to DOS. In addition to most DOS commands, ActaeOn has many other features, including:

- Graphical representation of directory tree.
- Find file/find string within files.
- Mouse support with scroll bars.
- Full context-sensitive help.
- Text editor.
- Multiple file Copy/Move/Delete.
- Fully configurable (Colors/Sort order, etc.).
- Add your own commands.
- 43–50-line support on EGA/VGA.
- Browse file in HEX/ASCII, Zip, LZH, GIF, or dBASE.

ChkFrag

This software program reports, on-screen, the extent of file fragmentation on a hard disk, and provides an ERRORLEVEL value usable within a batch file to call a defragmenting program. Executed without any of its optional parameters, ChkFrag produces a screen report indicating the number of files and directories, how many and what percentage of these are fragmented, and how many extra (noncontiguously stored) parts of files are contained on the currently logged drive. An ERRORLEVEL code from 0–100, reflecting the percentage of fragmented files (the /% parameter is the default), is also shown. An alternative drive (d:) to be analyzed can be specified, as can ERRORLEVEL return codes that reflect the actual number of fragmented files (the /N parameter) or of extra sections (the /E parameter).

The DOS CHKDSK command, specified with its /F switch, should be executed before running CHKFRAG. Lost clusters or cross-linked files are interpreted by CHKFRAG as an error condition. Drives created with the DOS ASSIGN or SUBST commands, and directories created by JOIN, are not checked by CHKFRAG. On-screen recommendations are to consider using a defragmenting utility when file fragmentation is in the range from 11% to 75%, and definitely to defragment disks that show more than 75% fragmentation.

Cache Test

This program helps a user evaluate the impacts of a cache, or the changes to an existing cache, on a drive's performance. Furthermore, with users increasingly using caches during a variety of tasks, it is important to find the optimal cache settings given the task at hand. Whereas a word processor might make heavy use of sequential files, a database might make heavy use of random access files. With this in mind, the memory requirements, sectors cached, write delays, and so forth have many possible variations.

By default, Cache Test creates a 10,000-record sequential file using 32-byte records and then a 10,000 record random access file that again uses 32-byte record lengths. Note: Even if the default is changed, any sample string length will be increased by two bytes by the addition of a CR/LF. You might want to compensate accordingly if you wish to be exact in your measurements. The sample record size indicator displays the TRUE sample string size.

This program uses BIOS to do all of its reads and writes, so any hardware or software cache should be able to get some good "hits." In an attempt to flush caches during the test, files are frequently opened and closed.

dCOM

The dCOM program is a full-screen, high-performance, interactive DOS shell, file handler, text editor, print spooler, macro processor, and menu system. Under the file handler, you can change directories, sort directories, execute, edit, print, copy, move, rename and delete files with a single correlating keystroke. You can selectively tag files to be copied, moved, deleted, or printed. Entire directory structures can be copied, moved, and deleted. Subdirectories can be renamed or hidden.

The copy command can span several destination disks when multiple files are tagged. Directories can be sorted in many ways. One of the sort modes puts executable files in alphabetical order before other types of files so you can find program files more quickly. Another sort mode categorizes files based on their extension first and then by filename, and, of course, the rest of the standard sorts using date/time or size, in ascending or descending order.

Hidden and system files can also be seen with the touch of a key. The file handler and the text editor use a command set that correlates with functionality, making it easy to learn and remember. All destructive commands prompt for confirmation before they proceed. The dCOM program features a powerful and flexible macro facility that allows each function key to be programmed as if it were a batch file.

The dCOM program also includes a high-powered print spooler that tunes itself to your computer and printer. It can be resized or disabled at will. The program dCOM is supported with a hot key that clears its contents and any internal buffer in the printer. When its output is configured to talk directly to the printer port, CPS rates exceeding 7000 have been noted during graphics or font downloads to high-demand laser printers having a fairly fast CPU attached.

Fully configurable mouse support is built in. With the mouse you can perform almost all of your normal file-oriented functions such as copying or deleting files. You can also use the mouse to change directories, run programs, execute macro keys, and invoke the visual tree. The dCOM program also provides a multiwindow, multi-buffer text editor that outperforms many editors twice the cost of dCOM. The editor in dCOM probably loads, saves, and searches as much as twice as fast as anything you're already used to. And, the searches are case insensitive. In a development atmosphere, the macro keys and the text editor soon become an indispensable tool.

DAAG (Disk-at-a-Glance)

DAAG (Disk-at-a-Glance) makes it easy to change directories (never type "CD\..." again). Overall features:

- Tree-structured directory map.
- Useful statistics relating to file and directory size.
- Shows hidden directories.
- Pie charts (requires EGA or VGA graphics).
- Contains a file BROWSE program.
- Handles up to 5000 directories.
- Is "mouse aware."
- Can switch across logical drives.
- Jumps to directory name by typing first letter(s).

ORG

This disk organizer has many features not found in other similar programs. In addition to the traditional function of unfragmenting your disk, ORG allows you to customize the sequence in which files are placed on the disk, lets you "lock" (leave unchanged) any file or group of files, and lets you sort files in any directory and delete temporary files on the fly, all in a single pass. Moreover, ORG does this as fast or faster than most commercially available unfragmentation software, as you will discover when you use the program.

You can run ORG on any valid drive in your system. This might be a floppy, a hard disk, or a logical partition on your hard disk. ORG should not be run on virtual (RAM) drives or network drives, and it issues a warning if you specify one accidentally.

If available, ORG uses expanded or extended memory for its tables and buffers. Extended memory requires the Microsoft XMS interface (HIMEM.SYS) to be loaded in your system.

QFiler

QFiler simplifies management of your hard disk. Two different, sorted directories are displayed side-by-side, allowing you to select both individual files and groups of files for QFiler to act on. Features:

- Copy and delete files.
- Rename files and directories.

- Pass files to an external browse utility or favorite text editor.
- Move files from one directory to another.
- Add and delete subdirectories.
- Run other programs from within the QFiler environment.
- Display files in sorted order.
- Set a directory mask so that only certain files are displayed.
- Back up files.
- Set Epson-compatible printer page length and send a form feed to the printer.
- Print a hard copy of a sorted directory. You can also set or clear file attributes, change file time/date stamps and rename the disk volume label.

Memory Management

The Last Byte Memory Manager

The Last Byte Memory Manager is a collection of software that can provide up to 384K of additional memory to your computer in the upper memory area between 640K and 1MB. It does this by using shadow RAM, existing fixed read/write (RAM) memory, or by mapping expanded memory (EMS) pages into the upper area.

Using The Last Byte Memory Manager, device drivers, terminate and stay resident (TSR) programs, and more can be moved up into upper memory, leaving more conventional memory available for your application programs. Depending on your hardware, you might also be able to extend the total conventional memory from 640K to as much as 736K.

This software package also includes several advanced utility programs that can move the master environment, DOS FILES, and DOS BUFFERS into upper memory. Other advanced utilities create RAM disks, print spoolers, command-line recall-(history)-buffers emulated expanded memory, and TSR "markers" (to facilitate TSR removal). Best of all, by the presence of the display buffer, unshadowed read-only memories (ROMs), and other adapter cards, these utilities can even use shadow RAM that has been disabled (referred to as *bank-switch memory*).

The Last Byte Memory Manager offers seven significant advantages over other MS-DOS upper-memory managers:

- It leaves no "stubs" in conventional memory.
- It works with any processor chip, even the 8088 used in the original IBM PC. Many memory managers require a 386 CPU or better.
- The Last Byte Memory Manager doesn't use protected mode. Unlike those memory managers that do, this software is totally compatible with any protected mode software, not just Microsoft Windows.
- It doesn't require extended memory. Some memory managers depend on the processor's ability to remap physical memory from above 1MB into the upper memory area.

- On motherboards that use one of the supported memory controller chips, The Last Byte Memory Manager, unlike all other memory managers, can use the unused shadow RAM normally disabled by the display buffer, unshadowed ROMs, or adapter cards. This provides the unique ability to use all 384K of upper memory.
- It does not slow down the performance of your computer because it does not incur the 5–10% execution overhead inherent in protected mode.
- On motherboards that use one of the supported memory controller chips, The Last Byte Memory Manager extends conventional memory, using regular full-speed motherboard memory, rather than using the EGA/VGA display buffer. (The display buffer memory of some EGA/VGA adapter cards can be as much as six times slower than regular memory.)

Menus

AutoLoad

AutoLoad is a graphical user interface (GUI) menuing system and file manager. Once installed and set up with all the application programs available on the system, AutoLoad loads and executes each program with a single key press or click of the mouse. Using separate menu data files allows for an infinite number of menu selections. The file management half of AutoLoad makes file and directory maintenance quick and easy. All the menuing and file management functions of this software can be controlled by either mouse or keyboard.

Also included in the AutoLoad package are four separate utilities used to extend the functionality and ease-of-use of AutoLoad. These utilities are:

- ALE, the AutoLoad text editor.
- LK, a file-locking security utility.
- BALLS, a graphical display utility used for screen blanks.
- FINDFILE, a utility to help locate files on the hard drive.

DOSMaster

DOSMaster is an easy way to get around DOS without having to remember all of the DOS commands. Instead, it puts all of these commands, plus a lot more, in easy-to-follow pull-down menus and windows. Also, instead of having to type DIR every time you look at something, it is already there in a menu you can scroll around in. Just press Enter on a program to run it, instead of typing its name. Also press Enter on a directory name to access it, instead of using CD all of the time. All of the file management commands are in the pull-down menus. So, all you really need to know about DOS to use the DOSMaster is the basic knowledge of how to manage files and directories.

Menu Master

This hard drive menuing system lets you run your applications from a simple menu. Menu Master lets you skip over ever having to use DOS at all, except when you load software onto your hard drive. From the menu, all you have to do is use the arrow keys or mouse to highlight the program you wish to run and press Enter, or click the mouse button. This way you can get your work done and not have to deal with DOS at all.

Menu Master also has password protection for each of your applications, as well as a main system password for completely locking other people out of your computer. A sample DOS commands submenu is already set up for you with the most common things you require from DOS in a simple menu. You can add your own options, or submenus, and you can also edit or delete the options in the sample submenu.

Medley

Medley is a menu program that displays a menu on the screen and lets the user select an operation from the menu. Menu selection is accomplished with a single keystroke, or by moving a menu bar to the desired operation. The purpose of Medley is to make computers easier to operate. It's great for DOS-impaired or typing-impaired people. Jim Button wrote Medley several years ago for use at ButtonWare. It quickly gained wide popularity. Best of all, Medley is free. ButtonWare asks for no payment for this program.

Employees can do their daily computer work without having to learn DOS commands. The program is lean and trim and is only 6K in size. (When it's running, it actually uses about 12K of memory.) So, Medley doesn't use up very much of your computer's resources. This makes it ideal for use on laptop computers, or wherever disk space is a concern.

Medley offers more nice features. There's an automatic screen blanker (so that your menu lines won't get "burned in" to your screen image). Users can be asked questions before a menu item is actually run. The answers to previous questions can be recalled to save having to retype them.

Is Medley for you? If you need to set up a complex menu system, with many levels of menus and dozens of menu items, Medley probably is not for you. If, however, your requirements are simple (one or two levels of menus), then you should take a close look at Medley. Its ease of use and simplified approach make it ideal for most people. If it fits your needs, please use it, share it with your friends, and enjoy. If you don't like it, toss it. Button makes his living by selling computer software, but he gets his kicks by pleasing computer users. He hopes Medley pleases you.

Fastmenu GOLD VGA

Fastmenu GOLD VGA provides fast one or two key strokes to your applications. Other features:

- Consumes no memory during applications' execution.
- Allows up to 40 folders, up to 20 applications per folder.
- Allows quick editing of the applications lists.
- Can be operated with a mouse.

- Has built-in auto-mouse detection.
- Has PassWord protection on any or all outside applications execution, module operations, and DOS exits.

Hard Disk Menu System (HDM)

Hard Disk Menu System (HDM) is the "push-button-easy" menu system you need. Once HDM is set up, you start programs and organize your files with a press of one or two keys. HDM is:

- Compatible.
- Fast.
- Powerful.
- Flexible.
- Network-ready.
- Affordable.

With a keystroke or two, you can start your favorite programs. This program accesses pull-down menus. However, you can also access the same menus from "shortcut" keys. The Hard Disk Menu System is a powerful yet flexible menuing system. It puts a friendly face on the disk operating system, or DOS. Yet, HDM is not a terminate-and-stay-resident (TSR) program. The Hard Disk Menu System doesn't require you to "speak" the DOS language.

HDM is an easy-to-use DOS shell (or menu system) that runs on top of DOS. This helps the person unfamiliar with DOS, without slowing down the more proficient user. Like a restaurant menu, HDM gives you a way to choose quickly what you want to do. HDM bypasses DOS, so computing is enjoyable and useful. HDM has virtually unlimited potential. You can have up to 1,000 menu files containing 26 menu pages each. Each menu page can have up to 10 menu entries. Altogether, HDM allows you to choose from up to 260,000 entries. You use the Hard Disk Menu System in two ways: You can simply use the menus that someone has already created, or you can design your own, either automatically or manually.

MIMenu

MIMenu (pronounced MY-MENU) was originally developed to fulfill a need for a user-friendly menu program for the Air Force. Features of MIMenu include:

- 10 choices per page, with a maximum of 10 pages in one set of menu pages.
- Virtually unlimited as to how many submenu page sets you can have.
- A quick screen system display determines your system's configuration.
- Password protection capability prevents exiting MiMenu.
- Password protection prevents unwanted editing of menus.

- Individual password protection prevents certain menu choices from being run by unwanted persons.
- Easy transferring or copying menu choices to new locations on any set of menu pages (even across different sets of menu pages).
- Built-in screen blanking capability, or you can run an external screen saver of your choice.
- Editing of menu choices and page names are all accomplished on the current screen without having to run an external editor.
- Mouse is supported throughout MiMenu.
- Option of returning to current menu choice after a program has been run.
- Full support to select different color choices for screen displays.

QuikMenu

Complex DOS commands are replaced by GUI-style, easy-to-use command buttons and pop-up windows in this program. QuikMenu, a graphical menu system, allows you to use your computer quickly, easily, and intuitively.

Running programs is as easy as pressing a single key or clicking the mouse. Best of all, you don't need to be a programmer to add an item to your menu. It makes using your computer system as easy and enjoyable as possible. You can use a mouse or your keyboard to access any menu function.

QuikMenu runs in all standard graphics modes and can operate on most networks that use DOS commands. When running QuikMenu for the first time, a dialog box appears, asking if you want a menu created for you. If you answer YES, QuikMenu searches your hard drive and automatically creates a button for any program it recognizes. Any programs not found by QuikMenu can easily be added at any time. Don't forget help. If you ever have trouble understanding an option while using QuikMenu, simply select the help command button from the top of the screen, or press the F1 function key.

Taskmaster

Typical menu systems make computers easier to use, which makes users more relaxed and productive. With a menu system, users interact with a friendly "front end," while the menu system deals with the unfriendly operating system. This effectively shields the inexperienced or infrequent user from the unfriendly aspects of the operating system.

Taskmaster does all this and much more. As the only programmable user interface product on the market, Taskmaster is unique. No single "front end" can satisfy all users. Taskmaster therefore allows far more extensive and flexible configuration than competing products. Typical menu systems allow limited configuration of the user interface; e.g., define the names, locations and parameters of your applications.

Taskmaster, however, is implemented as a high-level language. It comes with user menu tasks that permit you to be up and running immediately while, at the same time, maintaining flexibility. Yet you can also program Taskmaster for total control of the user interface. Everything displayed on the screen is under your control. You are not stuck with someone else's ideas; Taskmaster is the only user interface that you cannot outgrow.

If you can speak English, you can program using Taskmaster's intuitive fourth-generation language. Yet it offers you complete control over the computing environment. Features include:

- Supplied end-user tasks make Taskmaster suitable for beginners.
- Programmability makes it suitable for experts.
- It can launch even the largest of applications.
- Restricts access to system prompt, applications, and directories.
- Prevents other applications from providing access to a system prompt.
- Implements time-outs on menus, for security and screen saving.
- Optionally provides secure access via user names and passwords.
- Can return you directly to any previous menu level.
- Ideal for data capture.
- Able to pass data to applications.
- Data can easily be saved to disk for later restoration.
- No restrictions on complexity of each option.
- Is an ideal first language, easier than BASIC.
- Taskmaster is portable to other operating systems.
- Performs scheduling of periodic routine tasks such as backups, etc.
- Reports and acts on results from any application.
- Automatically records user activity in an encrypted catalog.
- Easily logs actions to a journal.
- Creates and reads ASCII text files—e.g., output from utilities.
- Directly reads the screen, closing messages, etc.
- Switches between video "screen pages."
- Gets and sets all file (and directory) attributes.
- Is intelligent, flexible, expandable, indispensable.
- Can be upgraded to Protean, a fuller programming language, LANs, UNIX, and CDOS systems.

In short, when your system grows, Taskmaster grows right along with it. It is upgradeable to interface with UNIX, Xenix, and CDOS micro, mini, and mainframe systems.

Virus

F-Prot

F-Prot is an antivirus program. The package consists of several files:

- VIRSTOP, the virus-intercepting program.
- F-Prot, the main program.
- SIGN.DEF virus signatures.
- TROJAN.DEF signatures for several Trojans.
- *.TX0 language files for various languages.
- VIR-HELP.* virus information in various languages.
- *.DOC information about viruses and the package. VIRSTOP intercepts known viruses; it prevents the execution of any program infected with any of the viruses it recognizes. The other program provides virus scanning, disinfection, and information.

Integrity Master

Integrity Master is the world's most powerful antivirus and PC integrity software. It provides function and performance far beyond any other antiviral or data integrity software, yet is easy enough for novice users. Integrity Master features:

- Recognizes known viruses by name, describes their characteristics, then guides you through their removal.
- Detects not only existing viruses, but also as-yet unknown viruses.
- Unlike other programs, it detects sectors and files that were damaged by a virus, not just those that were infected.
- Understands which files and areas on your disk are special and provides special, specific diagnosis and recovery if these areas have changed.
- Reloads system sectors on disks that are so badly damaged that DOS can no longer recognize them.
- Detects any form of file or program corruption, not just that caused by viruses.
- Provides easy-to-use menus with built-in help and the experience of a PC integrity expert through its Integrity Advisor feature.
- Is useful as an aid to PC security. If someone changes, adds, or deletes any of your files, you will know.
- Is useful with disk diagnostics.
- If you just restored your files from a backup, this software tells you if the files are really okay.
- If you just deleted *.BAT rather than *.BAK, Integrity Master tells you exactly which files you need to restore.

- If your hard disk is having problems and DOS will not even recognize it as a disk, Integrity Master can diagnose and then reload your partition and boot sectors to "fix" your disk.

Viruscan (Scan)

Viruscan (Scan) is a virus detection and identification program for IBM PC and compatible computers. Scan searches a PC for known computer viruses in memory, the partition table, the boot sector, and the files of a PC and disks. Scan also detects the presence of unknown viruses.

The program works by searching the system for sequences of bytes unique to each computer virus, then reporting their presence when found. This method works for viruses recognized by Scan. It also checks for new/unknown viruses by looking for viruslike code patterns, plus comparing against previously stored validation (checksum) data.

Chapter

4

Personal/Professional Advancement

This chapter is dedicated to your personal and professional advancement. I describe a number of programs that will help you find more challenging and rewarding employment, better manage your personal time, minimize stress, improve your health, take care of your personal finances, and better manage your home. For each topic, one or more programs are listed, along with a brief description of what each program accomplishes.

Vocabulary

People judge you by the vocabulary you use, both in verbal communications and in the words you write. So it pays to continue to improve your vocabulary for conversing, for better understanding, and for writing the memos, reports, proposals, and other documents that form a part of the everyday lives of scientists and engineers. This next program helps expand your reading and speaking vocabulary.

MaX

MaX enhances your vocabulary. It runs only once a day, gives you a short quiz on a certain word, and provides the definition of the word. The quiz reinforces the definition of the word in your memory. MaX keeps a record of your score and displays it if you choose.

The software checks each time that MaX is run to determine if it has already been run for the day. If it has, it informs you and exits. If not, it proceeds with the quiz, using the date your computer keeps. If your machine doesn't set the date automatically, you have to use the DOS DATE command to set the date before MaX runs. MaX is installed this way so you can run it from your AUTOEXEC.BAT and not have to remember to run it every day. Of course, this works best if you turn on your computer every morning. If you leave your computer on all the time, there is no easy way, except remembering to run MaX daily.

MaX also accommodates more than one user, keeping separate records for each. If you run MaX without any command-line parameters, it asks for your name. You can also tell MaX your name on the command line. The words and quiz data are read from a vocabulary file. MaX comes with a default vocabulary file. The first time the program is run for each user, the software sets up a record for that user, using the default vocabulary file. The vocabulary file that comes with MaX has 150 words in it. You are quizzed on each word twice before the file is exhausted; that means if you run MaX every day, the file lasts 300 days, or almost a year, before it runs out of words.

Resumes

Resumes are usually the first item your potential employer sees from you, whether you are applying for a job with a new company or being considered for an advancement or a new position within your own company. A resume is a very important document, so you should devote considerable care to creating, checking, and publishing it. On the basis of hundreds (maybe thousands) of resumes I've personally reviewed, here are some musts to observe when you compose a resume:

- The text must be neat, not crowded. All pertinent data must be included. Don't try to get too fancy with clip-art, special borders, etc.; they merely distract from the presentation. Be plain and neat.
- Account for any blanks in continuous employment with some logical explanations.
- Check, double-check, and triple-check the spelling and capitalization of all the words and terms. Carelessness here might indicate to a potential employer that you're prone to carelessness in your day-to-day work.
- Print your resume with a laser or other high-quality printer. If you don't have access to such a printer, bring your disk to a printer shop or service bureau and have them print it out and make the copies as well. This is an important investment and is no time to economize.

With these caveats, let's look at some of the software programs that are available to help you with this vital document. Try several of these programs and see which one best meets your specific needs.

Digital Resume

This program creates an interesting interactive resume on a disk for an on-screen presentation. In addition to giving potential employers more information about yourself, it accomplishes it in a fresh and innovative way. This software program approach announces that you are not afraid to explore and find new and better ways to solve common problems, so it gets you noticed in a positive manner. Resumes can also be printed, if desired. This program requires a word processor.

Disk O' Resume

Today's job market is tough and is getting tougher. With all the competition for jobs, you have to find a way to get noticed to get hired. Disk O' Resume gives you a unique way to get noticed and get the upper hand on other job applicants.

Every business worth working for has a computer. When your prospective employer puts your Disk O' Resume into his or her computer and runs it, you are sure to get noticed, and remembered, not only for its content, but also for your creativity.

This program is not meant to replace your paper resume, rather to supplement it. You can give it to your future employer, in addition to your paper resume. And with its nice looks, it not only leaves you looking attractive to your prospective employer, but it also makes sure he or she remembers you above all others.

Resume Best

This program provides an excellent tutorial that guides you through all the steps of preparing the most professional resume possible. Resume Best represents five years of extensive experience in preparing more than 5,000 resumes. It even includes important information on getting prepared for an interview.

Covered are such subjects as one-page versus two-page resumes, basic components, action verbs, style, format, proper identification, and wording of objective sections for "general" and "targeted" resumes.

It also covers the proper presentation of qualifications, education, training, experience, military experience, salary, civic affiliations, personal and business-professional references, etc., as well as the type of paper to use, how to prepare a cover sheet, cover letter, and an interview thank-you letter. It even includes hints for proper mailing of your resume package, and much more.

Resume Master

This resume generator prompts you to enter information about yourself, your past work history, your education, etc. It then takes this information, formats, and prints it out to create a professional-looking, one-page resume. You do not need to know anything at all about creating a resume. All you need to do is enter the information that is requested of you and Resume Master does the rest. Additionally, this software allows you to create your choice of either a chronological-type resume or a targeted-type resume. Thus, you can choose the resume type that works best for you, based on your own background.

Even though Resume Master is simple to use and straightforward in its approach, a help system is included if you get stuck and don't want to refer back to the manual. You can get help at practically any point in the program by simply pressing the F1 key.

You can also send a copy of your completed resume to a disk file. This allows you to customize your resume by reading it into a word processing program and making minor changes or additions there. Menu-driven, it has exceptional output to LaserJet and dot-matrix printers. Resumes can be saved and later recalled. The output is very professional looking with the use of bold characters and underlining to highlight key information.

Resume Professional

This software helps you organize your resume information by storing your personal, educational, work, and reference data in different modules. Once the data is entered, you cut-and-paste the information to create and print the resume. A full-screen editor helps you add the finishing touches. A second editor window can be opened up to create that all-important cover letter. You can save, recall, and edit your work. Resume Professional is fast and easy to use and accommodates mouse or keyboard operation.

Resume Shop

Formerly called One-Page Report, Resume Shop has the novice-user in mind and provides menu-driven help that prompts you to enter every item you need to construct your resume, including your name, address, job objective, education, work experience, and references. You can create a number of resume files tailored to specific employer requirements. The result is a concise, one-page presentation of your job background and credentials. A user guide offers advice on improving your resume writing skills. The program also includes a sample cover letter to be mailed with your resume.

Job Search

Along with a good resume, you also need a program that helps you keep track of your employment contacts, your letters, your notes about the company, the names and positions of the people you met, the results of your interviews, and similar information. This is especially true in today's highly competitive job market where you often might have to submit a large number of resumes and need to keep track of all of the contacts you have made. The following list of programs will help with this vital task.

Apply

Apply monitors your applications for jobs, grants, contests. Apply does your busy work and frees up time for you to search for the best opportunities. It allows you to focus on personalizing each application. Apply writes your letters and:

- Merges the routine parts of your applications.
- Allows you to edit your boilerplate.
- Uses employer's buzzwords to prompt you.
- Stores codes to remind you of special requirements for the position.
- Asks if you need to order dossiers and reminds you whether you have.
- Prints your letters and envelopes.
- Prints your resumes.

This program stores data about your earlier applications and searches this information for your reference before you apply again. Apply lists those to whom you have applied currently, those who have rejected you earlier, and those who rejected you in previous years. The program tallies your decisions. Apply prints several lists and summaries.

Job Hunt

Job Hunt is an interactive guide to marketing yourself. It helps you develop a job search plan with a timetable and specific activities, and it reviews and analyzes your time and financial resources. Job Hunt provides step-by-step guidelines and criteria for developing a list of target companies and gives guidance in preparing a detailed employment history.

The program also has step-by-step instructions for preparing industry-specific resumes and cover letters. You can contact companies in groups, or individually with high-quality letters and mailing labels that look professionally prepared. Job Hunt keeps track of your phone contacts, cover letters and replies, company responses, interviews, appointments, and more. It prods you along with a week-by-week list of tasks, activities, and milestones.

Job Search

This program was developed to help all job seekers, whether they are college graduates, recently unemployed, or seeking a career change. It helps locate a job in less time by using dedicated modules that make the job search more organized, better structured, and more disciplined.

The first module is a database used for tracking companies that can be searched by industry or geographical preference. The second module is a word processor, with a spell checker for creating resumes, cover letters, and follow-up letters. Thirdly, the time manager module helps you plan daily job-hunting activities. The fourth module keeps you on course by tracing historical notes, to-do notes, and expenses. Also included are a pop-up phone dialer, calculator, timer, calendar, and area-code list.

Personal Scheduling

Schedules are as much a part of our professional and personal daily lives as are eating and sleeping. We usually rise, leave for and return from work, eat, and go to bed at about the same time. Because we all have many tasks to accomplish and only a limited amount of time each day, it makes sense to make the most use of it, rather than racing from one late and uncompleted task to another.

A personal schedule is meant to be your slave; you are not going to be a slave to it. It is your helpmate. A schedule is simply an orderly, organized method of accomplishing tasks. It can serve as a security blanket, giving you the confidence that you have your life under control. It might be also thought of as an always available prompter of what you have to do. It reminds you what should be done, gives you a rough idea of how long each activity will take, and the order in which your tasks should be accomplished.

Still, any schedule is only a collection of educated guesses. (Some people use dart boards for estimating; some even use blindfolds.) So a key factor in any schedule you decide to use is its ability to reschedule, reorient, and reorganize activities that don't get accomplished on time. Probably the most important features of any schedule are flexibility and ease of change. You must be able to continually monitor and update your schedule to adjust to the real world. Often many of the events we schedule cannot be completely controlled by us. We are often at the mercy of others, of unplanned events. So remember, whenever you create a schedule, accept the fact that it's going to change. Even so, accommodating changes is much better than the uncertainty and helter-skelter, fatiguing existence of working with no schedule at all.

For the following schedules, try out a few of the ones that appeal to you. You probably won't find a schedule that suits you perfectly, so the best way is to put one into practice and see how closely it meshes with your activities. And when you find a schedule to use permanently, be sure to register it with the author. You'll probably receive an updated version of the schedule, plus perhaps a manual and the privilege of consulting with the author.

Personal Information Managers (PIMs)

Active Life for DOS

This program is the most powerful system known for planning, managing, and tracking active business and personal lives. Flexible schedules manage work flow effectively. All types of recurring activities need be entered only once. The Active Life Notebook provides an indexed text/graphics database manager, featuring telephone auto-dial. Active Life also includes pop-up calendars, alarms, and week-at-a-glance. Printed schedules provide portable reference and notation. It was developed so you can plan time for everything you want to do.

The Active Life package contains all the software necessary to run Active Life for MS-DOS. You can also run Active Life under Windows. With Windows, you can take advantage of additional features of the Windows environment.

You can run a number of applications under Windows at one time and easily switch between them, creating an integrated work environment. You can transfer data between Active Life and other standard DOS applications, as well as other Windows applications. From the Windows environment, you can easily access all Windows and non-Windows applications, files, directories, and disks, and control all DOS-related tasks such as managing directories or files and formatting disks.

Calendar Mate

This appointment calendar program produces daily or monthly calendars that can be printed in English, French, German, or Spanish. Printers supported include Epson and compatibles and HPLJII/IIP.

Features include the ability to compute the date for Easter, compute the lunar cycles, and print when a full moon and new moon occur in the month. The program can print events in compressed format on a monthly calendar, display future occurrences of a repeating event, automatically reschedule events, print a single calendar or a range of calendars, print a yearly calendar on one page, and more.

Chronolog

This easy-to-use personal information manager lets you enter data about your job (or any other facet of your life) that is important to you and organize it. Its primary purpose is to make sure you are kept aware, and are reminded of, future events for which you have a responsibility. Chronolog is excellent for managing resources, scheduling people/things, and minimizing conflicts.

The Daily Calendar

This very useful program tracks projects, phone calls, letters, clients, and accounts. It lists appointments, prints calendars, and features extensive record-keeping ability. You can add any amount of text to the calendar entries, log conversations, keep records, and track projects and correspondence. Text can be imported and exported, and the excellent TEXT SEARCH feature lets you search for keywords and phrases, then assemble them into a document. The handy TO-DO-LIST manages lists of items by category, priority, or date.

If you have a modem, the DIALER calls numbers for you from within the calendar. The NAME/ADDRESS DATABASE keeps names, phone numbers, birthdays, and other important information. LIST MANAGER produces phone, address, and other lists. The FORM LETTER utility creates form letters with mail-merge for all or part of a list. LABEL MAKER produces mailing labels in a number of styles for any part of a list. The FILE MANAGER manages your hard disk files, allowing easy import and export of text. Multiple databases can be created and maintained. Printer support is provided for 100 different printers.

Day Master

Day Master lists all the events, birthdays, anniversaries, and appointments for the date, as well as a quote for the day. Any event or birth/anniversary that you enter has a small arrow displayed next to it; appointments flash. The year of the specific event is displayed, along with what was significant about it. Other features include the ability to create multiple data disks and a reminder system for any range of dates.

PC Tickle

This program is an easy-to-use organizer and checkbook balancer. PC Tickle organizes your schedule, displays a daily agenda, maintains a history log of tasks accomplished, and prints a copy of the schedule to be used as a checklist. You can view the schedule for a previous date or for future dates, balance your checkbook, track and total your expenses, track and total your calorie intake and output, edit your calendar file, etc.

Reminder for Windows

Reminder for Windows helps you manage your appointments and tasks with a comprehensive time management system. Items are sorted by date and time, with everything needing immediate attention placed at the top so you can "see" what's next on your list. Reminder alarms can be set to display warning messages and even execute programs in your absence. Detailed time management reports show time usage. Also included are a built-in calendar and automatic telephone dialing utilities. ASCII text files can be imported and exported via the clipboard. The program supports dynamic data exchange on both client and server.

Schedule

Schedule is an appointment calendar, reminder, and memo program. It works well for a single user, but it also supports multiple users on a network by recording a name and access code for each. The appointment scheduling part of the program allows repeating entries for holidays, etc. Memos can be made in month, week, day, and subject files. A week-at-a-glance is offered, and a monthly calendar can be printed, but it does not display appointments.

Diet and Health

Better Diet Analyzer

Have you been advised to keep an eye on your calorie, fat, cholesterol, and/or sodium consumption? Or, are you the lucky one who monitors someone else's diet?

The Better Diet Analyzer handles up to approximately 9,000 days worth of dietary summaries of the day's calories, fat, cholesterol, and sodium consumption, plus your weight, if entered.

CalCount

CalCount, the Calorie Counting System, is designed around three files, FOODCAL.DAT, MEALCAL.DAT and PERS.DAT. FOODCAL.DAT has more than 550 different food items (165 in Demo) and contains each item's caloric value per a specific unit of measure. MEALCAL.DAT is an empty file in which you record "meals" that are created using CalCount. PERS.DAT is also an empty file in which you record personal data including weight, blood pressure, and caloric intake on a daily basis.

Diet Balancer

Need to lose weight or reduce your sodium and cholesterol intake? With The Diet Balancer you will be able to determine your appropriate weight loss goals according to sex, age, and exercise levels. The Diet Balancer can:

- Chart your dietary performance during a 60-day period.
- Analyze your diet for 26 different nutrients, including calories, fat, sodium, and cholesterol.

- Save meal plans and have instant access to the previous day's diets and meal plans.
- Look at the nutritional contents of any foods in the database, including brand names and fast foods.
- Add any new food to the database.
- Analyze your personal recipes for nutritional content and save the information in the database for inclusion in future diets.

Doctor on Disk

This computerized medical book lets you quickly determine what illness is suspected when you or your family start showing symptoms of sickness. The index of symptoms contains more than 490 entries, arranged in alphabetical order. It can be searched by using the Up/Down arrow keys, PageUp/PageDown keys and the Home/End keys, or by pressing the letter on the keyboard that corresponds to the first letter of the symptom listing.

The index of illness also contains more than 490 entries arranged in alphabetical order. It can be searched by using the similar procedure discussed for the index of symptoms. You can select as many as six symptoms at a time. You must select a minimum of two symptoms. The more actual symptoms you select, the more accurate the diagnosis will be. If you select six symptoms and the computer determines that a large number of illnesses are common to those symptoms, you can narrow the diagnosis down by selecting three or four of the first selected symptoms, then select one or two different symptoms.

Headache Free

This software discusses the different types of headaches. It compares migraine and cluster headaches and discusses possible cures, including acupuncture, hypnosis, vitamins, and others. Many of us have learned to head right for the bottle of aspirin when life's little aches and pains occur. This might not be good for your health and it is not always convenient. What do you do when you don't have a bottle of medicine handy? What can you do to get quick relief? How can you avoid side effects?

This program helps you overcome minor aches and pains. It instantly shows you how to apply a combination of techniques including Shiatsu, Swedish massage, foot reflexology, accupressure, and others gleaned from a variety of eastern and western sources. When you or your family and friends need relief, the applicable knowledge will be there.

Leaner

Are you tired of the diet syndrome, only to get bored and eventually gain the weight back, and maybe more?

Leaner is a collection of behavioral modification strategies and motivational techniques that turn your PC into a motivational tool for weight control. No traditional diary method can provide so many different ways of reviewing your entries without forcing you to do the plotting and charting. Leaner keeps you motivated by graphing, reporting, charting, and reviewing your progress.

With more than enough chart space to keep a large family's (or a small company's) progress neat and tidy (20 charts/directories and virtually unlimited directory space), Leaner ignores the small stuff while you stay focused on getting into shape. Leaner calculates your body fat, gives tips on boosting your metabolic rate, designs custom forms to use as food diaries and behavior modification tools, and turns the information you give, based on daily activities, into charts and graphs for seasonal evaluation. (The program accepts either Metric or English weights and measures and converts freely between them.)

Leaner uses the food exchange system developed jointly by the American Diabetes Association and the American Dietetic Association. The program helps you change your eating and exercise habits so that over a period of time not only will you reach your desired weight, but you will also have adopted a healthier and more satisfying lifestyle.

Personal Finances/Small Business/Accounting

Business Accounting System

If you have a small consulting business, this accounting software package might be just what you've been looking for. It's not intended to replace your accountant, only to make his or her job easier at tax time. It saves you money in professional fees, while at the same time it makes your bookkeeping job easier and less prone to error. This system gives you information about the financial status of your company at the touch of a key.

Business Accounting System tracks and reports revenues and billings, accounts receivables, and business expenses for a small service/consulting type of business, for both the current and previous tax years. Client billings are recorded and maintained via an accounting menu option available from within each client screen.

The invoicing mechanism includes the ability to perform point-of-entry editing, thereby allowing you to make corrections up to the point you instruct the software to save the invoice to the live system files. Once saved, the invoice can then be immediately printed, or you can defer printing to a batch mode if you so desire.

If necessary, invoices can also be reviewed and edited from the accounting menu. You can review, edit, void, or reprint any invoice, so long as it was created within the current year. Also from the accounting menu, payments and credits can be posted or applied toward any open invoices, or they can be set up as prepayment deposits for future application to invoices.

Checkbook Management Program

Checkbook Management Program's primary functions are to: enter information on transactions, figure and display the running balance, correct any information on transactions, reconcile a bank statement, delete transactions, generate reports, and print checks. Transactions include checks, deposits, autoteller transactions, bank charges, and interest from NOW accounts. Reports generated include:

- Check Register—listing all transactions within specified dates.
- Checks Only—listing all checks within specified dates.

- Deposits Only—listing all deposits within specified dates.
- Automatic Teller Transactions—listing all automatic teller deposits or withdrawals within specified dates.
- List Bank Transactions.
- List All Budget Categories, plus many more features.

Checks can be printed from within the program. Because the printing format is completely under user control, no special checks are required. Up to 50 budget categories can be maintained for reports and pop-up selection. Categories are not defined in advance, but they are added to the list as they are entered on transactions.

Checkbook+

This software is a checkbook register and (for lack of a better term) an accounts-payable system. Unlike most systems available, it looks and works like your check register, rather than the check itself. It simplifies the task of maintaining a checking account. This includes keeping track of your balance and reconciling your bank statement. In addition, Checkbook+ also tracks your regular bills, such as charge accounts, utilities, rents, loans, etc.

For charge accounts, CB also lets you enter information about your charges, such as what you purchased, who you bought it from, a reference code, and how much it cost. This is how CB keeps track of your account balance. It also provides you with an historical record of your purchases that you can keep as long as you like. To make a payment on an account, all you have to do is write them a check and enter the account code (or select one from a menu) in place of the payee. CB automatically applies it to your account.

Checkbook+ is not a program for accountants. There is no general ledger, financial statement, or balance sheet. If you want a financial program, there are plenty on the market. Personally, they seem much too complicated and clumsy.

FastBucks

FastBucks is a fast and "user-friendly" finance manager for home or small business use. It supports checking, charge, savings, miscellaneous expenses, cash, other assets, and other liability accounts. It uses a recursive budget model to generate a relatively painless and workable budget for your home finances. FastBucks also has a financial calculation section to produce loan amortizations, savings, and depreciation calculations.

FastBucks supports three pin-feed check forms and two laser-printer check forms. It also has a user-definable fiscal year for both personal and business accounts, and it supports international currency and date formats. It has 39 expense categories each for personal and business accounts and an additional 15 income categories for personal and business accounts, which can be edited "on-the-fly" while entering data into the program. It also supports 63 recurring payees so the program "fills-in" frequently used entries for you with a couple of keystrokes.

Home Budget Management System

This software is an easy-to-use financial record-keeping and reporting system. Users have complete flexibility in the way they design their accounts, allowing them to track not only checkbooks, but also any income, expense, asset, or liability. You can find out exactly where the money went (and how much is left). Budgeting is also allowed and compared. It has a simple and fast data entry method, and transactions can be split into 14 different accounts.

Home Budget Management System provides extensive screen reports and 10 printed reports. (Printed reports can also be directed to a file.) Some other features are transaction editing and the ability to save recurrent or similar transactions to a "template" with optional reminder flags (due dates) for automatic entry.

Easy Bank

EB lets you track checks issued, withdrawals and deposits made, automatic transactions (monthly payments and deposits), and funds transfers between different accounts. All expenditures (checks, withdrawals, bank drafts) are categorized in up to 98 different user-definable expense (tax) categories. The software makes a transaction record of each of the above, which can be edited, voided and cleared (posted), or viewed in the reports section. EB provides statements of income from all, or specified, sources for any given time period. EB can supply a record of all, or only one, given expense for any specified time period.

EB has advanced features such as backup and restore, a pop-up calendar and calculator, optional user-file maintenance, and user configuration.

Personal Financial Planner

Personal Financial Planner helps individuals conduct their own total financial situation analysis. Subject areas include: income and expenses, assets and liabilities, net worth, economic assumptions and tax rates, education expense, retirement needs, capital needs (life insurance), income protection (Disability), and the impact of taxes and inflation on rates of return. The accurate completion of this program helps identify any financial problem areas where adjustments might be needed. Users can monitor and update information as necessary to know their current status and ensure they are on track or find out what adjustments are needed to achieve their identified financial goals.

The program is easy to use, even for those who do not have extensive experience with computers. The program menus encompass all that is needed to perform a total personal financial planning analysis. In addition to the regular program mode, there is an automatic demonstration mode and a regular demonstration mode to assist in using the program.

Financial Manager With EZ Budget

This software is a personal and small business financial program with features and ease-of-use that meet or exceed the best commercial or shareware programs. It maintains and balances a checkbook, records payments and deposits by category codes of your choice, and generates reports of monthly, year-to-date, and average monthly payments. The software reconciles bank statements and summarizes income and expense data for financial planning and income tax preparation.

The EZ Budget part of the program provides a simple but effective way to prepare monthly and yearly financial plans. Estimated expenses and other payments are compared to anticipated income so that adjustments can be made to meet financial goals. Actual payments are later compared to the budget. Practice sessions are available for new operators—a convenient way to try to learn the program without updating files. Test data is provided to test and review the program.

Financial Manager is easy to use, with automatic menus, on-screen prompts, and help messages available throughout the program. Help messages are specific to the operation being performed. The beginner or experienced user can get the help needed when he or she needs it. No lengthy operating instructions are required.

Files are loaded and saved as needed without operator intervention; the program is intuitive. The software automatically adjusts the display for a color or monochrome monitor. You can also specify a laptop computer for high-resolution on a liquid crystal display.

PCBudget

This is a simple and organized budgeting tool for your computer. PCBudget assists, simplifies, and organizes your home financial records. You can use it as a simple checkbook manager or as a detailed record-keeping-and-tracking home-accounting system. You can keep track of your bank accounts, credit cards, expenses, incomes, assets, liabilities, loans, and bill payments. You can use PCBudget to track budgets, plan your future needs, measure your progress, and alarm you of problem areas.

PCBudget uses a double-entry accounting system. All transactions affect at least two accounts. For example, when you use your checking account to purchase groceries, two accounts are affected. Your checking account balance decreases and your grocery account increases. As more and more transactions are recorded, the better you understand the "big picture" of where your money came from, where it is going, and how to keep it under control.

Mailing

Bulkmail

This software is a specialized system of computer programs designed to make the job of preparing U.S. domestic bulk mail as fast and as easy as possible. Bulkmail is not a "database management system" or a "list management system" within the meanings most widely accepted nowadays for those terms.

Bulkmail is based on experience with large quantities of bulk mail and is the latest in a series of mailing-list programs. Because of this, you'll find that Bulkmail is full of features found on any bulk-mailer's wish list.

Bulkmail has several features to speed up entering records into a mailing list and accelerate the job of bundling mail. The speed with which it runs in your computer is one of those features. Bulkmail was written in the C programming language, a language that produces machine code so efficient that it's used to produce most modern operating systems and compilers. Bulkmail's "template" lets you repeat fixed information without having to retype it for each record. The template lets you update it on the fly. When you print labels in zip code order, it marks the labels so you'll know, without counting, when to tie off a bundle of letters.

This program maintains all records in a mailing list in both name and zip code sequence. The touch of a function key takes you directly to the first record, last record, previous record, or next record in either sequence. An efficient binary search lets you find any specific record in an instant. Bulkmail's records can be resequenced at any time on any field. Resequencing can filter the records to include only those currently of interest. Out of a list of local professionals, for instance, you might produce a list of doctors sorted by telephone number.

Bulkmail contains all the usual features that will let you configure your printer output to a specific label format. In addition, it lets you see your label format graphically on the screen while you set it up.

Pony Express

This software program is for all businesses who use the postal service, especially smaller businesses. As businesses and individuals, we all use the postal service.

For individuals or small businesses, this program gets you out of the lines at the post office and back in your office. For large businesses, it streamlines your mail and shipping rooms. Pony Express is a sophisticated program that knows what the window clerk knows. We call it "a civil servant in a box."

The U.S. Postal Service is the largest delivery system in the world. The system works well and works hard. Even on holidays when you don't see your letter carrier, internal processing continues and never stops. Every day the system moves more than 450 million pieces of mail through 29,000 domestic post offices. They are at every address in the nation almost every day.

And U.S. postal rates are the lowest in the world. Junk mail is analogous to a commercial on television in that it helps keep all rates lower. No system is perfect, but if you agree that to make the changes we might desire is a feat similar to changing the rotation of the earth, then it makes good sense to accept the postal service and learn more about how the system really works.

First, let's take a minute to take the mystery out of the postal rate types. For important stuff: first class—letters; priority mail—packages (first-class zone rated). For unimportant stuff: third class—letters; fourth class—packages (parcel post).

The previous explanation is an easy way to mentally keep the services straight, but it is somewhat oversimplified. While we might think in terms of letters and packages, the postal service really just looks at weight. Don't worry; Pony Express thinks the same way. For my example, I use the term "letters" because first- and third-class rates refer to small weights.

There are also options that you can use with these rate types. "Special delivery" gets delivered to you immediately upon receipt at your local post office. "Registry" (that's how the Fed moves the nation's gold supply) means that the item is signed for and secured every step of the way. A "return receipt" can be returned to you showing when, where, and to whom the item was delivered. "Certified mail" gives you proof of mailing, and, of course, there are others. The beauty of this system is that you pay only for the options you need. In addition, the rate structures are based on the distance the item travels, rather than a single fee to send anywhere. This is referred to as "zone-rated."

Pony Express figures zones, rates, option fees, and automatically applies discounts. It also blocks out services that don't apply to your objective. Remember, all you need to do is affix the proper postage amount to pay for this transaction. Maybe now you can see why Pony Express helps sell a lot of postage machines. If you don't have one, stamps work just fine.

The unsung hero of the postal service is priority mail. Priority mail gets delivered to major cities within two days, and to outlying locations in three. For example, two pounds sent by priority mail costs only $2.90. You can even use it for letters, and it also delivers on Saturday.

Don't give your outgoing mail to letter carriers when they deliver yours. They have a whole day ahead of them before they get back to the post office. You can easily add a day to the delivery by making this mistake.

Drop off mail at a drop box, but only one with a posted schedule. Pay attention to the schedule because they are strictly adhered to. Different boxes have different schedules, so either shop the neighborhood or call the post office for schedules if you need flexibility. Do not use the mail receptacles inside the post office. These do not have posted schedules, so they are emptied as time allows. Use the drop box outside. Use the morning mail for important items. Most drop boxes have a morning schedule, so you might consider batching morning mail and sending it out with someone from your office (early lunch, delivery person, etc.)

To get the most from the Pony Express program, experiment with different options on the rate screen. For options or services that are new to you, just use the help screens. It's really easy, and you'll see not just how smart the program is, but also how much more you know. Once you've tried it, use the Pony Express program to figure your postage. Apply the postage, make sure items are properly marked, and use the drop box or the caller window at the post office. Stay out of the lines, save time, and save money; it's that simple.

Retirement Planner

Retirement planning is becoming more and more complex because of numerous changes in tax laws and the different types of retirement benefits available. Social Security, personal savings, and business retirement programs are usually what compose the retirement benefits a person earns. Because people live longer now, retirement benefits are of prime importance to most individuals. A big fear is outliving your income benefits.

To add to the problem, most retirement programs, including IRA, 401K, and other company benefit programs, do not have a defined retirement benefit (not even considering the impact of increases in the cost of living), but they pay benefits dependent on the uncertain results of the investment performance of the program.

Future availability of Social Security benefits is also a question. Individuals must take more responsibility to make sure they have sufficient assets to support themselves in retirement. This program helps you reach your retirement goal. And it's never too early to start. You need to know where you are and if what you're doing leads to your goal. By using this program, all these questions and more can be answered. The program allows you to monitor your progress as often as you like.

Taxes

This is everybody's favorite subject. I will say no more.

qTAX

The qTAX program is divided into four modules, or sections:

Main module. You automatically enter this module when you type qTAX. It allows you to name your tax return and provide name, address, and social security information for you and your spouse. You also can list the names of all tax forms and worksheets available from the software.

Printer module. You must access the printer module at least once to identify the printer you use. The printer module establishes the print codes that qTAX uses for printing tax forms and worksheets. You have the option of selecting from a list of printers or entering your own codes from your printer manual. Once you identify your printer, save it; there is no need to access this module again, unless you need to select another printer.

Worksheet module. This optional module lets you enter wage, income tax, social security, interest income, dividend income, itemized deductions, and dependent and child care information. An advantage of this module is that it provides you with an excellent checklist of supporting information to make sure you have not forgotten anything. All data is automatically carried forward to the appropriate tax forms.

Taxform module. This module generates a tax return. All tax forms are completed and printed out of this module.

Charge Accounts/Credit

Charge Account Management Program

Charge Account Management Program's primary functions are to:

- Enter information on transactions.
- Figure and display the current balances.
- Correct any information on transactions, delete transactions, and generate reports. Transactions include charges, payments, credits, and interest.

Reports generated include a listing of all:

- Transactions between dates.
- Charges between dates.
- Credits between dates.
- Payments between dates.
- Interest and service charges between dates.
- Budget categories currently in use.
- Transactions between dates for a specific budget category.
- Transactions between dates for all budget categories.
- Charge account names currently in use.

Annual Expenditure Report. A table of expenses totaled by month and budget category, with totals for each month and category.

Summary Report. Summary of total charges, payments, and interest for a set of accounts by month, for a year.

Current Status Report. Summary of total charges, payments, interest, and current balance for all current accounts. Reports can be generated for a single account or for combined accounts. Up to 50 budget categories can be maintained for reports. Categories are not defined in advance, but they are added to the list as they are entered on transactions.

Credit Fix

This program gives you the inside secrets of credit repair. It is definitely possible to have negative items legally removed from your credit report. Bankruptcy, judgements, foreclosures, tax liens, repossessions, late payments, etc., can be permanently removed with an easy-to-use technique. This is the same technique that America's best credit lawyers use (and charge up to $3,000 or more). Now you can do it yourself with your PC.

Credit Fix was developed by a former loan officer. It reflects years of experience in dealing with credit bureaus and the hidden loopholes in the credit reporting system. These loopholes were created when Congress enacted The Fair Credit Reporting Act. In response to angry taxpayers' (and voters') complaints about inaccurate credit reporting practices by almost all credit reporting agencies, Congress created a series of hard-line credit reform bills. As with most bills designed by Congress, it is full of inconsistencies and loopholes. It didn't take credit attorneys long to figure out how to use (and abuse) these laws to make it possible for any negative remark on a credit report to be removed permanently.

These same powerful techniques, and the required forms, are available in this software package. Also contained in this package is a step-by-step plan to rebuild or establish your credit. In addition to repairing your credit report, this software package also orders your credit report and applies for a secured credit card from one of several major banks of your choice. It also rates your credit history, using a scoring system utilized by many banks that issue credit cards.

TRW, Inc., has announced that consumers can receive one free copy of their report each year. TRW's decision came as Congress was considering changes to the Fair Credit Reporting Act, which governs credit bureaus. To request a free report, write to: TRW's National Consumer Relations Center, 12606 Greenville Ave. P.O. Box 749029, Dallas, TX 75374-9029. Or call 214-235-1200, ext. 251.

Checkbook

CheckMate Plus

This program helps you easily and accurately maintain your checking accounts (including simplifying regular monthly transactions and printing out your checks). You can "predefine" checks that you write often, and even have CheckMate Plus "remind" you when they're due. Print the checks you want, or write them by hand. (You can freely mix the two methods.) When your statement arrives, the software even helps you reconcile your account. You can split each check twelve ways, attach notes, even "customize" the check stub. You can print reports at any time.

Because CheckMate Plus is a real double-entry accounting system, you can generate balance sheets, income statements, and budgeting reports. And don't forget the high-resolution color graphics. Whether you're tracking a family budget or launching a business, you'll have complete financial control effortlessly.

The best feature? Ease of use. CheckMate Plus sports pull-down menus, pop-up windows, and context-sensitive help. Features:

- Real "double-entry" accounting that's truly easy enough for home, and powerful enough for business.
- Up to 200 accounts.
- Checking accounts support checks, deposits, withdrawals, debits, credits, and ATM transactions.
- Can attach "notes" and "stubs" to each transaction.
- Prints checks (computer form, laser, or right out of your checkbook).
- Helps reconcile your checking accounts.
- Many powerful and flexible reports (including standard accounting reports).
- Powerful "search-and-filter" functions.
- Reports can go to the printer, disk, or the screen.
- Data export to "WKS" format for Lotus 1-2-3, Microsoft Excel, Borland Quattro, Trius As-Easy-As, or virtually any spreadsheet.
- Financial calculator (loans, deposits, and compounding) with full amortization reports.

- High-resolution color graphics.
- Multiple ways to view your data ("full screen" and "browse" modes).
- Pop-up, context-sensitive help.
- Pull-down and pop-up menus throughout.
- Pop-up calculator with rolling tape.
- Password protection.
- "Budgeting" capabilities.
- Automatic transactions (pick from a menu or have CheckMate Plus "remind" you when they're due).
- "User-defined" Reports.

Express Check

This program's simple goal in life is to help you manage your checking accounts. Because the various screens look like your paper checks, deposit slips, and check registers, you will find the program extremely easy to learn and use. But the program goes far beyond the usefulness of a check register. Every check and deposit can be recorded to various budget codes that you set up. This makes it possible for you to see at any time during the year where you stand in regard to your spending. How much do you spend on food each month? How much have you given in tax deductible donations? Car payments, electricity, gasoline, entertainment?

Without ExpressCheck, you either guess at where you spent all your money, or you spend long hours with a calculator. With ExpressCheck, you can find out in just a few seconds.

WinCheck

Wincheck manages any number of checking, savings, and credit card accounts. WinCheck's features include a check form designer to build check templates for printing on virtually any preprinted check form (dot matrix and laser checks are both supported).

The custom report maker allows you to view and print a variety of supplied reports, such as cash flow reports, or you can create custom reports tailored to your own needs. With the custom report maker, you can print an entertainment expense report, a credit/deposit history, or all tax-deductible transactions for the year to date.

WinCheck's transaction classifications, or "categories," help you keep track of the money earned and spent on categories such as business and entertainment. WinCheck's budgeting facilities can be used in combination with these categories to plan a monthly budget. Add recurring transactions to a menu or have it remind you to make transactions such as mortgage payments and paycheck deposits. It also supports Windows Dynamic Data Exchange (DDE) and sports an internal batch language.

Home Management

Home and Business Legal Guide

You actually receive four programs in one with this software. The first is an educational program to read, review, and learn your legal rights by reviewing, on screen, the full, annotated text of laws and governmental regulations. The annotations, prepared by an attorney, carefully point out the parts of the statute that are very important for your legal rights.

Legal forms. The program features an artificial intelligence document-generation program that prepares legal forms. In many instances each form has several different variations that you program by answering on-screen questions. At this time, the program generates more than 226 different forms drawn from more than 193 categories. (A complete list is contained in the manual.)

Tutorials. This program also features tutorials, which are instructional texts that explain the nuts and bolts of various legal problems in an easy-to-understand way. In addition, the tutorials feature an online dictionary and the ability to examine both the full text of the laws and the explanation on screen.

Word processor. Use this online processor to alter forms once completed.

All of the programs feature an online note pad and online help. Also provided is a DOS shell that allows you to perform system functions or run other programs without exiting this legal guide.

Home Maintenance and Repair

Don't you just hate it when you have to call a repair person, but you think he or she might take advantage of you? Wouldn't it be nice to have some basic knowledge of your home and auto systems just so that you could ask the right questions and at least sound like you know what you're talking about? If you don't happen to have a brother-in-law to double-check with, you can learn basic home and automotive maintenance and troubleshooting techniques from Home Maintenance and Repair.

Home Maintenance and Repair has maintenance tutorials and troubleshooting charts for air conditioning, home electrical systems, heating, plumbing, appliances, and automobiles. It gives you tool lists, illustrations, and schematics of the systems. It allows you to print maintenance forms. It offers step-by-step support for doing regular maintenance on your home and car—the simple kinds of tasks that help to keep those expensive repair people away from your door.

And when something does go wrong, look to Maintenance and Repair's troubleshooting charts for a suggestion of what the problem is and what you should do about it. In both the troubleshooting guides and the general maintenance tutorials, the program takes special care to alert you to all dangers, gives cautions and advice on working safely, and lets you know when it's time to call in a professional.

Depending upon your interest and skill level, Maintenance and Repair can be a beginner's tutorial and guidebook for dealing with mechanics and repair people, or a ready guide for hands-on maintenance and repair. If you are among the thousands of people who don't really understand much about plumbing, wiring, how your furnace works, or just what you're supposed to do to keep your car in good condition, Home Maintenance and Repair does know.

Home Manager

Home Manager manages the details of operating a home and its finances. It can make a database for your assets and liabilities, budget and track your home finances, print addresses on envelopes (laser printers only) or labels, calculate various payments and interest for your finances, schedule and track maintenance on your home, create a general database of things to do, and track your vehicle maintenance and make databases of lists of general items, names, addresses and phone numbers, and your important records. You can print all of these records. There is also a built-in calculator and calendar to view each month of any year.

Some of the chores performed by Home Manager are:

- Make a database to track your assets and liabilities.
- Budget and track your home finances.
- Print addresses on envelopes or labels (laser printers only).
- Calculate various payments and interest for your finances.
- Schedule and track maintenance on your home.
- Create a general database of things to do.
- Track your vehicle maintenance and make databases of lists of general items, names, addresses and phone numbers, and your important records.

Home Management II

This package of five programs helps you keep track of your money, plus a whole lot more. Unlike other home finance software, this program is uncomplicated, easy to learn and use, and written so that you need learn only those portions of the software you will actually use.

Different people want to accomplish different things in financial planning. Some only want to balance their checkbook; others want to track stocks, and still others want to plan their family budget, etc. But not everyone wants to do all these tasks. With the five programs in Home Management II, you can do all of this and much more. Because each program is self-contained, you need learn only what you want to use. This was kept simple because it is frustrating to waste time learning (and paying for) parts of a program that are useless for you.

Many financial management software programs try to be all things to all people, both in the home and business. As a result, they do little useful work for anyone. This program takes a different approach. Instead of incorporating all tasks into one giant program, five different programs are included in Home Management II. This makes the sum of all the parts more flexible and more suited to the individual needs of the home user.

Whether you're a homeowner or a renter, there's something for you in Home Management II. All five programs in the package are designed to help you out with many of those big and little chores that get in the way of what you really want to do.

Point and Shoot Home Manager

This program arose from dissatisfaction with available programs to manage the details of operating a home. Point and Shoot Home Manager encapsulates, within a single program, the primary data management tasks needed in the home. It is a collection of dedicated databases for complete home budgeting, tracking of home maintenance, home inventory, important record tracking, name and address tracking, and vehicle maintenance record keeping. A full-featured financial calculation system is included. This is not simply a home budgeting system like so many others. This is a comprehensive home management system.

Home Buying

Homebuying and Refinancing

This is much more than just another loan amortization program. It is a comprehensive financial analysis of buying or refinancing a home, or prepaying an existing mortgage. It helps homebuyers and homeowners save thousands of dollars. Prospective homebuyers will find indispensable the program's analysis of buying a home versus renting. Real-estate agents can increase the level of service they provide to their clientele by using the program to create quick and accurate homebuying reports.

Homebuying and Refinancing is the premier home purchase analysis program designed for the home buyer, homeowner, or real-estate agent. For the home buyer, the program determines total yearly and monthly costs (PITI, points, and expenses), after-tax costs, and present value costs for several types of loans, including fixed and adjustable rate. The program also rates a home buyer's chances of loan approval and compares the cost of buying a home versus renting. For the homeowner, the program helps evaluate the financial consequences of making prepayments or refinancing an existing loan. Refinancing results include calculation of monthly payment savings, total present value savings over loan life, and break-even times for a wide range of interest rates and points. For the real-estate agent, the program produces a professional-looking package of clear and straightforward results with a customized cover page.

Homebuying and Refinancing is an easy-to-use, menu-driven program with pop-up pick lists, context-sensitive help, and an online users' guide. Many of the program's data items have default values that the user can accept or easily change. Although the program contains the most recent IRS tax parameters, the user can change these items so the program's tax calculations stay accurate in future years without the need for yearly updates. Additionally, a user can change the program's loan qualification criteria to account for variations between regions of the country and lending institutions. The program also allows a user to save and recall cost analysis data to disk.

The Home Inspector

Will your new home be a lemon? Probably not, but before you buy, check it out using The Home Inspector. A new home is probably the most important investment you will ever make, so protect that investment. This program was designed by professional home inspectors to help a prospective home buyer objectively evaluate a home's condition.

All of the major sections of the home are covered, including electrical, plumbing, heating/AC, basement, roof/attic, garage, general interior, and much more. Many times, the buyer looks at the surface of the home (Is it painted? Will my furniture fit?, Is there closet space?, etc.) without digging deeper, possibly missing important and costly details.

Organize Everything

Household Register (HR)

This is a fast, easy-to-learn-and-use software package that helps you maintain a detailed list of your possessions. If you are an average person, you probably have no accurate inventory of your household possessions. That's what HR is for. By recording your possessions using HR, you'll have a permanent record of your possessions that can be used in case of loss or damage by fire, flood, windstorm, burglary, vandalism, etc. Even if these events do not occur, using this software is ideal for warranty purposes. For example, if you require repairs or servicing for warranty work, having the model number, serial number, and the warranty expiration date helps facilitate service.

HR also provides the total value of your possessions, which gives you an accurate figure to help determine the amount of insurance to carry. You can compute the value using purchase cost or replacement cost. HR is the best inventory program available. If you use it to record your possessions and keep it up to date, it eliminates much trouble and anguish in case of disaster. Once you record your inventory using HR, be sure to store a backup data disk and one of the inventory listings in a secure place, away from your home. You might want to give a copy to your insurance agent or company.

Household Register features: Holds 17 fields of information describing each possession, category, item, quantity, owner, location, manufacturer, model#, serial#, description, date purchased, where purchased, purchase price, total price, warranty (Y/N), warranty expiration date, replacement price, and total replacement price; has twelve easy-to-use functions from the main input screen (Next, Prev, Find, Top, Last, Edit, Add, Copy, Del, Srch, Report, ChangeMenu Help).

It prints seven types of reports: Total Value-Sm, Index Cards, Index Card Format on regular paper, Rolodex Card, Warranty, Total Value-Lg, and Summary. Reports can be sent to the screen, printer, or disk text file. You can print your entire inventory or partial inventory based on the selection criteria of your choosing. The reports can be indexed by category, owner, location, or item.

Automatic indexes. Each possession is indexed by category, owner, location, and item. By using the main screen FIND function, you can quickly list your possessions indexed by category, owner, location or item.

Pop-up menus. The category field has a 24-choice menu, the location field has a 24-choice menu, and the owner field has an 8-choice menu. Each menu can be customized to fit your needs, and menus can be turned off if desired. Search on any combination of fields. The command-line menu can be switched from the bottom to the top of the screen.

College

Financial Aid Profiler

This program matches a student's individual background and educational goals with a large database of more than 200,000 listings for scholarships, fellowships, grants, and loans. This database represents more than ten billion dollars in private-sector funding and is constantly being updated. Without this program, you could spend months in a library researching your opportunities for funding from the private sector, or you could waste time researching the information and still miss thousands of funding resources.

Small Business

Business Adviser

This unique expert program puts the tax, legal, and business knowledge of a Harvard lawyer and "big 6" CPA at your fingertips. This is a real "no-brainer" to install and use—even for computer novices. It gives your small business instant and current tax, legal, and other business-related help on a vast range of tax, legal, and other subjects, both federal and state (for whichever state you select). Business Adviser covers IRS, OSHA, ADA, ERISA, EEO, labor laws, workers' compensation, environmental laws, and hundreds of other easily accessible topics.

This program creates a multipage checklist of business start-up requirements in two minutes or less, customized for your business and the laws of your state. It has: built-in worksheets/outlines to create a business plan and other documents, easy-to-use income and estate tax calculation/planning modules, and consulting sessions in which the program asks you detailed questions to help you make key business decisions such as whether to incorporate, choose Sub S status, etc. There's nothing else like it at any price.

Chapter

5

Word Processing

"The man of science appears to be the only man who has something to say just now—and the only man who does not know how to say it." SIR JAMES BARRIE

The most-used application of the personal computer is for word processing, so a huge number of word-processing programs are available, ranging from free or low-cost to expensive commercial programs. To help refute the late James Barrie quoted above, this chapter describes a variety of word-processing and related programs—software that can make literary geniuses out of men and women of science.

Included in this chapter are several free to low-cost versions of word-processing and related programs of various sizes and complexities. Many are comparable to, and several even exceed, the capabilities of commercial programs. Also included to further enhance the literary abilities of scientists and engineers are creativity/brainstorming programs, outliners, integrated programs, text editors, spellers, dictionaries and thesauri, grammar checkers, foreign-language word processors, presentations, desktop publishing programs, fonts, and screen-capture software. If this excellent collection of free and low-cost software doesn't make literary geniuses out of all of you, I'll be sorely disappointed.

Creativity/Brainstorming

Engineers and scientists are probably the most creative group of people in the entire world. Whether you consider yourself to be creative or not, there are several software programs you can use to improve your innovative endeavors. These creativity/brainstorming programs are highly subjective in their approaches, techniques, and results, so you should sample as many of them as possible to ascertain which functions best for you. Remember, they're available to be tried out for only a modest cost. If you find some that work best for you, be sure to reward the authors for their genius, register the program, and continue to use the software.

Creativity Package

One of the best of the genre, Creativity Package, consists of three disks (if the program is not archived). The package consists of Thomas A. Eaton's book, *Think Thunder! and Unleash Your Creativity* and two programs, Versifier and Thunder Thought. Eaton's excellent and well-written book explains what creativity is and how to use it.

Thunder Thought is not an outlining program; nor is it a think-tank program. It is computer-aided brainstorming that facilitates the creative process by automating its initial stages. It is a program for anyone who needs ideas. With it, you no longer need to worry about how to come up with novel combinations of words, images, or ideas. Thunder Thought accomplishes that for you, leaving you with the task of finding or recognizing sense in whatever combinations of words and ideas it produces. Essentially, Thunder Thought works by throwing dice. You give it a vocabulary of up to 800 words or short phrases by filling in four lists of nouns, verbs, adjectives, and adverbs. When you then ask it to help you brainstorm, it picks words at random from the appropriate word lists and plugs them into fill-in-the-blank sentences.

How can such a process as Thunder Thought possibly generate sensible sentences? You provide the vocabulary, generally to go with a specific, defined topic. Random choices from such restricted sets of words can easily seem nonrandom, especially when those choices are arranged to look like sentences. In addition, the human mind has a way of imposing order on disorder. In the process, it exposes thinking that is going on beneath the surface. This is the key to psychological tests such as the ink blots so beloved by psychiatrists. Like the ink-blots, Thunder Thought gives your subconscious a chance to work. That is the secret.

Versifier is a creativity-enhancement program for poets. The program combines words and phrases from a vocabulary you supply into different kinds of poems, which you then can save and edit. It allows you to specify vocabulary, small or large, and is optimized for generating poetry, not for brainstorming in the broader sense of the term. Versifier requires more user input before the computer accomplishes its random magic.

Writing poetry with Thunder Thought or another "poetry machine," such as Versifier, is a matter of using what the computer produces, not as poetry, but as the raw material for poetry. Edit and elaborate it, just as when you use the computer as a source of ideas for other kinds of creations. Then, impose suitable poetic form. The computer program's randomness guarantees unusual originality of image, thereby helping you avoid triteness.

DynaMind

You can use this idea processor or thinking tool to help solve a complex or important problem, or to create innovative approaches to a situation by making tradeoffs. DynaMind stimulates your creativity and imagination. It helps you examine problems from many angles to make sure nothing is overlooked. DynaMind helps you clarify and sharply focus your thinking about all aspects of a problem. You identify and prioritize your goals and use them to understand your motivation. Once the situation and the goals are defined, the program helps you create strategies for achieving your goals. It generates ideas through a variety of techniques that have proven to stimulate creativity and imagination. Finally, it helps you evaluate your alternatives, then select the best one for implementation.

DynaMind can be used by just about anyone with a problem to solve, whether for business, education, or personal use. In addition to helping solve problems, it helps you communicate with others about them. After analyzing a situation, you can provide a DynaMind file to another individual, who can then see exactly what factors have gone into your decision.

The program can also be used for joint efforts at problem solving. After one person has worked with a situation, someone else can work with the same file and create additional options, or evaluate the ones already presented.

Idea Tree

This "thinking tool" takes a general concept, a "parent" or root idea, and provides the mechanism to visually list and separate the concept or main idea into its basic components, called "children."

Most ideas worth thinking about have some degree of complexity; a main idea exists with several lesser or interdependent ideas. For example, if you decide to design a new computer peripheral, the overall design task must be divided up into many sub- and sub-sub tasks that are interdependent and which must be organized into some logical order.

The objective of Idea Tree is to organize this complexity by identifying its component pieces and their vertical position in the hierarchy (main ideas at the top, subideas at lower levels). This structured approach to developing ideas is presented as a graphic tree with branches and subbranches, much like the tree in your computer file directory structure.

Outliners

Outlines are an essential element of writing and should be created before any document longer than a page or so is written. Even short documents can profit by making a rough or scratch outline of your thoughts before you begin writing. An outline gives you an overall view, or "view from the top" of what your document is going to cover, and the order and subordination of the topics.

While your ideas are still in this malleable outline form, they are much easier to evaluate, manipulate, add or subtract to, and organize, than after you have translated your random thoughts and ideas into disorganized writing. It's just like taking a trip without planning where you're going and ending up futilely attempting to compensate for your inadequate forethought.

Outline Plus!

This program helps you develop and use computerized outlines. This is essentially the same document-planning tool we learned to use when writing essays in school. However, computerized outlines increase the power and usefulness of ordinary outlines, in much the same way that word processors improve on typewriters for developing complete textual documents.

An outline processor is a computer program that simplifies the development and management of outlines and other structured text. It shares many features with word processors, but both tools have their distinctive strengths. While word processors are tools for the entry and development of complete, formatted texts, outline processors are tools for creating a well-organized and comprehensive structure for a document before you begin writing the text. Outliners are effectively used for rapidly shaping a document at its inception, while it is still in a malleable form. However, word processors are better at taking a shaped document to completion.

In general, use Outline Plus! whenever you organize ideas or concepts into a formal structure. It lends itself to classification tasks, with its built-in decimal and classical labelling notation that automatically adjusts to changes in the structure. It also is an acceptable text editor. The size of your outline is limited only by available memory.

PC Outline

This outlining and planning program lets you work with ideas as you would deal with words. You can randomly enter information of almost any type (thoughts, plans, ideas, etc.), then organize them into a hierarchical structure. Once in the hierarchical structure, you can view the information in many powerful ways for review and analysis. For example, using the hide/unhide feature, you can view down to any level of detail in the outline that you choose. With a single keystroke, you can jump from viewing the lowest level of detail, to a view that shows you only the highest, the most important topics.

With another few keystrokes, you can print the outline, send it to a file compatible with your word processor, copy it into another outline, or paste it directly into your word processor. PC-Outline is particularly appropriate for list making. Making to-do lists or planning a project are ideal applications of this program.

PC Outline features: automatic numbering; easy, pull-down menus; automatic paragraph formatting; hanging indents; search and replace; block operations; sorting; freely mixing outline and text; customizable printer drivers; and much more.

Word Processor Utilities

AnyWord

Formerly called Total Recall, AnyWord is a general text indexing and retrieval program. It consists of two parts: Indexer and Finder. When Indexer is run against a set of text files, it produces an index that's much like the index of a book. These index files can then be read by Finder to quickly locate any word or combination of words in the text.

Again, like a book index, your text is untouched. All of the indexing information is carried out separately. Your text files are not modified or rearranged in any way. This means that you can "play around" with AnyWord as much as you want to without worrying about making an inadvertent change.

You supply information about your files through the profile file used by both Indexer and Finder. Here are some AnyWord features:

- Search by any word or combination of words.
- Display selected text.
- Extract selected text into new files.
- Search with simple Boolean logic, i.e., AND, OR, NOT.
- Search with "wildcards," e.g., comput? for references to computers, computing, etc.
- Exit to DOS and return without disturbing selected text or search results.
- Review previous search results within a session.
- View all of the words that have been indexed.

Hytext

Hytext brings hypertext—the latest development in computer learning and information retrieval—within the means of every interested author, teacher, or electronic publisher. *Hypertext* can be defined as the nonlinear, hierarchical presentation of information. In contrast, a book represents the most well-known linear system for information storage and access. You usually read a book from beginning to end. The most sophisticated references in a book are the table of contents and the index. A concept on one page has no direct reference to another concept on a different page.

This is not true with hypertext. A page can contain hot words that are directly linked to other related topics anywhere in the document. These hidden topics can be viewed by selecting the hot word in question. Any topic can contain one or more hot words, so the thread of reference continues.

This technique provides a treelike arrangement of information, where various branches can be cross-linked like vines running and connecting among the branches. Instead of accessing information in a linear mode, you can skip between several different pages and find all the information related to the topic selected. You can control the level of detail with which information is presented, and the information retrieval becomes more rapid and direct.

Index Maker

Give your document files that professional touch by adding an index. Manually creating an index to a document is a time-consuming process. Index Maker greatly reduces that investment of time by having your computer accomplish the work. Load the file to be indexed, create a word list for the index, then sit back and watch. Index Maker indexes up to 500 words, with a maximum of 50 page citations per word.

This program is menu-driven for ease of operation. Index word lists can be saved on disk, then imported when needed. Add or delete words from existing lists. Once created, the index can be viewed on screen before being printed. Index Maker can also be used to locate keywords as markers in large document files to locate information quickly and easily.

Iris

This program lets a writer create an electronic book that can be read online with many sophisticated features. This technique might be the next evolution in books. Unlike printed copy, electronic books do not need to be read in a linear fashion. With this hypertext-type software, instead of citing page numbers, authors cite topic names. Readers select the names directly from the screen, or from a menu. Electronic books can ask questions and respond to the answers. Readers can jump from one topic to another with the flick of a key. Windows can change color and size.

If you already have material stored in an ASCII text file, converting it to a virtual book is easy. Internally, Iris uses a format similar to the well-known DOS batch file. The major difference is that "unmarked" lines are displayed, while "command" lines use a special one-character prefix.

ProCR

ProCR stands for Professional Optical Character Recognition. Here are the basic specs for this optical character recognition software:

- Reads 8- to 30-point mono and proportional fonts.
- No font selection required.
- Up to 260 words per minute.
- Supports HP ScanJet directly.
- Supports other scanners producing TIFF or PCX files.
- Selectable resolution, including 200 (fax) and 300 dpi.
- Preview and online-correction modes with graphics adapter.
- Real-time viewing of text during processing.
- Continuous scanning if auto document feeder is attached.
- Misrecognitions flagged with selectable character.
- Menu-driven or noninteractive mode from DOS command line.

ProCR performs Optical Character Recognition and runs on an XT or AT; however, an AT is strongly recommended because of the highly CPU-intensive nature of the program. A graphics adapter is not required for basic character recognition, but is needed for the preview and online-correction functions. If a graphics adapter is used, it should be Hercules, EGA, or VGA.

Proindex

If you need to create an index for the back of your book, thesis, report, or software manual, this is an easy way to index any word-processed document. Proindex takes your text file, along with a list of keywords or phrases, and looks for matches between the two. It produces an index in any one of numerous different styles, including those suggested by *The Chicago Manual of Style*. There is no limit on the length of your text file, or the number of keywords in your list. The resulting index can be saved as a text file, or sent to a printer.

The index produces an alphabetical list of names, subjects, references, etc., together with the position where each appears in the text. This index usually appears at the end of a book or publication. The exact format or style of this index can be selected by the user.

Proindex creates an index from almost any kind of text file. All that is needed is your original text file and a list of keywords or phrases. The software compares this reference list with your original text file, then marks each keyword it finds in an index. Utilities are provided to automatically prepare a reference list for you if you don't want to create your own. The resulting index can be paged by a utility.

Writer's Heaven

This program is a power-packed add on for the popular word-processing program, PC-Write. A new set of command keys, in an ingenious arrangement, offers you unsurpassed editing capability. With a single keystroke, you can move the cursor or delete by character, word, line, sentence, or paragraph—in either direction—without moving your hands from typing position. Writer's Heaven can double your editing speed—yet it's easy to learn and use.

Writer's Tools

This software contains three separate shareware programs that can be used by writers: Datakeeper, The Comedy Generator, and Writer's Dream.

Datakeeper is a super-easy-to-use, very free-form database to help you keep track of your research, notes, recipes, or any text information at all. Datakeeper fills a huge niche in the computer world that has existed for a long time. For business people, scientists, students, CEOs, homeowners, and everyone, there has always been a need to quickly jot notes into a computer, then have fast and simple access when that information is needed. This program is a fast and easy-to-use free-form database.

The Comedy Generator was originally written for speech writers and performers, but has been found to be useful for anyone working with comedy. This software can also be used for many noncomedy operations, such as helping you write poetry, generating fictional character names, coining new phrases, and creating ideas. Here's how it works: You type a sentence, then the computer returns your sentence with one or more words changed—substituted from one of many carefully designed word lists. The resulting sentence might or might not be humorous, so you continue to experiment until you achieve the desired results.

Writer's Dream is important for anyone who would like to be published, which should include most engineers and scientists. This magic tool allows you to write for 7.2 million readers. To reach this audience, you will not have to query a publisher, deal with editors, pay a vanity publisher to print a stack of your books, or sell yourself or your book to anyone.

Text Editors

Software in this general category is used to edit straight ASCII text either as prose, or for programming. Some of the programs have many of the capabilities of word processors, such as block move, search and replace, etc. These text editors are easy to use, require little RAM, and are fast in operation. Some emulate WordStar commands; others use pull-down or simple menus.

Boxer

This highly regarded text editor has mouse support; EGA/VGA support; multiple files and windows; block functions (sort, total, average, auto save); context-sensitive help; search and replace across files; and much more.

A fully functional text editor, Boxer, features: 512-level Undo, full mouse support, keyboard reconfiguration, 43/50 lines for EGA/VGA, multiple files and windows, column marking, macros, color, pull-down menus, word processing, and context-sensitive online help. Also available are: forward and backward searches with wildcards, EMS swap for DOS shell, graphic drawing, auto indent, find mating parenthetical characters, fill, print, invert, and detab.

Boxer also features 26 clipboard areas and a pop-up clipboard manager, support for multiple compilers with auto-select by file extension, support for DESQview, a clever synchronized scrolling feature for easy file comparisons, auto-frame of marked rectangles, and more than 75 other new features and options.

pE

Perfect Editor (pE), is a text editor with:

- Full, carefully thought-out, mouse support.
- Easy menu-bar operation or Alt command keys (WordStar optional).
- Multiple files.
- Windows (up to 20) with easy cut and paste.
- Line, column, or stream blocks.
- Block copy, move, delete, read, write, and drag (with mouse or command key).
- Find and replace (forward, backward, ignore case).
- Regular-expression search and replace.
- Brace match (and parend) forward or backward.
- Undo to recover unintended deletions.

- Key macros with repeat and chaining.
- Utilities to browse disk directories.
- Count words.
- Full DOS access: subdirectories, DOS commands, shell.
- Variable number of text lines and columns displayed on EGA/VGA.
- Programmable function keys and many user-configurable options.
- Line and box drawing with mouse or key commands (9+ styles).
- Multimath calculator.
- Hex mode for non-ASCII files.
- Powerful sort command.

QEdit Advanced

QEdit Advanced is a text editor, as opposed to a word processor. This means that most of its features are geared toward creating and maintaining program source code and other text files. However, the software can also be used as a word processor to write documents or letters, since it includes word-wrap and paragraph reformat capabilities.

QEdit was developed to operate quickly, especially in the following areas: initially loading a file for editing, displaying/updating the screen, moving through the editing window, searching for text, and saving a file back to disk.

QEdit features very fast operation; it can be customized and can easily be configured the way you like it. It utilizes all available memory, and you can edit files as large as your computer's memory allows. The number of files you can edit is limited only by available memory. You can view files through up to eight "windows." You can have multiple views of the same file, or view a different file in each window. It provides up to 99 scratch buffers and exit to DOS, is easy to use, can use macros, is compact, and much more. Even with all these features, QEdit requires less than 50K of disk space.

Text Editor

This excellent, public domain, full-screen ASCII text editor uses commands similar to those used in WordStar and Sidekick. You can easily create/edit batch files, generate forms, and edit files captured by telecommunications programs. You can edit files as large as available memory, and pop-up help is available. Other features include word wrap, DOS shelling, insert, delete, split, join, block commands, and more. You can print a file/block to LPT1–LPT3, enter any ASCII code, find/replace a phrase, and set left/right margins and page length. Because the program is public domain, you can incorporate its text editing capability into your own software programs.

Sail

Sail is a text editor, a software tool for creating and modifying programs, documents, and other text files. Because of its power, Sail is ideal if you do lots of text processing—be it electronic publishing, program development, etc. While very powerful, Sail is also very easy to learn. Easy-to-use help screens help you quickly find a command. Its command structure is mnemonic and easy to remember.

For beginners learning Sail as a first text editor (or for advanced programmers learning Sail after having experience with other text editors), Sail's simplicity, speed, and power make it an attractive way to do editing tasks. Here is a summary of its features:

- Edit several (>20) files at once.
- Support * and ? DOS wildcards.
- Open several windows.
- Record and playback key macros.
- Interpret commands in its embedded language.
- Capitalize entire blocks of text.
- Edit via pull-down menus.
- Undo text deletions.
- Cut and paste.
- Escape to DOS and return.

Easy-to-Use Word Processors

For those of you who do not wish to invest the time and effort needed to learn the fine points of some of the more complex word processors, but you still need to turn out plain-vanilla documents, here are several word processors that are easy to use, yet they perform many of the key functions of the more complex programs. These programs usually require less RAM and less disk space than the heavyweights.

Breeze

Breeze is an easy-to-use word processor/text editor that does everything you ever wanted from a text processing program, and more. This program was designed with one thing in mind . . . pleasure. You will find the simplicity of Breeze a breath of fresh air, compared to the oppressive and boring alternatives. Features include:

- Pop-up hypertext help system.
- Attractive multimenu environment.
- Text to .EXE converter (converts documents into self-displaying executable programs with menus, printing, and text search features).
- Multiwindows/files (multiedit).
- Clipboard.
- File selection menu.
- Exploding windows.
- Handles giant files.

- RAM dictionary for fast spell checks of whole documents or as you type.
- Create your own personal dictionary if you wish.
- Print spooling (lets you print and edit at the same time, saving hours of time).
- Mail-merge/form letters.
- Scientific calculator.
- ASCII table.
- Analyze text readability.
- Sort paragraphs or lines.
- Statistical analysis.
- Pick list of recent files.
- Mouse-supported.
- Fully configurable.
- Automatic line and box drawing.
- Fast find and replace.
- Text styles.
- Word wrapping.
- Auto paragraph reformatting.
- Autoindent.
- Autosave.
- Restore line function (undo).
- Columns.
- Comprehensive print options.
- Menu-operated printer control.
- Full block operations.
- Powerful macro functions.
- Index words.
- Operate in condensed line mode on EGA and VGA systems.
- DOS shell.
- Insert date into text.
- Center line.
- Execute other programs and any DOS commands from within the editor.
- Go-to line and page functions.
- Set place marker.
- Duplicate lines.
- Auto screen save.

Flex

This program features the usual editor functions, plus:

- Online help.
- Background printing.
- Keystroke macros for repetitive tasks.
- Auto smart block indents for programming.
- Flexible tabbing.
- Optional ruler line.
- Case change: character, block.
- Undo by line or block.
- Supports 27 popular printers.
- Optional autosave to guard your work.
- Adjustable margins.
- Word wrap or long lines (up to 999 characters).
- Shift entire blocks left or right.
- Save file under different name.
- Justification, left, right, or center.
- Up to 10 windows onto the same or different files.
- Cut and paste from one window or file to another.
- Customizable colors.
- Hide blocks for clarity.
- EGA/VGA 43/50 line support.
- Optional nonflashing block cursor.
- Edits files as large as available RAM.
- Fully DesqView-aware video and keyboard functions.
- Quick comments for Pascal programming.
- Lightning-fast function-key block moves.

Flex works with DR-DOS, 4DOS, and Command Plus.

Galaxy Lite

This easy-to-use word program is a fast, RAM-based word processor that offers a wealth of features designed to make the program powerful yet extremely easy to learn and utilize. Users are offered a choice of using menus or quick keyboard commands for all of the major program functions.

With this program, you're never stuck with a cumbersome menu system for commands you already know, and you're never stuck with a confusing command-based system for commands you use less often. It has spell checking, macros, multiple document printing, mail merge features, and it edits within dialog boxes. It understands WordStar commands and handles WordStar files.

Galaxy Lite has a very short learning curve. Its logical, easy-to-use-and-learn commands resulted from considerable research into how people really use word processors. If you have experience using any full-featured word processing program, you'll be up and running with this program in no time. If you're new to computer word processing, or to IBM-compatible computers, you'll be pleasantly surprised at how quickly you can learn all the advanced features of this program, which offers a complete pull-down menu system for entering commands. These menus allow you to use the full power of this software almost immediately, without having to memorize keyboard commands.

KwikWrite

KwikWrite (Executive Word Processor) allows you to create documents such as letters, reports, and manuscripts. The editor uses ASCII files, so it can also be used for programming and creation of other text files. It is ideal for creating E-mail. KwikWrite allows merging of information from dBASE III database files to provide easy mail-merge capabilities. It is also suitable for writing source code (Fortran, etc.), for writing batch files (.bat), creating data files, and for creating memos and documents to print or merge with other programs. Features of the program are:

- It's easy to learn and is suitable for professionals, executives, office workers, students, and others doing a moderate amount of word processing.
- It fits on one disk and functions with a bare-minimum installation.
- Resulting documents are simple ASCII files, so they can be merged into other systems and other word processors with ease. The software can handle files of any length.
- Software has enough features so users can grow into more sophisticated uses such as mail merge, programmable printer options, external text entry, and other applications.

Mind Reader

This unique program is a text editor with an address book, glossary, and a calculator. It is unique in that it uses a patented artificial intelligence-based machine to learn the way you write, and to whom you write. The more you use the program, the smarter it gets. The program suggests words and phrases as you type, even before a complete word is actually typed.

The idea behind this program is to help busy professionals create or modify a document and produce final copy without having to use a secretary or editor. This program also can create standard letter and memoranda formats, store text that is used repeatedly, and provide the user with a handy way to avoid the frustrations of a screen littered with typographical errors, misspelled words, and runaway sentences.

Mind Reader has an expandable dictionary containing the words most frequently used for business letters and memoranda. You can add your special technical terms, names, etc., to this dictionary. The address book stores name and address blocks so you can input them in your correspondence with the touch of a key. It keeps telephone numbers handy and creates mailing labels. The calculator provides standard mathematical functions.

PC-Write Lite

If you want a word processor that doesn't gobble up memory, consider this slim version of the widely acclaimed PC-Write. This trim version has been designed for beginners and pros. Easier and faster to use, it only needs 384K of RAM, or only 256K if you leave out the spell checker (great for laptops), and includes a great new formatting menu and preview mode. Lite is PC-Write file/keystroke compatible and handles files larger than 60K.

Lite has retained such features as word and character count, WordStar conversion, page-jump, page-preview, bookmarks, date stamp, search and replace, tab/space conversion, split-screen editing, auto save, auto file-lock on networks, common ruler margins and controls, basic page-layout commands, footnotes and endnotes, insert file, and automatic page numbering.

Full-Featured Word Processors

This section covers some of the more complex word processors, programs that are suitable for writing anything from a memo to a full-length book. Most have spellers; some have thesauri built-in, and several have many of the features you find only in desktop publishers, such as columnar arrangements, headers and footers, graphics import and editing, multiple window operation, mouse support, and macros. The full-featured processors require a little more learning to become proficient in using them, but most have pull-down menus and extensive context-sensitive help.

Black Magic

Word processors, databases, and spreadsheets all manipulate and structure information in specific but different ways. Hypertext systems, such as Black Magic, enable you to expand information in another way—by dynamically linking keywords, graphics, and blocks of information to each other.

For example, you are creating a document and need to define a concept or word, but the detail of the definition would distract the reader from the main text. You can tag a keyword that indicates that the definition is available. An area of the screen opens up, and you type the definition or explanation. To explore the definition, the user triggers the tag and the reference is displayed. This is a simple illustration of hypertext.

Black Magic shells to DOS and links and transfers information between hypertext and other applications. You can build context-sensitive help systems, link specific items in text documents to technical drawings or illustrations, and expand obscure references with detailed explanations. A screen grabber is available to integrate graphics into Black Magic. The extended ASCII character set is supported for the use of international characters.

If you are currently using a document generator for live electronic documents such as computer-based training guides, foreign language tutorials, interactive catalogs, or educational courseware, then you should try Black Magic hypertext writing. You will find Black Magic to be more than you ever imagined.

FreeWord

FreeWord uses all available memory for text editing. It features macros, mouse support, and can open up to six windows/files at once, with on-screen sizing, multiple font support, and a maximum line length of 999 characters. FreeWord is menu-driven. This means you need to learn only a few commands to use this program. Many users can utilize the program without reading the manual or trying the demos. Online help service is available from the main menu.

FreeWord supports a variety of ways to move around in your document. You can, for example, move the cursor by character, word, line, screen, or by page. A block of text can easily be marked to be deleted, moved, or copied. You can search and replace the text of your document for a word or phrase. With this program, you can right justify (left margins are always justified) your text so your document will have that formal look, or you can select to have the right margin "ragged" (so it looks as if it were typewritten). You can print up to nine copies of any document. Printing goes on in the background, letting you edit the document you are printing, or another document, while printing. The software also has a "typewriter" mode for sending special printer codes to your printers, for addressing an envelope, or for filling out forms.

Nova

Galaxy has gone Nova, presenting a unique word processor that is easy to learn and use, yet has all the powerful conveniences that today's users demand. Features:

- A modern interface similar to Windows and OS/2, using pull-down menus, dialog boxes, multiple windows, mouse support, and a keyboard.
- Work in up to ten windows that are on screen at once.
- Size, overlap, or zoom any window to full screen in a single keystroke.
- Logically structured keyboard commands that are color-highlighted to help you learn shortcuts. Context-sensitive online help.
- Powerful editing features such as cut and paste between documents, style sheets for easy, consistent formatting—or create your own formats, and more.
- Headers and footers, micro justification, and automatic reformatting.
- Multiple fonts, proportional spacing, and laser printer support.
- Edit files as large as eight megabytes, using EMS expanded memory or hard disk.

PC Type

This major word processor has many advanced writing tools. The following is a summary of the program's features. MAIL-MERGE allows you to use keyboard input, PC-file databases, ASCII files and comma-delimited (WordStar) files, as well as PC-Calc data. FAULT FINDER catches misspelled words and offers substitute suggestions. GRAPHING lets you take your tables of data and create bar and pie charts or line graphs and scatter diagrams. MULTIPLE WINDOWS let you work with up to ten different files in up to four different windows. HELP is context sensitive and provides page references to the manual. COMMANDS are available for power users. The last ten commands are saved for later use. CALCULATE equations as commands or values in your text. DRAW boxes and lines. PC-Type automatically inserts the intersection characters.

KEYSTROKE MACROS memorize your keystrokes, and they can be saved and modified. TABS, MARGINS, and INDENTS can be redefined within your files. You can MOVE and COPY lines, paragraphs, sentences, words, and best of all, blocks or columns of text within or between files. You can SEARCH and REPLACE data, and ALIGN and SORT columns of data. DOS commands are available from within the program. The WHOOPS key is available to undelete modified or deleted lines.

DATABASE and SPREADSHEET DATA can be pulled in from PC-File, dBASE, and PC-Calc+ (or even from ASCII or comma-delimited files). PRINT features abound. GRAPHS created can be incorporated into your text. DATE and TIME are easily inserted into your documentation. PC-LABEL is included so you can print mailing labels in a single or multicolumn format. COMPATIBILITY is provided among PC-Type, PC-Calc+, and PC-File.

PC Write Standard Level

This easier-to-use version of PC Write Advanced (listed below) is the adaptable word processor from Quicksoft. Context-sensitive help, online tutorial, pull-down menus, mouse support, and auto-repage make it easy to use. The Adapt feature lets you add or remove feature sets depending on your needs. You can start with core features, then add more as needed. The pull-down menus change as you adjust feature sets. You can even read about features in help and instantly adopt them. Also included are *Roget's Thesaurus*, easy margin setting, up to 10 windows, and import/export from WordPerfect, DCA-RFT, and ASCII. It supports microspace justify, auto-reformat, auto-number, mail merge, spell check, and more. Great for both formatted documents and ASCII files, PC Write supports more than 900 printers; it's tops for LaserJets. You can set more than 200 customization options with the Adapt feature and Customizer menus, in areas such as editing, keyboard use, screen display, and file handling. It's easy to tailor PC-Write to your preferences.

PC-Write works well on laptops. Here are some useful tips:

- You can install and run PC-Write with only a single 720K disk drive. All PC-Write files fit easily on a 1.44M disk.
- If your laptop screen is hard to read, two customizer options are available to help.

Pc Write Advanced

This top-of-the-line, outstanding word processor has context-sensitive help, an on-line tutorial, pull-down menus, mouse support, and auto-repage features that make it easy to use. The Adapt feature lets you add or remove feature sets depending on your needs. You can start with core features, then expand its repertoire and add more as needed, "learning as you go." Pull-down menus change as you adjust feature sets. You can read about the features in help and instantly adopt them.

Added features include *Roget's Thesaurus*, up to 10 windows; import/export from WordPerfect, DCA, and ASCII; microspace justify; auto-reformat; auto-repage; auto-number; multiple column and left/right page layout; and multiple-file updates of index, table of contents, and endnotes. Additional features include mail merge with conditionals, spell check, customizable button bar, and more. It's great for both formatted documents and ASCII files, and it supports more than 900 printers. It's tops for LaserJets.

You can set more than 200 customizing options with the Adapt feature and Customizer menus in areas such as editing, keyboard use, screen display, and file handling. You can easily tailor PC-Write to your preferences. The optional Wizard's Kit tells how to set many more options and how to change the menus, print control files, and internal program notes.

Super WP

This WordPerfect look-alike from Norway is a world-class word processor with features you won't even obtain in high-priced commercial programs. It boasts 117 equivalent WordPerfect commands, full file compatibility, a much easier-to-use menu system. Super WP can also be configured to "act like" several other well-known word processors on the market, such as WordStar. The shareware version has a default configuration of WordPerfect and can read (and optionally write) WordPerfect file formats. The excellent help function is more advanced than many tutorials. Initially, more than 160 functions are available in two screen pictures.

Super WP also contains a built-in cardfile system for keeping track of addresses, client information, phone calls, activities, etc. The Personal Manager is a PIM function that contains "today's work," planning, calculator, date and time setup, as well as access to the cardfile. The program can also read ASCII files. It has a print preview function, spell checker, calendar, calculator, and an auto-dialer. Super WP has a 120,000-word spell checker and supports more than 60 dot matrix and laser printers. However, Super WP cannot import or handle clip art or graphics.

Word Fugue

Word Fugue is a heavy-duty word processing program intended for a variety of word processing and editing tasks. As installed, it is ideal for the touch typist. You can keep your hands on the home row to activate any of the functions. The program is easy to learn because it comes with context-sensitive help. For novice users to try out the various functions, there is also a tutorial on the program disk. The program comes supplied with industry-standard pull-down menus to simplify learning. In addition, each command on the pull-down menu can optionally display the command sequence needed to activate it, so novices can learn the commands.

Document files are ASCII text files. Up to eight editing windows can look at different files or the same file. Printer support includes laser printers, and you can customize the printer drivers. Printing also occurs in the background, so you can continue other word-processing tasks. The program comes with a 100,000-word dictionary and an update program to add or delete words from the dictionary. In addition to the main dictionary, you can set up auxiliary dictionaries in each directory to contain words that are specific to particular projects. Main dictionaries are available in American, Australian, and British versions.

The program can generate indexes and tables of content. Both of these files can be customized by using options embedded in the text. Automatic hyphenation is also available. Tables are supported, as is word wrap and justification of each table column independently. A pop-up calculator with 16 functions is included; the calculation result can be pasted into your text. Word Fugue comes with support for box and line drawing. Additional macro files contain European and Greek characters.

Snaking columns of text is supported. Newspaper-style columns (not to be confused with tables) have text wrapping to the next column when the bottom of the current one is reached. There are hundreds of other sophisticated features like insertion of current date or time (in different formats), the ability to mark changed text in the left-hand margin, offset page number on odd or even pages, print page numbers as Roman numerals, and so on. The installed commands make it perfect for laptops, while the cursor can be changed to a hard block of color.

Wordmaster

This full-featured ASCII word processor has an extensive pull-down menu system and online help. The program provides: multiple fonts, macros, windows, search and replace, cutting/pasting text, headers/footers, automatic pagination, the ability to customize colors and command structure, and more. The documentation includes examples of using various functions to help you work easier and faster.

Spellers

Spellers not only help us spell difficult or unusual words, they also do an excellent job of catching our typos, which are often difficult to spot even during a careful proofread. Most spellers can add to their dictionaries special words, names, and acronyms that you commonly use in your writing. But spellers are not magicians; if you use a word such as "pane" and meant to use "pain," your checker won't catch it.

MicroSPELL

This software is a compact, efficient, fast spelling checker for use both as a stand-alone program and with MicroEMACS on many different computers. Currently supported computers include MS-DOS and OS/2, Atari ST, Amigas, and Unix machines of various types. MicroSPELL is supplied both as IBM-PC executables and as C source code suitable for compilation by most C compilers.

MicroSPELL has tools to mark misspellings, scan for replacements, suggest single words, and maintain user and main dictionaries. MicroSPELL is normally used in conjunction with MicroEMACS to scan text files, stepping through misspellings and making suggestions and corrections.

ShareSpell

This is a "stand-alone" spelling checker. With it, you can check the spelling of text files that you have written. ShareSpell is completely interactive. It displays the text as it goes through the file, checking for spelling errors. At each word it thinks is misspelled, you are given the option of replacing the word from a list of suggested correct spellings, manually correcting it, adding it to the dictionary, marking it for later correction, or ignoring it throughout the rest of the document.

Currently, ShareSpell has a 112,000+ word dictionary, and you can add words that you use all the time, such as your name and words that are specific to your field of writing.

SpelCheck

SpelCheck extracts words from an input file, or several input files, and checks them for membership in a superimposed code dictionary. Any words not found in the dictionary are written to an output file, one per line. The program recognizes a number of options for:

- High-order bit stripping.
- Appending additional information to the output word list.
- Defining the characters composing a "word."

WinSpell

WinSpell is a Windows-based spelling supervisor that checks your spelling three different ways.

- WinSpell acts in real time, as you type. While typing, it notifies you as soon as you misspell a word.
- WinSpell spell checks text files such as Windows Write ASCII.
- WinSpell checks the spelling of information copied to the Clipboard.

These three modes of WinSpell allow you to check your spelling in virtually any Windows application, including DOS-based applications running in a window, spreadsheets, telecommunication packages, etc.

Grammar/Style Checkers

Even though you might think you remember all of the unbreakable rules that Miss Agatha Coldbottom hammered into your nonreceptive head, there is a higher source to which you can appeal to make sure your prose is readable. Although these programs will not take the place of Miss Coldbottom (the software only suggests, it doesn't insist like Miss Coldbottom did), they will help you avoid some cliches and dangling participles (I'm still not sure what they dangle), use shorter sentences, and improve your overall grammatical construction.

Cliche Finder

Cliche Finder checks your writing for old, hackneyed, trite, and overused words and phrases and produces a report listing the "cliches" found in a document. Hundreds of cliches are supplied in the program's database. You can view and change the information in this database. The document you check has to be in ASCII format.

While there are valid reasons for using cliches or jargon in your writing, their overuse might indicate that you are writing like a literary robot and have given little thought to the subject matter, and to its readability. To keep your prose interesting, you need to develop your own distinctive style and not rely on banal expressions that lack originality.

FogFind

FogFind uses the Gunning Fog Index to measure the readability of your writing. To use the program, the text should be in ASCII format, without any special control codes. The software analyzes your writing from the standpoint of the school grade level of education that would be required to understand your prose by checking sentence length, the number of polysyllabic words, etc. It provides a running tally of your writing and stops either at the end of your writing or after it has checked fifty sentences or longer prose.

PC Style

PC Style is an easy-to-use program that "has a way with words." This program helps you to:

- Put "punch" into what you write.
- Reduce the number of long words in your prose.
- Add a personal touch to your writing.
- Make your writing more readable.

PC-Style helps you write more effectively by looking at what you have written, then it provides a report that you can use to analyze the strong and weak points in your text. Several concepts of style and readability are incorporated, including the use of shorter sentences, action verbs, and personal pronouns. You can use the program to check various types of writing, including reports, letters, books, manuscripts, speeches, and technical writing. The program shows you both graphically and in percentages the strengths and weaknesses of your writing. You can print out the report and use this information to improve your writing. PC Style works hand-in-hand with PC-Type+.

Proscribe

Proscribe checks your writing style and helps you make your letters, reports, speeches, ads, etc., more readable. The software scans your writing for the kinds of writing mistakes we all make, the traps we fall into at least once in a while; these are called word wasters—words or phrases that are wordy, confusing, often misused, or just plain unnecessary. There are five categories: wasted words (WW), misused words (MW), tongue waggers (TW), verbs-to-nouns (VtN), and passive voice (PV).

Proscribe helps you eradicate the jargon and "excess baggage" from your writing. And you can customize this feature: Include your own "pets"—jargon or phrases you want to avoid.

In general, your ideas are handicapped if your writing has an overall grade level that's higher than 8–10. Many people can deal with text written at higher levels, but they have to work harder and strain more to grasp your ideas. Avoid words with more than 1.5–1.6 syllables per word, on average. Long, complex words are too abstract to create pictures in your reader's mind and do not convey the proper ideas. Use 15–20 words per sentence, on average. Long sentences tax the memory; readers often forget what you said at the beginning of a long sentence.

Business letters, memos, and reports often have grade levels of 13+; their sentences average 25–35 words. Often they are formal, dull, and stiff, lacking a conversational style. In technical writing, grade levels of 16+, and 30–40 words per sentence are prevalent. Yet, Bell Labs found that good technical writing had grade levels of 10–12. Complex ideas can and should be expressed in easy-to-read writing. Popular newspapers and magazines, however, usually show grade levels of 8–10, with 17–19 words per sentence. Some sentences are very short, some very long, but they consistently average 17–19 words.

Readability Plus

Readability Plus is the ideal complement to any grammar checker such as Grammatik or RightWriter. Unlike grammar checkers, this program assumes that what it's been asked to analyze is mechanically correct. Rather than look for errors, it looks for sentences and words that aren't used appropriately for the writer's intended audience. Comparing the text to any of nine writing style models, Readability Plus guides the writer in "aligning" the style of the draft with the style model chosen. For example, by using the magazine style model for company reports, writers produce more interesting and easier-to-understand documents. Readability Plus reads documents produced with WordPerfect, WordStar, Microsoft Word, and all ASCII (text) files.

Style checkers

This group of checkers, including Count, counts lines and words in an ASCII text document.

- Critic, a writing analysis system, gives a full report of word and sentence length, index of readability, active and passive verbs, repeated words, and it catches some grammatical errors.

- FogFind computes a graded reading level for documents.
- Phrase points out sentences that are too long or run-on.
- Proof Reader counts number of words, long words, number of sentences and length, and gives a readability level.
- Punct checks ASCII text for punctuation preceded by spaces or not followed by spaces.
- Pro-Scribe is an extensive program that finds wasted words, misused words, passive voice usage, and complex words.
- WC tells which words are used the most, and it counts words, gives average word length, and a count of letters.

Thelma Thistleblossom Grammar and Style Checker

This software puts your venerable English teacher to work. She checks WordPerfect documents for passive voice, long sentences, subject-verb agreement, possessive pronouns, cliches, punctuation, and spelling. To top all that off, she helps you write for your appropriate audience level.

WStyle

The Wintertree Writing-Style Analyzer—WStyle for short (pronounced "Double-you Style")—helps improve your writing style in two ways. First, it examines your writing for common problems such as redundant words, wordy phrases, and awkward sentence structure. Second, it analyzes your writing style and presents statistics that tell you where your writing is weak and where it is strong. Although WStyle is not a substitute for an independent review, it helps you to write more clearly and directly and avoid common writing-style errors.

Dictionaries/Thesauri

JORJ

Spelling checkers are great for checking spelling, but what do you do when you need to know the meaning of a word? Pick up a book? Not in this modern computer age. You won't have to with JORJ, a computerized dictionary.

JORJ is a 58,000-word dictionary that you can run as a stand-alone program or as a memory-resident utility inside your favorite word processor (or other program). If you don't know the spelling of a word, or are unsure about its exact meaning, just pop-up JORJ. Type in the word, and JORJ displays its spelling, the spelling of popular forms of the word (-ance, -ous, -ly, etc.), and its meaning. You don't even have to know the correct spelling; JORJ takes your phonetic spelling and locates the correct word. For instance, if you type in "dayrigur," JORJ correctly answers with "de rigueur."

A special feature is JORJ's ability to find related words. Scan for the word "president" and get a list of presidents. Look up "planet" and get a list of the planets. Next time you're at a loss for words, let JORJ come to your rescue.

Thesaur

Searching for the right word to say what you really mean? This computerized thesaurus should help you become both more stylish and concise (make that succinct) in all your writing and prepared speeches. Add panache to your letters, eloquence to your term papers, and a Phi Beta Kappa impact to your speaking endeavors by spending just a few minutes with Thesaur. When you call up a word, you are provided an array of synonyms and related words. Set the pointer over your synonym of choice, and you get a definition to confirm its aptness.

There are more than 10,000 main words and close to 50,000 synonyms in Thesaur. The program also allows you to add your own words as synonyms, and even delete words that you think are inappropriate.

Thesaur Plus

Thesaur Plus is a pop-up thesaurus for use whenever you need a different word or a different shade of meaning than the word that comes to mind. You can pop up Thesaur Plus and get the word you wanted but couldn't quite remember.

Thesaur Plus was designed with simplicity in mind. You only need to remember your hot key (the key combination to pop up Thesaur Plus). You can look up the word at your cursor or call the thesaurus and type a word to reference.

Technical Word Processors

MathText

MathText is a full-screen word processor that's especially designed for technical typing. It manipulates lines of text, where a line consists of up to 40 overlapping rows of characters. It works with five character sets: the standard characters, italics, and three special sets, each special set containing 94 characters, all of which are user-definable. A special feature is a set of 94 internal macros, each of which can store up to 49 keystrokes for fast entry of common expressions and symbols. There are no complicated commands to embed in your document. The text is completely and continuously formatted on the screen. To enter a character as a subscript, you simply place the cursor (which moves by half-lines), and type the character, be it a Greek V or whatever. In addition, most features of standard word processors are in MathText.

The program uses a graphic screen to display the 470 possible characters, which can appear vertically by half-lines. In general, the technique used by the printer program is to print the standard characters using the font internal to the printer, and to print special characters using graphics. Thus, printing of its files involves mixing standard characters with special characters, using the graphics capabilities of the printer. Superscripts and subscripts are easy to create. Large characters are obtained by putting together graphic pieces.

RGB TechWriter Scientific Text Processor

This special software program is an easy-to-use word processor designed specifically to make typing complex symbolic expressions fast, easy, and accurate. Equations in technical documents require characters with special attributes, including superscript, subscript, and Greek and other special symbols not found on a standard keyboard. This program handles this situation by using the color of the character as it appears on the screen to convey the additional information, so there is an exact correspondence between columns on the screen and on the printed page. This simplifies lining up the numerator and denominator of a complicated expression, whereas other word processors insert control characters that make alignment very difficult.

Many standard word processor features are provided, including page numbering, headers and footers, search and replace, line centering, bold printing, underlining, right-justification, and the ability to move blocks of text. This program is also a good editor for writing source code in structured languages such as Pascal.

Individual files are limited to 62,768 characters, which is about 20 pages of single-spaced text. The software requires an 80-column color monitor and works with CGA, EGA, MCGA, and VGA. Files having a standard DOS ASCII format can be created by and imported into this program. Printer support is extensive.

Foreign Language Word Processors

Chinese Word Processor

NANJIXING. Type on your standard keyboard, and Chinese ideograms appear on screen, beautifully rendered. This program features an "intelligent" input method where, as you begin to type in the Chinese, a numbered list of possible ideograms shows up. You find the character you are looking for and type the number.

KDPlus

The KDPlus system is a collection of programs that you can use to display, edit, and print Japanese Kanji text on any EGA/VGA-equipped, IBM-compatible personal computer, mostly using existing software (i.e., software not written with Japanese-language information processing in mind). For example, the ancient EDLIN editor can edit Japanese texts (but there is also an editor in the system, JWRITE, which is much more suitable for this purpose). Probably the most important application of the KDPlus system is accessing Japanese-language bulletin boards and databanks, using existing communications programs such as TELIX or PROCOMM.

Intext

Whether you're a linguist, translator, student, or work for an international company, you're going to be glad you heard about Intext. Intext is the first shareware word processor created for worldwide use. Intext lets you create documents in English and any of the following languages: Albanian, Arabic, Croatian, Danish, Dutch, Farsi, Finnish, French, Gaelic, German, Hebrew, Italian, Macedonian, Norwegian, Polish, Portuguese, Russian, Serbian, Spanish, Swedish, and Turkish.

Intext supports all the functions you would expect to find on a full-function word processor: search and replace, text block commands, text formatting commands, etc. What makes it unique is that all the special characters and symbols for the languages are displayed on the screen and printed out on either an Epson FX, LQ, or IBM Proprinter. On-screen help is provided, and the programs support both 9- and 24-pin dot matrix printers.

Specialty Word Processing Functions

Bannerific

Bannerific is a program that lets you produce large banners on your computer printer. All IBM PC keyboard characters can be used. The size of the characters produced on the printer is variable and can be as large as the width of your computer paper. Banners can be printed horizontally or vertically. You can choose the size of your banner. Default values for the size and placement of the letters are chosen by pressing Enter at the prompts. Otherwise, enter an appropriate number for the questions: 1. The height of the letters. 2. The width of the letters. 3. The left margin on the printer, and what the height, width, and margin options do.

Gothic Banner

This software prints a Gothic banner on your line printer. Upper case, lower case, and numerals are supported. The banner can be up to 78 characters (and spaces) long.

When first run, Banner allows you to enter control codes to be sent directly to your printer to set it up for compressed print, bold face, etc. You enter whatever your printer requires to perform that function. Banner assumes that your printer can print 132 characters in a line. Most printers cannot, without being configured for compressed print. Sending an ASCII 26 to most printers configures them to compressed print. Check your printer manual. After the configuration, simply enter the line of text to be printed on the banner. Entering nothing ends the program.

Rockford!

If you need a business card, Rockford!'s your program. Using Rockford! with Microsoft Windows, you can design professional-looking business cards in just a few minutes. It's far easier to design your cards in Rockford! than to follow the traditional path of going to your printer and trying to pick out a business card design from a book.

There's even more. Using Rockford! you can print sheets of your customer's designs, then copy them onto card stock. Just cut the cards, and you can offer one-hour business card service. Rockford! doesn't take the place of a true print shop, but being able to offer your clients a quick set of 100 cards while they wait can mean the difference between a sale and a walkout.

Rockford! isn't limited just to business cards. Clubs can use it to produce membership cards. Used with readily available pin-back card holders, Rockford! produces ID cards. Many users also use it to create humorous or gag cards. Use your imagination. You can create just about any style of card with Rockford! Here's a list of some of the features of the program:

- WYSIWYG design screen.
- Horizontal and vertical card designs.
- Use any font available in Windows.
- Line, box, and circle drawing.
- Bold, italic, bold-italic, and reversed type.
- Four border types.
- Adjustable margins and border offsets.
- Text alignment tools.
- Split lines of type.
- Automatic bulleted lists.
- Insert clip art on card.
- Preview clip art before loading.
- Scale and position pictures to .01-inch accuracy.
- Insert character formats inside lines.
- Print card layouts with 200%-sized camera-ready copy.
- Print sheets of ready-to-cut cards (9 or 10 cards per sheet).
- Print four-up sets of camera-ready cards (preferred by some printers).
- Print cards on precut card stock.
- Save completed cards to disk.
- Easy-to-use interface.
- Online help.

Integrated Programs

This section covers a genre of software that has not only word-processing capability, but additional functions such as databases, spreadsheets, etc., so that interchanging data within the integrated program is much easier than when using separate programs.

SSI

This integrated program combines a word processor, a list/data manager, communications, a midsize spreadsheet (64 columns by 255 rows), and a file and directory browser. To make the program even more comprehensive, pop-up aids are a keystroke away anywhere in the system. These include a calculator, an aid for graphing data, file/directory viewing, DOS shell interface, and a tiny utility to make macros or smart keys.

Surefire

This is a word processor with a fairly new concept called "smart document processing." This allows you to build databases and spreadsheets within your word processing document to the point where it's hard to know if you are still in a word processor. The word processor contains all the standard functions: find and replace, headers, footers, page numbering, bold, underline, line drawing, foreign and special characters. The databases are dBASE III compatible, can be searched, and can have up to 254 characters per field and up to 128 fields per record. The spreadsheets are supported by 70 arithmetic, date, time, and conditional functions, along with absolute and relative cell addressing.

It has online help and a pop-up calculator, and documents can be linked to create larger databases and/or 3-D spreadsheets. Dot-matrix, laser, and daisy-wheel printers are supported.

Presentations

BitFlick

This Windows application shows bit maps in a slide-show fashion. You point BitFlick to the proper directory, and BitFlick flicks through all of the bit maps that are there. Its options include show speed, autosizing window, and a removable start/stop button. You should put a collection of .BMP files that all have the same resolution into your show directory. Bundled with BitFlick are two .BMPs that have a resolution of 150 × 200 × 256. Please note that you will have to have an SVGA to see these properly.

Blackboard

Blackboard was carefully designed to suit the largest possible audience by using the high-resolution CGA mode so that most IBM-compatibles can display your creations. Blackboard works with VGA, EGA, CGA, Tandy, laptop, and other computers. Blackboard is a multimedia program incorporating sound effects, music, pictures, and text, all in ASCII vector graphics, so large amounts of information can be stored on a single disk. Features of this ASCII-vector-graphics drawing and tutorial-creating program follow.

Authors can combine text and graphics easily on-screen in a WYSIWYG environment. Graphics files are very small. On-disk tutorials can contain more than 100 graphics screens on one 360K floppy disk. Using ASCII text files to store graphics commands and vectors means the end-user can search for any text within any picture—an automated index. Sound effects and music composition are easily incorporated into any presentation.

This is a multimedia package. Tutorials can include text, graphics, music, sound effects, and some animation. Full screen is available for drawing; the menu is another full-screen overlay. No scrolling is necessary. Some tools from sophisticated drafting programs are included, such as "grid" and "coordinates." Again, because of the ASCII file format, minute adjustments in finished products can be made.

Especially for Writers

AVScripter

If you write scripts that require separate audio and video columns, AVScripter is for you. Unlike programs that graphically create two columns on your computer screen, AVScripter works with full-screen ASCII text files created with your word processor, using simple dot commands. Working with the whole screen, you see more text on the screen at one time, you don't have to continually jump back and forth from one column to the other, and you need not worry about keeping the appropriate audio and video lines opposite each other. AVScripter does it all for you automatically. AVScripter does the following to your full-screen ASCII text file:

- Sends two-column output, single-spaced, double-spaced, or single-spaced video/double-spaced audio to your printer, your screen, or a disk file, automatically syncing audio and video lines.
- Allows variable-width audio and video columns.
- Automatically numbers and renumbers your scenes and pages.
- Looks for the best place to start each new page, then adds the appropriate "continued" where necessary.
- Automatically centers your title lines.
- Prints a header line at the top of each page.
- Lets you switch between full-width lines and two-column format within your script.
- Allows underlining anywhere, even in headers and title lines.
- Sends printer commands to your printer.
- Prints selected pages and multiple copies.
- Chains files for processing.
- Generates prompter/narrator scripts from the audio column.
- Generates scene lists and logs.

Fame

Fame is a database program for poets and writers. The program keeps track of the poems/stories that you write, the publishers that you send them to, and the status (result) of the submission. It also lets you know which poems/stories are available for submission to a particular publisher, factoring in whether or not the publisher accepts simultaneous submissions. It even writes submission, withdrawal, and query letters. Fame handles 198 poems/stories and 162 publishers.

In addition to allowing you to update the status of your poems/stories, the program has a series of preestablished queries that provide many types of summaries. The program is menu driven and very useful. Program options are:

- Add poem name.
- Change poem name.

- Delete poem.
- Add publisher name and related data.
- Change publisher name or related data.
- Delete publisher.
- Change poem status (initialized to "not submitted").
- Change date database maintenance utilities.
- Queries (display option listed first, print option after /).
- List of poems with dates written and status (sorted by date).
- Alphabetized list of poems with status of each.
- Alphabetized list of publishers.
- List of poems submitted to selected publisher, with status.
- List of publishers that were sent selected poem, with status.
- Summary of poems for each status.
- Status of selected poem for selected publisher.
- Publishers for whom poems are outstanding.
- Overall status summary.
- Candidate poems to send selected publisher.
- Create publisher mailing label file or print letters.

During initialization of the software, you can choose to have the word "story" appear in all places instead of the word "poem."

GScript

The Gadfly Script Formatter (GScript) is a suite of styles and macros for Word for Windows to help edit and format a screenplay. In the olden days, writers hired script services to prepare their manuscripts for submission. With the PC, screenwriters either settled for the best their word processor could do or purchased a second application to create a properly formatted print version of the script. It greatly reduces the effort required to enter, edit, and format a screenplay. GScript formats a screenplay with proper indents, font selection and size, and strings inserted at the top and bottom of the page (when the script is formatted for printing).

It is used in two stages: A template is a specialized Word for Windows file that contains style information, macros, glossaries, menu, key, and toolbar assignments to be associated with a class of documents (like screenplays). When you create a Word for Windows document, you base it on a template. GSCR is one such template. During the creation of the screenplay, GScript simplifies editing with specially tailored styles, key assignments, toolbar assignments, and macros. After a draft of a screenplay is complete and you wish to format the screenplay for delivery, with page breaks and continuation marks inserted, it does all the hard calculation and work automatically when you run a single macro: FormatScript.

The resulting layout conforms to the standard established ages ago (before the dawn of the personal computer) by the Writer's Guild of America and the studio system. This format is based on the formula you've heard a zillion times: One page equals one minute of screen time. That might seem like an arbitrary producerish concern, but it does possess a general truth that can be extremely helpful during the writing of the script. The font used in this template is Courier 12 point. If you are accustomed to measuring indentations in terms of characters (rather than in terms of inches), Courier 12 point is a 10-"pitch" font; there are ten characters per inch. Font: Courier 12 point, 10 pitch (that is, 12 points high, 10 characters per inch).

This is the same font you might remember as being standard on the antique IBM Selectric typewriters. Margins: left—1.25 inch; right—1.0 inch; top—1 inch; bottom—1 inch; character name: 2.5 inches (26 characters) in from left margin; dialogue: 1.2 inches (13 characters) in from left margin; parenthetical—1.9 inch (20 characters) in from left margin (the first line has a hanging indent of one character); transition 4.5 inches (46 characters) in from left margin.

You can cheat and make your screenplays appear shorter by increasing the margins for action and dialogue; you can use proportionally spaced fonts (such as the Times Roman 12 point). The reasons against doing so are twofold: No one is fooled. A script that comes in at 120 pages of 10-point Helvetica is, we all know (even producers know), actually 132 pages of script. Secondly, it helps the actual composition of the screenplay if it is formatted according to these styles and indents. The flow of the script, the tempo, and timing are simply easier to see and feel if the script is formatted in Courier 10 pitch with proper margins. You can sense how much description and dialogue is on the page.

What GScript does that other formatters do not: It makes no changes to your document that cannot be easily reversed. It does not tamper with the disk file. In fact, it doesn't even save its additions to your file unless you specify that it should, or do so manually. All other script formatters that I know of require some kind of conversion from your word processor to a format they can read and understand. Even if they can convert in the other direction, back to your word processor, something is inevitably lost in the translation.

Not so with GScript. More specifically, because GScript is integrated with Word for Windows, it possesses several advantages: Even in "draft" mode, the styles in GScript are such that you will never get "stranded" character names at the bottom of a page, and dialogue will never be broken. You don't have to create an ASCII file for formatting. GScript is a unified system capable of creating and printing both "draft" scripts and fully formatted "presentation" scripts. Other formatters either ignore, strip, or choke on some of the most useful tools found in powerful word processors. With GScript, because you have complete control over what styles are hidden, you can fill your screenplay with notes, outline levels, footnotes, annotations and the like, and still create a formatted print file without changing a thing. There is no need for two copies of your screenplay (i.e., one draft, one presentation), since GScript strips out all of the added strings and all of the altered styles.

Muses

Muses is a program for authors and their agents to organize the circulation of manuscripts to publishers. A very specialized database, it helps keep track of publishers' responses, as well as current mailings. Muses tracks correspondence with publishers and gives reminders when it's time to follow up on a submission. It also mail merges the address in the files with appropriate form letters.

Muses prints manuscripts with several kinds of headers, footers, and footnotes. It requires ASCII text. Muses also keeps a separate bibliography of published work.

Orpheus

Orpheus is for writers of online books, manuals, tutorials, or any other form of electronic documentation with a hypertext component. It consists of two programs: OH.EXE, the authoring program itself, and OHREAD.EXE, a reading program for distribution with compiled electronic books.

Unlike other authoring systems, Orpheus lets you write hypertext quickly and intuitively, without having to think about the mechanics of linking. There are no special characters to insert, and no program codes; you always see what your intended reader will see.

In most hypertext systems, the reader is expected to scroll through long text files. This makes navigation difficult because linkwords might be just offscreen. It also makes for a boring, unattractive presentation. Orpheus focuses on the screen as the medium of communication; during the authoring phase, each card has a 50-line workspace, but only the top 24 or 25 lines go into the finished work. (If you choose 24-line cards, the Orpheus Reader displays your work with a title bar at the top.) With a plethora of formatting and text-drawing tools, Orpheus makes it easy to create eye-catching, informative screens. End-users find it easy to navigate through your hypertext network because they can always see at a glance what paths are available. Thus, your end-product looks fantastic and is a pleasure to read. Orpheus offers an unbeatable combination of power, ease of use, beauty, polish, and flexibility.

Documents can contain up to 32,000 linked cards and graphics. A text card can hold up to 250 links to any combination of text cards and graphics. There are 7 link types, of which one can be reserved for the author's notes and excluded from the compiled version, while two others are intended for programmable features that will enable interactive tutorials.

PenName

PenName is a lighthearted aid to assist writers in selecting character names or pen names for themselves. The program allows you to select the category of name you want (you can pick different categories for first and last name), and PenName allows you to try different combinations.

You can also save names you like in a list of "possibilities" to be examined later. The program is distributed with a few hundred names supplied, but you will probably want to add more names to the database. Don't worry about duplications; there is a built-in duplication preventor. If you wish to change the category of an existing name, simply enter it as a new name and the categories you select will replace the previous categories.

ScreenWright

ScreenWright is a text formatting program specifically designed for screenplays and teleplays. Special care has been taken to keep the software simple to use, so you can concentrate on your writing instead of a stack of computer manuals. To use the program, you type in your screenplay using almost any standard text editor (e.g., PC-Write, WordStar, Perfect Writer, etc.). Then the program types the manuscript on your printer in perfect screenplay format. Examples are given of how you can use the software with your scripts.

Word processing for screenplays is different from word processing for many other types of writing in that an unusual and rather inflexible format is required by the film and television industries.

ScreenWright handles these special requirements quickly and easily. To use the program, you need to think of a screenplay as being made up of only a handful of basic elements:

- Numbers for new scenes.
- Directions and other descriptive matter, including setting, lighting, action, etc.
- Names of characters (speech headings).
- Speeches of characters.
- Parenthetical or descriptive remarks about a speech.
- Camera angle.

You note the beginning of each of these sections in your text with a simple notation, called a command. A command tells ScreenWright how to properly format a portion of your screenplay. Once your screenplay (or draft) is entered into a file, you can use ScreenWright to format it.

Writer's Dream

The software developers of this program have found out that most of their customers are good writers. Most also know something very useful and possibly unique. Perhaps you are a retired computer pro, or maybe you could teach people how to cook nutritious foods. Perhaps you could write a course in advanced electronics or elementary geology. Maybe you would be good at writing science fiction or children's stories. Some think that becoming a published writer is nearly as difficult as becoming a rock 'n' roll star and that self-publishing costs money and requires marketing skill.

But there is another way to be published that works for absolutely everybody. Yet, few know about it. Using this technique costs almost nothing, but literally millions of people will read your writing. You can make unlimited money. The Writer's Dream is a way to put your information on IBM-compatible disks. If you can operate any word-processing program, you can do this.

Let's say you have written a disk about programming. You release it as a shareware product. This means you send copies to some shareware distributors. There are 72 million IBM-compatible computer users in America. Almost all of them are frequent shareware buyers. Some of them will get your disk and make copies, which will, in turn, be read and copied by others. Eventually, one person out of every ten is likely to read your disk-based programming book. You, with your own little computer, will be an author with 7.2 million potential readers. That's not all. You can put a commercial on your disk that inspires some readers to send you money.

Because you can put text on a disk and therefore create information of interest to others, shareware is only one means of distribution. You might like to use it for in-house training or disk-based conferencing within your company, or you might like to create and sell a retail disk-based book product.

ALMANAC 1992 is one of the hottest-selling titles in shareware. It is not a video game or a spreadsheet. No, it is just text files. The author has also written several nonshareware electronic books that sell very well through mail-order distribution. Even though he is a programmer, his single best-selling "program" is not a program, just a text-based tutorial, an electronic book. He is not a professional writer.

Write about anything in which you have some knowledge, anything in which you are interested. Release it as shareware, and watch what happens!

Desktop Publishing

This section covers the epitome of the word-processing genre: desktop publishing (DTP). DTP differs from word processing in being able to mix graphics and text on the same page, use multicolumn layouts, cover more than one subject on the same page, and continue with the different subjects on separate columns on subsequent pages. Some of this software can only accommodate a single page; the more complex programs can accommodate an entire book, with graphics and limited color.

CopyFitIt

First, where does copyfitting fit in? Usually before copyfitting is started, you need a layout sketch—a drawn plan of how the final printed piece will look—so you'll know, for one thing, how much space each of its typeset elements take. The copyfitting process itself is described later in this section.

After copyfitting, you write your decisions regarding type size, etc., onto your typewritten text (you "spec" the "copy"). Then you're ready to have your text typeset. A professional typesetter uses a highly sophisticated machine to prepare what's often called a galley, a copy of your words set in the typefaces (alphabet styles) and format you have requested. You need to proofread the galley for errors and have any corrections set.

Then you make a pasteup, joining together these typeset words with any borders and other artwork called for in the layout sketch. Finally the printer makes a printing plate from this pasteup and uses it on a printing press to print numerous copies for you.

Essentially, copyfitting is the process of figuring out what typesetting format enables your words to fit the space allowed by the layout, using a typeface (or typefaces) you have chosen from those available at your typesetter.

Generally speaking, most copyfitting involves multiplying how much text there is, by how big a selected size of the typeface is, then comparing the result to the space available in your layout. If the result is too big, you recalculate using a smaller size of the typeface, or if it's too small you try a larger type size. The various CopyFitIt copyfitting procedures provide a number of specific ways of doing this process, and they make it faster and easier.

Introduction to Desktop Publishing and Design

This excellent book on disk is an introductory course on designing documents for desktop publishing. The course was originally a series of messages to all on the City2City network in the DTP conference, of which the author was the conference host. It was also uploaded to the NAnet DTP conference hosted by Mr. John Kerr. The course gives good, solid basics for those new to DTP who want to improve the quality of their output.

Envision

This outstanding, full-featured WYSIWYG desktop publishing program is comparable to PageMaker and Ventura Publisher. It has page layout and composition features. Text can be scaled, rotated, bolded, italicized, kerned, or filled with any fill pattern. Its modern icon menu lets you work on your document and display it just like it's going to print.

You have complete control over text blocks, text columns, line leading, and line justification (left, right, center, and full justify). You type text directly on the page or import it from a file (the program supports many word processors). A complete set of drawing tools for line, box, circle, ellipse, polygon, polyline, Bezier curve, with various line styles, colors and fill patterns are available. You can import PCX images and scale them. This program supports laser, dot-matrix, inkjet, and PostScript printers. All graphics monitors are supported.

LookBook

LookBook is an electronic book and reference system for publishing and creating a variety of disk-based documents such as newsletters, program manuals, catalogs and depositions. It also works with ASCII data files like print report files. LookBook also includes a robust complement of options and a versatile TSR utility that enables documents to support other applications.

Although the system makes quick work of informal ASCII text files, LookBook's forte is automating the "paged" text of books and manuals. It does this by creating a unique page location map that establishes "hypertext" links with a traditional table of contents and "back-of-the-book" type of index list.

LookBook makes document preparation more of a word processing task than a database task. The traditional format using ASCII text contributes to this. The traditional format also enables LookBook to present users with an intuitive interface, while the ASCII text gives users a clear idea of the text's organization.

LookBook's performance characteristics function to increase the utility of its documents through effective floppy disk use. The small space requirements—documents can be compressed to 60% of original size—pack approximately 150 pages for each 360K of disk space. The program's capacity eliminates the need for hard disk installation, except for very large and frequently used documents.

This software is a complete reference system that includes a note editor for taking page-linked notes and for editing bookmarks that remain from session to session. LookBook also provides for point-and-shoot document selection, with automatic detection of file type. For authors, LookBook provides "video aware" color enhancement through simple-to-use commands.

The Rubicon Publisher

Rubicon is a batch processor that is used to publish text files that have been previously prepared in plain ASCII format using any suitable word processor or text editor, or any other software that generates plain ASCII text files. It is a serious desktop publishing program, not really designed for beginners; but if you can use a word processor, you can install and use the publisher successfully, provided that you read and follow the instructions and pay reasonable attention to detail when preparing text files for publication.

This comprehensive desktop publisher, modeled on traditional typesetting concepts, has more than 120 features. This Postscript or LaserJet system has two LaserJet font families (Trajan and Renner) and supports many others. The batch processor is controlled by tags embedded in the input file. No mouse or graphics capability is required to run Rubicon Publisher, but an ASCII word processor is necessary. The LaserJet fonts require 1.4MB of disk space.

It enables you to produce a variety of documents—forms, notices, advertisements, newsletters, reports, financial statements, leaflets, brochures, book-length projects, etc.—with typesetting of professional quality. In short, it's a very useful program, and it's also more fun than most video games because its creative possibilities are limited only by your imagination.

Typesetter PC

This page layout program gives you WYSIWYG capability in designing a single page and prints it out in the highest resolution achievable on a 9-pin dot-matrix printer. Fonts and graphics can be mixed anywhere on the page. Fonts can be italicized, inverted, made bold, and more. A full-featured graphics program including such tools as pencil, k-line, circle, rectangle and spline, is part of Typesetter PC. There are many on-screen help menus that list these functions.

Typesetter PC is not geared to use the intelligence of more expensive laser printers; instead, it fully utilizes the computer's CPU to push low-cost dot-matrix printers to limits not realized with other products.

Most desktop publishing packages operate on systems that consist of a scanner, expensive laser printer with some type of page definition language, plus a software package that goes for just less than $1,000. These systems are capable of remarkable results. However, the total cost for such a system is often in excess of $13,000. Still, considering the capabilities of such systems, there is certainly a big market for them.

On the low-cost end, most software products either produce signs and cards or newsletters. Typically, these products are easy to use, produce attractive results, and are priced at less than $100. However, all of these products produce output that is obviously generated with a dot-matrix printer. Most of these products either provide the only artwork that can be used with the product, or they force you into purchasing additional graphics or artwork. Fonts usually are of limited resolution and varieties. While these products are certainly useful, they cannot be considered substitutes for even entry-level systems to desktop publishing.

Typesetter PC produces a product that can generate multiple fonts and graphics without relying on any other products. The most important capability of Typesetter PC is its ability to produce output that in many cases can be substituted for pages created with an expensive desktop publishing system. All that is required to achieve this capability is any MS-DOS compatible system with 512KB RAM, a graphics card, and an Epson graphics-compatible printer.

Typesetter PC creates custom-sized printouts so they can be used to create labels, business, index, and Rolodex cards. Two different printout resolutions are provided. The standard printout lets you print with 960 × 695 pixel resolution. This is greater than a half-million pixels. A double strike option produces a darker output.

One of the best features found in Typesetter PC is its ability to generate near-laser output. If the near-laser output option is selected, Typesetter PC produces a page with 1920 × 1392 pixels. That's 2,672,640 pixels. This is the absolute limit of the Epson FX 80 and compatible printers. This translates to 245 × 145 dots per inch. Other products claiming to be entry-level desktop publishing systems costing three times the price of Typesetter PC only drive the Epson to 72 × 72 dots per inch. Typesetter PC utilizes a mathematical algorithm to rescale the image in memory to produce the highest printer output. At the end of each printed line, the computer processes the graphic image to smooth out the curves, increase the resolution, and rescale the image to the printer. An additional capability found in all Typesetter PC products is that you are provided with a WYSIWYG environment.

DTP Tools

This program provides, right on your desktop, two things that are often needed for desktop publishing: a conversion between inches, picas, and points, and a quick and easy way to preview a font.

To use this program, you need the Visual Basic runtime library, VBRUN100.DLL and SPIN.VBX, which should have come in the ZIP file along with this file and the EXE file. Make sure VBRUN100.DLL and SPIN.VBX are in your path. It's best to keep them in your \Windows or Windows\System directory.

Conversions. By setting the value of the units that are known, the closest value of the other units is displayed. This value is rounded up or down accordingly. Inches are to the nearest 1/32, picas are to the nearest 1/4, and points are to the nearest whole point. Use the up and down arrows to set the known value. The rate at which the values increase or decrease can be changed by using the speed button to toggle between fast and slow.

Fonts. By choosing fonts from the options menu, you can preview a font (ASCII 33 to 126) by selecting the font name, font style, and font size in standard point sizes from 8 to 48 points. The font names are the printer fonts available. Be aware that the font shown on the screen might not be exactly what your printer prints on the paper. WYSI not always WYG. The font display window can be closed by double-clicking on it.

Rulers. These were added just for the fun of it, but they can be extremely useful, especially with Paintbrush and other drawing programs where rulers and grids aren't available. Units can be set to inches, picas, or points. The ruler windows can be closed by double-clicking on them.

Fonts

ANSIIKey

This small Windows program lets you see all your fonts, with the characters as they will appear on your keyboard. This includes the difficult characters used by symbol and dingbat fonts, so you can quickly view the characters and know which key to press to get the character you want.

You can also press any key and see a blown-up image of that character for a detailed view. Plus you can use a scroll bar to move through all the ANSI characters, from ANSII 33 to ANSII 255, to see those characters that you can't type. Naturally, the ANSII code will be right on your screen.

Then, you can use the <Alt>-number pad method to put the characters into your other Windows applications. But there's an even better way: Turn on the automatic clipboard feature, and any character that appears in the zoom window is automatically pasted into the clipboard for even easier transfer to another application.

Finally, you can save any character's image as a .BMP file for use as clip art in any Windows program that accepts .BMP files. You can keep ANSII Key Layout minimized, then pop it up whenever you need a special character. It's easy, fun, and educational.

TrueType

In most digital type systems, whether implemented on a computer or in a printer, there are just a few basic steps that the system goes through to display characters on the target device. The essential steps are:

1. Load the font outline and feed it to the rasterizer.
2. Scale the outline to the correct size.
3. Fill the outline with pixels, creating a raster bitmap.
4. Transfer the raster bitmap to the display device.

Expanding upon this list gives a more detailed look at how TrueType works. Here are the steps in order:

1. Load the font and feed it to the rasterizer.
2. Scale the outline to the correct point size for the correct resolution of the device.
3. Apply the hints to the outline, distorting the contours to build what is known as a hinted, or grid-fitted outline.
4. Fill the grid-fitted outline with pixels, creating a raster bitmap.
5. Scan for dropouts (optional).
6. Cache the raster bitmap.
7. Transfer the raster bitmap to the appropriate display device.

Step #1: Loading the font. In a printer, the loading of a font is usually accomplished by the printer's BIOS when it has to locate the font in ROM, or as a downloaded version of the font stored in local printer RAM. In a computer, this is normally accomplished by the operating system when it has to locate the font file on disk. In Windows in particular, TrueType fonts are stored as .TTF files in the system directory of your Windows directory. These .TTF are what is known in the Macintosh world as the raw "sfnt" (don't try to pronounce that if you have dentures) resources. The remainder of this disk covers all of the other steps in detail.

Screen Capture

Whenever you have a computer screen display that you would like to capture and incorporate into a report or proposal, this software captures and saves it as an image that can be converted to any of several formats.

Capbuf

This software is a memory-resident program that lets you capture any text screen and save it for later processing or printing. It is quite useful for programs that do not otherwise provide means of printing their results. Use it for intermediate results of games or collecting copies of messages while on a bulletin board. Remember that many elaborate program screens are line graphics. They, too, can be captured, then brought into an ASCII editor. Capbuf can be removed from memory once you are finished capturing screens. A second program, Filter, eliminates extra lines or line-drawing characters (especially useful if your printer does not support these characters).

GR4AI

GR4AI is an alternative version of the Grabber screen capture program. It is used exclusively to capture high-resolution 640 × 480 and 1024 × 768 graphic images from PC systems that are compatible with the IBM 8514/A adapter interface, and which are running in "real mode" under MS-DOS. (GR4AI, like GRABBER.EXE, does not capture screens from Windows or Windows applications.)

The Adapter Interface, or "AI," is a programming interface that lets application programs work with your high-resolution video adapter. The AI usually is loaded as a memory-resident (TSR) program in your computer's memory. When present, it allows GR4AI to capture certain types of high-resolution graphics.

GR4AI can only capture a screen if the foreground program that is generating the image is actually using the AI to do its screen work. Note that this is not always the case. For example, if you have an IBM 8514/A adapter, or another manufacturer's card that is compatible with IBM's 8514/A at the register level (such as an ATI 8514/Ultra, Graphics Ultra or Graphics Vantage card), then it is possible for some application programs to work with the screen by writing directly to the adapter's hardware registers and ignoring the AI completely. GR4AI cannot capture any screens from such applications because, unless the AI is being used for screen output, GR4AI has no way of knowing that the adapter is in use at all.

Therefore, if you have an application from which you want to capture screens, and the application's installation or setup allows you to choose between AI-compatible and 8514/A register-level compatible video, then you should choose the AI-compatible option.

Grabber

This screen capture program lets you capture text and graphics to a COM file which, when run, redisplays the screen. A memory-resident utility captures character-based text screens directly to plain ASCII files.

Grabber supports most video systems, including MDA/EGA/VGA and Hercules-compatible mono-graphics adapters, in the text mode. Grabber also converts captured images to PCX, GIF, and PIC formats and has hot key support.

Professional Capture System

This program has two capture programs. WinCapture captures a Windows-defined area, full screen, windows, or client area. DosCapture captures standard and extended VGA text modes and standard VGA graphics modes.

Chapter

6

Engineering and Science

This chapter covers the technical software spectrum, ranging from elementary to highly advanced engineering and scientific applications. You can pick and choose these programs as they apply to your specialties, and don't overlook the elementary programs covering your specialties. Reviews help you remember the facts you might have forgotten, and refreshers often present new topics you might never have learned completely.

General

Accurately Aim Your Satellite Dish

To use this program, you need to know the geographical coordinates for the location of the satellite dish. The better the accuracy, the more accurate the results. If you don't know the coordinates, they can be found on any topographical map. The coordinates of the nearest airport can also be used if it's within a few miles. This program has been tested throughout the northwestern hemisphere and has been found to be quite accurate. Although it can calculate dish aiming for the other hemispheres, no guarantee can be made of its accuracy. Also, the Canadian-U.S. satellites are aimed at the northern hemisphere and might not be viewable elsewhere.

This program displays the aiming coordinates for all the geostationary satellites currently in the North American section of the Clark belt. If hard copy is requested, the printout also contains areas so that technical values of each system can be recorded.

World

World provides a rapid, graphic representation of the area surrounding any location on earth. Rapid graphic location determination by several methods is provided. Tables are provided for U.S. state capitals, national capitals, and locations provided by the user. The location can be entered as a latitude and longitude. The user can select a location by typing the name, or a portion of the name, and selecting from the possibilities presented. New locations can be added to the database.

The continents of the world can be displayed. Move a pointer to a location on a projection of the world, and a new projection centered on that location is created. Distances between selected points can be determined. Storm or other element tracking throughout the world is provided. Data can be entered within the program for future viewing, and single storm or other track plotting is provided.

General Engineering

CC-SURVeyor 4

This program is a geometric design program for civil engineers and land surveyors. It includes traverse, inverse, sideshot, bearing, and distance intersection routines, area, inscribe curve, rotate/translate/scale, lot summary, radial stakeout, predetermined area, curve solutions, triangle solutions, and perpendicular and parallel offsets. A set of topo utilities allows reduction of field notes for EDM or total station topo, or stadia topo.

Road design features include vertical curves and grades and intersections, cul-de-sacs, and knuckles. Screen graphics include assignment of points to user-named layers; these might also have different line types, and they will screen plot in different colors. EGA or VGA graphics is required. CC-SURVeyor plots on Houston Instruments or Hewlett-Packard pen plotters, and on Epson-compatible 9- or 24-pin dot-matrix printers. It includes plot rotation without having to rotate the coordinate base, graphic plot preview, and pause for pen change by layer.

COGOWare

Here are two hot programs for engineers, drafters, or all who are interested in technical drawing on their computer. COGOWare is a coordinate geometry surveying program for plat designers. The main use of the program is to calculate coordinates of each point in the plane of the survey and of each point calculated by the plat designer. Each pair of coordinates is assigned a point number by the designer as the calculations progress. These designated coordinates are saved in batches of 100 points to a disk file. Any designated point can be recalled and the survey extended as the design progresses. All calculations are carried out in double precision.

PLOT

This program is a set of quick, easy-to-use tools for technical plotting. It handles rectilinear, polar, and vector plotting. After plotting, you can review a tabulation of all plotted points (except for the vector program).

Loci

This public-domain program by Greg Jackson accomplishes Root Locus design. It doesn't require graphics capability, but the program uses it if it's available. All major graphics systems are supported, including Hercules and AT&T monochrome graphics.

Root locus design is fundamental to control theory and is used by chemical, electrical, and mechanical engineers. Basically, it solves for the roots of the characteristic equation in a transfer function, then plots them in the complex plane. If this sounds like absolute nonsense, then you might as well quit right here. Features: Transfer function descriptions can be stored to disk, read from disk, and edited. It's menu driven and very simple for first-time use.

Solutions in the form of calculated poles, zeros, and loci can be stored in ASCII format for use with other programs. Hard copy of root locus plots is available for Epson or IBM dot-matrix printers. Plotter interface through TPLOT (another public-domain program) is available. Limitations: Maximum order of the polynomial is 20. Maximum number of loci is 100 × order. Root Locus analysis was developed by W.R. Evens in 1950. The program uses Laguerre's method and linear deflation for root calculations. The root calculation code is taken directly from Borland's Numerical Method toolbox.

Mechanical Engineering 3 and 4

This collection of programs is designed for computer-aided instruction, problem solving, and analysis in mechanical design and includes programs that:

- Compute circular and rectangular duct sizes.
- Analyze the forces on a continuous beam over simple supports.
- Analyze trusses and frames.
- Calculate weights of various metals, etc.

PC Stran

This is a 1-, 2-, or 3-dimensional structural analysis program. PC-Stran analyzes 2- or 3-dimensional truss/frame structures or 1-dimensional continuous beams. PC-Stran employs a matrix method of structural analysis, specifically the direct stiffness method of solution, which is sometimes called the displacement method. A truss/frame is defined as a structure whose members can be attached to joints (or nodes) where only axial forces are transmitted (a truss with frictionless swivel connections) or where bending and twisting moments are also transmitted (a frame with rigid connections).

Pendulum

The pendulum is one of the simplest, yet best known, and most used mechanical devices. Understanding in detail how it works, and why it behaves the way it does, is an important step in understanding more complicated systems. In learning about the pendulum, you also start to appreciate the need for more advanced mathematical tools to describe such systems.

With this program you can create and modify computer-simulated pendulums, move their suspension points, and see how they react. You measure the length and the oscillation periods and compare your results with calculations. You actually perform computer "experiments," but it is very important that you also perform experiments with real pendulums to acquire a good feeling for mechanical systems and for experimental methods.

One thing computer simulations allow you to do is to create conditions that you couldn't easily achieve in reality. For instance, this program allows you to change the gravitational acceleration, g, and therefore you can see what a pendulum would do on the moon without the need for space travel. You can even find out what would happen if g could become negative!

Other aspects that are not so easy to study with a real pendulum are the behavior for very large oscillation amplitudes, or for cases when the friction is very large or very small, etc. These "extreme" conditions are not only fun to observe, but they also often give you a better understanding of what is going on under more "normal" circumstances.

A section of this program shows a moving pendulum with attached force and acceleration vectors, which is a big improvement over the static diagrams you find in books. You also see a dynamic graphical representation of position, velocity, and acceleration plotted as functions of time, and you can stop the action at any instant and read all the values. Carefully examine these moving diagrams and graphs to gain a better understanding of how and why the pendulum moves as it does.

Finally, the program contains problem generators and assignments in which all the numerical values are changed for each run. You will solve interesting problems requiring measurements and calculations and see examples of how pendulums can be put to good use. You enter your answers, and the computer tells you whether they are right or wrong.

Survey 1 and 2

These civil engineering programs include a Lotus 1-2-3 template for calculating cut and fill quantities. Also included is PlatChek, which checks acreage, closure, and errors in simple surveyor's plats and legal descriptions and shows a graphical layout of plat. To use Road Model, you enter the coordinates of a setting-out station, bearing and distances, etc:, along the road. The program generates coordinates along the centerline at any regular chainage interval or at specific points.

Survey Land Yourself

You can survey land yourself—quickly, easily, and inexpensively. This simple compass and tape surveying system can plot deeds, find lost property corners and lines, calculate acreage, and write land descriptions so that you can find your boundary again at a later time. You can also establish test plots, subdivide property, lay out building foundations and gaming fields, hide and recover buried treasure, or even map a cave.

The surveying instructions (print them from disk files) are in plain English. They assume only a high-school education. No magic. The catch is that the precision is only about 99%. That is the precision you can draw on graph paper with a protractor and ruler. With BASIC you can calculate, and with CGA or Hercules, you can draw maps on the screen.

The intended audience is anyone who needs a good preliminary survey. Obvious users include buyers and sellers of land, and landowners who want to know what they own. Land use planners, developers, artists, landscapers, architects, foresters, geologists, prospectors, ecologists, hydrologists, pollution control engineers, and industrial espionage engineers also estimate with this method. And so can you, when you need to talk sense with these kinds of people.

Speleologists and other brands of -ologists use this method to measure points, lines, areas, and volumes on, above, and beneath the earth. For 2% of the expense and trouble, they get 98% of the information.

Electrical Engineering

A-FILTER

This program calculates the resistor and capacitor values for various op-amp active filters. The filters covered are: low pass, first order; high pass, first order; low pass, second order, equal component value Sallen-Key; high pass, second order, equal component value Sallen-Key; low pass, third order, equal component value Sallen-Key; high pass, third order, equal component value Sallen-Key; low pass, fourth order, equal component value Sallen-Key; high pass, fourth order, equal component value Sallen-Key.

Calculations simultaneously include seven response curves for each filter (greater than first order), best delay, compromise, flattest, slight dips, 1 dB dips, 2 dB dips, and 3 dB dips. These filter configurations can be cascaded to make bandpass filters.

AC Analyses

This program covers small signal analysis to determine the ac response of a circuit to filters, capacitors, resistors, inductors, transformers, op amps, and transconductance amps.

A Collection of Electrical/Electronic Formulas

This software contains a collection of basic electrical and electronic formulas that can be found in any electrical engineering handbook. The software was developed to alleviate the necessity of looking up individual formulas each time they are needed. This program includes the most commonly used electrical and electronic formulas. The software was designed to include the use of a math coprocessor, if one is present. The coprocessor is emulated if one is not found in your system.

The electronics section of the program calculates various electronic-related formulas including Ohm's law (dc and ac), coupled inductance, impedance, resistance total, capacitance total, resonance, admittance inductance total, susceptance, mutual inductance, and conductance.

The electrical section of this program calculates various electrical-related formulas, including: efficiency, fuse and wire size, power factor, conduit and junction box size, amperes from horsepower, wire resistance and voltage drop, amperes from kilowatts, kVAR capacitor sizing, amperes from kVA, supply system conversion, kilowatts and kVA, and a FLC and wire ampacity chart.

The Number Systems evaluates and changes a quantity in one number system to an equivalent value in another system, including decimal (base 10), binary (base 02), octal (base 08), and hex (base 16). The World Wide Voltage section provides a current list of the world's electric power supplied to consumers. The list covers a wide range of countries and includes frequency, system (wye/delta), single-phase voltage, and three-phase voltage, when available. Two types of calculators are available: an algebraic and an RPN (reverse Polish notation) style.

AutoSkem

This general-purpose drawing program has been optimized for schematics. Drawings are constructed from three basic objects: lines, text, and symbols. All lines are straight lines between any two points in the drawing area. Text can be scaled to one of five sizes and placed anywhere in the drawing area. Symbols are any collection of up to 30 lines, which can be defined by the user. They can be scaled in size, rotated in 90-degree increments, and placed anywhere in the drawing. The program comes with about 40 predefined symbols and a sample drawing that shows them.

Electrc2

Electrc2 performs the most common electrical contracting calculations such as:

- Power/voltage/current/resistance of resistive loads.
- Lighting fixture calculations based on zonal cavity method.
- Determine conduit size for combination of different wire sizes and types.
- Calculate voltage drop for various loads, voltages, and run lengths.
- Calculate required wire size based on allowable voltage drops.
- Determine the cost to operate various loads at different kWh charges.

The operation of the program is pretty much self-explanatory. All of the subfunctions can be called from the main menu, and after each individual calculation, you are prompted to choose the next course of action.

The Electrical Reference Database

This database gives electricians, engineers, designers, and panel shop operators a place on their computers where they can find those little notes that only experience produces. In this case a professional has given you his experience and created a forum for you to add your own knowledge. Databases include:

- dc motors.
- Single-phase motors.
- Three-phase induction motors.

- Three-phase wound rotor motors.
- Three-phase synchronous motors.
- NEMA data.
- Motor frame sizes.
- Three-phase conductor ampacities.
- Single-phase conductor ampacities.
- Maximum conductors in conduit.
- Three-phase transformers.
- Recommended copper wire and transformer sizes.

The present data is referred to as "seed" data just to get you started. You can add and delete data as you like. Each database can handle a maximum of 2.1 billion entries.

Electrical Engineering 1, 2, and 3

These programs cover a wide range of applications: display loudspeaker frequency response curve, perform delta-wye conversions, make common electrical calculations, display and manipulate spectral data, cover dc circuit analysis, and troubleshooting of an audio amplifier.

Electron

This software is great for electrical engineers, technicians, and hobbyists. Electron solves electrical equations, performs various electronic and electrical calculations, and displays many different diagrams of motors, transformers, etc., on a graphics monitor. Electronic calculations are available for dc, Ohm's law, parallel resistor, resonant frequencies, decibels, and resistance for ammeters. Electrical calculations are supplied for true power, watts, ohms, rpm, and frequency. Various tables, such as a standard wire size table, are also provided.

Eval

This program is a mathematical expression evaluator that's capable of evaluating virtually any formula, including those with trigonometric functions and user-defined variables. Formulas are typed in at the command line, and the formula and outcome (and optionally the variable values) are immediately displayed on your screen or can be sent elsewhere via standard DOS redirection.

Ferrite.COM

The Ferrite.COM program calculates the inductance of windings placed on an iron-powder toroidal core. Inductance-versus-turns data is presented in tabular format on the screen and can be printed if the printing option is selected. FERRITE.DOC and CHART1.TXT are text files that provide online documentation and a reference chart of maximum windings that can be wound on the core.

Once all parameters have been entered, a table of inductance versus turns is calculated and displayed on the screen. Also, the printer prompt appears on the screen, and a printed copy can be obtained.

Network

Network is an electronic circuit analysis program that analyzes ladder networks. Ladder networks are combinations of components that are "chained" together in a "ladder" format. Many circuits, such as filters and matching networks, can be represented as a ladder topology.

Network has been optimized for use by working RF engineers. For example, many circuit analysis programs provide output data in the form of voltage and current. While this might be useful in a general sense, it might not be in a form that is desirable for RF designers. Network, on the other hand, provides the following output data in either normal or scientific notation:

- Insertion loss (dB).
- Phase angle of insertion loss (degrees).
- Return loss (dB).
- Voltage standing-wave ratio, VSWR.
- Reflection coefficient, rho.
- Real component of the input impedance, Zin(R).
- Imaginary component of input impedance, Zin(I).

In addition, you can tabulate this output data either to the screen or to both the screen and your printer. You can also plot the data graphically to the screen. If your disk operating system (DOS) includes the Microsoft program GRAPHICS.COM, you can dump the resulting high-resolution plots to an Epson-compatible graphics printer.

PCBreeze

This program is a sophisticated, yet easy-to-use, tool for designing printed circuit boards (PCBs). It lets you create, view, modify, and output any one- or two-layer PCB of up to 300 square inches (using a 50-mil grid system). It has two text layers for each PCB layer and a common silk-screen layer. Most operations can be performed directly through the keyboard, or by pop-up menus. Microsoft-compatible mice are also supported.

Simply specify the component pads and use the cursor keys to draw the traces. Pan and zoom to any part of the drawing. Move, copy, or delete any section of the artwork. Sections can be saved to disk to create libraries to use repeatedly. Design is simple with the use of auto-routing facilities that allow an entire board to be routed by PCBreeze alone, or interactively edited and auto-routed to obtain precise control over the artwork generated.

Finally, the results can be output to a plotter or Postscript printer with the included plotting utility. PCBreeze also produces prototype-quality plots on an Epson FX/LQ or compatible printer.

PC-Schematic

This software lets you draw electrical schematics, create printed circuit artwork, draw electronic timing and clock diagrams, block diagrams, and more. After starting the program, you are shown an empty drawing board. To start drawing, simply use the cursor keys to draw lines or select any of the standard characters or electrical components from the menus.

Use the standard components (336 are included) or create your own; PC-Schematic lets you store up to 4096 components that can be used again and again. You can also copy, move, and automatically repeat components. You can even work on two drawings at one time.

When you're finished drawing, you can save to disk or print to a standard dot-matrix printer. Ample documentation and on-screen help get you started; you decide where to stop.

PC-ECAP

This is an easy-to-use ac circuit analysis program that analyzes circuits consisting of resistors, capacitors, inductors, transformers, transistors (both bipolar and FETs), operational amplifiers, and transconductance amplifiers. Totally integrated in one package, it contains a full-screen text editor for preparing analysis input, an electronic circuit analyzer that calculates your circuit's frequency and phase response, a high-resolution display plotter that supports all common graphics adapters, and a high-resolution plotter that's compatible with most printers. Features:

- Calculates magnitude, phase, group delay, impedance, VSWR, and return loss.
- Completely menu-driven and very easy to use.
- Can analyze circuits with up to 40 nodes—no limit on the number of components.
- Built-in full-screen editor with online help function.
- Free input format—common abbreviations such as P, U, N, M, K, MEG, and G are recognized. For example, if your circuit has a 47-pf capacitor, all you need to enter for its value is 47 pf.
- Supports CGA, EGA, VGA, Hercules, MCGA, AT&T, and even the monochrome display adapter.

All calculations are accomplished in 15 digits of double precision. High-resolution plots can be printed on IBM/Epson 9- and 24-pin compatible printers and HP LaserJet+. The program can be configured to send its output to LPT1, LPT2, LPT3, or a disk file. With the bipolar transistor model maker, just enter your transistor's operating parameters, and PC-ECAP creates a hybrid-pi model for your transistor. PC-ECAP will also utilize a math coprocessor if available, but it does not require one.

Professor PC Laptop!

Professor P.C. Laptop! is a comprehensive tutorial system. It provides laptop, notebook, and portable computer owners (or those interested in buying a laptop or notebook) hundreds of powerful tips, tricks and reference articles. Tutorials include:

- Travelling with laptops and notebooks.
- Clever equipment-purchasing strategies.
- Battery care and maintenance.
- Telephone and manufacturer support numbers.
- Software selection.
- DOS lessons.
- History of computers.
- Hard drive secrets.
- Laptop batch files.
- Modem use.
- Little-known CompuServe resources for laptop owners.
- Custom reading list.
- Computer virus protection methods, and more.

It's an amazing collection of laptop and notebook computer reference material.

The program contains a high-speed search system, the capability to turn sound on/off, tutorial printing, help screens, and color or monochrome light-bar menus. This software has no special requirements: It works on floppy or hard-drive-equipped computers, color or monochrome, and all screen resolutions (Hercules, CGA, EGA, and VGA).

Calculators

CMPCALC3

This program converts your PC temporarily into a programmable calculator for complex numbers, having real and imaginary parts. It uses reverse Polish notation (RPN), like Hewlett-Packard calculators. Complex numbers and functions become as easy to handle as real ones.

In addition to addition, subtraction, multiplication, and division, (+,–,*, and /), CMPCALC3 has complex exponential, complex logarithmic, complex trigonometric, complex hyperbolic, and a complex square root function.

It also has 9 labeled storage areas, which are shown on the screen, and 41 numbered storage areas, which are not. All stores can be saved in a .CLC file. A labeled store, whose label begins with POL, displays its data in polar form. CMPCALC3 can also be used very efficiently for repetitive calculations because it memorizes programs of up to 50 commands and saves the programs in .CLC files.

DOS-Calc

DOS-Calc is a fully operational desktop calculator program for your computer. It's actually easier to use than a standard calculator. DOS-Calc has many of the functions a high-priced calculator has, but best of all, it has a tape printout. This allows you to go back and review your previous calculations. Because DOS-Calc is just like a desktop calculator, you probably won't even need to read the manual. DOS-Calc is a program that you will probably use on a daily basis. A comparable calculator would cost many times the price of this program.

DOS-Calc has a unique unit conversion feature. Not only are there more than 400 unit conversions already established, you can add nearly 800 more. The program handles up to 1200 conversions in all. It also has a built-in table feature.

This software handles up to 100 tables. Each table can have a total of 1200 entries. The table feature is similar to the six constant register values DOS-Calc maintains, except that the table feature allows a description to accompany the value.

A number of tables are included with DOS-Calc and are directed towards technical interests. You can create tables to suit your own personal needs. For example, tables for a budget, price lists, cost standards, performance data, accounting information, etc. This feature allows you to exit to a DOS shell so you can run another program without leaving DOS-Calc. An added bonus to this feature is that you can specify a particular executable file name in a text file, and that program automatically runs when you select shell.

Reliability Calculator

This program was developed so your PC could assist in the lengthy calculations required to play "what if" calculations with such things as temperature acceleration factors, number of days required to demonstrate an MTBF (mean time between failures) with a given number of units, the number of units required for a given number of days, and what the MTBF of a product is, based on the actual number of units tested for a specific number of days at a given temperature. Menu selections are:

CALCULATE PRODUCT MTBF. This option calculates a product MTBF based on the number of units tested, number of days tested, test chamber temperature, actual number of failures (N + 1), and chi-squared confidence level.

CALCULATE NUMBER OF DAYS TO TEST. This option estimates the number of test days required to demonstrate a product's MTBF when the number of units to be tested is known, the MIL-HDBK-217 calculated MTBF is known, and test chamber temperature, estimated number of failures, and the desired confidence level are known.

CALCULATE NUMBER OF TEST UNITS. This allows you to estimate how many test units will be required to demonstrate a product's MTBF if you have a limited time to do so. Required information is: calculated MTBF (MIL-HDBK-217), number of test days available, temperature of test chamber, expected number of failures, and confidence level.

CALCULATE ACCELERATION FACTOR. This option uses the ARRHENIUS equation to calculate the acceleration effect on components as a function of temperature.

Stathelp

This software consists of three independent programs: WATSTAT, a clever guide to what statistical procedures are appropriate for your data; LOOKUP, a probability calculator for z, t, F, chi-square, and binomial tests; and RANDO, a pseudorandom number generator for uniform and normal distributions plus exotica-like lottery numbers, random binary, and random integers.

All three programs are menu-driven with context-sensitive help. WATSTAT also has extensive documentation and a terrific tutorial that leads you through the steps toward choosing an appropriate statistic. A great package for statistics students and teachers.

RPNCalc

RPNCalc is a program that allows your computer to function conveniently as a scientific calculator using reverse Polish notation. It can be used in two ways. As an ordinary "transient" program, it can be invoked from the DOS prompt. It can also be installed as a terminate-and-stay-resident (TSR) program, and it can subsequently invoked from any context by pressing the hot-key combination. (As a TSR, it occupies about 40K of memory.) RPNCalc has the following features:

- Scientific notation or fixed decimal display.
- Trigonometric functions and inverses (degrees or radians).
- Logarithmic functions and exponentials.
- Additional mathematical and stack functions.
- 6-byte floating point (about 11-digit internal accuracy).
- Pi key (3.14159 . . .).
- TSR can be removed from memory without rebooting.

XACT

XACT brings you a trio of programs emulating three top Hewlett Packard hand-held calculators—the HP-11C Scientific Calculator, the HP-12C Financial Calculator, and the HP-16C Programmer's Calculator. All three programs have approximately 70 function keys laid out in the standard HP format. The prefix key usage is the same on the computer and the calculator.

Zyac

Zyac performs all common math functions, from addition and subtraction to exponentials, square roots, and logarithms, to trigonometric functions including hyperbolics, in radians or degrees. Statistical functions (mean, standard deviation, population variance, etc.) and metric-to-English conversions are also part of its functionality.

Zyac uses the entire 80 × 25 computer screen and has online help and a menu, so it is much easier to read and use than a normal hand-held calculator. The system has two memory registers and two programmable function keys, making it one of the most comprehensive calculator programs available.

General Mathematics and Statistics

Al-Jabr

Al-Jabr is a multipurpose mathematical program that can be used as a sophisticated calculator with multiple-graphing capabilities. The program was written with college students in mind and others who would appreciate some assistance on their mathematical problems but cannot afford the more sophisticated symbolic packages.

Al-Jabr supports all of the common arithmetical operators: addition, subtraction, multiplication, division, exponentiation, modular operator (a mod b divides a by b and returns the remainder), integer division (a\b [or a div b] returns the largest integer that is not greater than a/b), a unary minus sign (as in —51) simply returns the negative of the value to the right. Most of the common mathematical functions are supported as well: sin x, cos x, tan x, sinh x, cosh x, tanh x, sqrt x, log x, ln x, exp x, abs x, asin x, acos x, atan x, and x.

Curvefit

This is a compiled BASIC program for computing the goodness of fit of 25 equations to sets of X, Y data (up to 255 data pairs). The coefficients of the equations are computed, along with the Coefficient of Determination for each of the applicable equations.

Routines in Curvefit allow easy additions, changes, or deletions to data. The program selects the curve of best fit based on the value of the Coefficient of Determination. Display of residual values (entered value of Y minus value calculated for selected equation) is supported. Data can be stored to disk (X and Y data pairs). Also, the predicted values for Y and the associated X values can be stored to disk. Data are stored to disk in a format that can easily be imported by 1-2-3 or compatible programs.

DAN

This software allows entire data files to be used as variables in formulas—a great idea. You then can easily list, print, or plot the "manipulated" data in a fraction of the time it takes by conventional means.

DAN comes with 21 sample algorithms, ranging from the mundane (gas and electric bill data), to the sophisticated (biorythms, three-dimensional equations, and electrical filters). This is a top-flight, well-documented package for math students and teachers, engineers, and scientists. You can combine math operators, including log and trig functions, with constants and data file names in algebraic expressions. Graphs can be plotted in linear, log, or polar coordinates. A wide range of printers is supported.

DataPlot

DataPlot is a general-purpose, two-dimensional data plotting system specially designed for scientific and engineering applications. Its simple menu-driven user interface allows a quick and effective method for scientific data analysis and presentation, yet is powerful enough to produce publication-quality graphs and documents. Here are some of the basic features of the plotting system:

- All functions and commands are menu-driven.
- Full-function graph setup editor provided to define axes, grids, scales, title, labels, and plot definitions.
- Supports many graph types, axis styles, and plot styles.
- Graph setup can be saved on disk for later use.
- Graph types include linear plot and single- and multiple-cycle logarithmic plot.
- Pt styles include scatter, continuous, step, and needle.
- Puts up to 10 curves on a single graph.
- Supports most common graphics adapters and HPGL-compatible plotters.

Differential Equations and Calculus

These programs visualize and solve high-level mathematical equations, conveniently graphing them on your EGA monitor. Great for students, scientists, or anyone who just enjoys math.

The Differential Equation program can solve a variety of both ordinary and partial differential equations. You can also integrate factors of an equation and create a graph of power series solutions.

Use the Calculus program to find the derivatives of an expression in x and/or y. Other features include graphing functions of y = f(x), graphing first and second derivatives, solving and graphing parametric equations, finding the integral of simple functions, and finding the surface contour of a function z = f(x,y).

Epistat Statistics Package

This package contains a set of routines for use in the analysis of small data sets and is meant to be used by a person well-versed in math and computer operations. This is a fairly complex disk and is not recommended for the casual user. But for those with an interest, it can be of great help.

E.Q.Filer

This program is for anyone who uses equations. It provides quick and easy access to formulas with clear and concise prompts. Before using any formula, the system displays the relevant equation and any descriptive notes. E.Q.Filer then goes on to accept data, with prompts that are input by you, and solves the equation. Any variable can be changed with instant results after the initial solution.

Equations are input in single line format and checked upon entry for obvious errors such as unbalanced parentheses or missing operators. After accepting an equation, E.Q.Filer accepts prompts, which are shown when the equation is executed, for each variable in the equation. An allowable range can be specified for all input variables, and a fixed decimal can be established for the resultant. Using simple or highly complex equations was never this easy.

Equator

The Equator is an equation-learning aid. This product helps you learn the function, use, and application of various general-purpose formulas used in the fields of math, finance, and science.

This product consists of 35 equations, divided equally among the three subjects. Each equation appears on its own screen or worksheet. The Equator lets you experiment with each equation as much as you like. The more you work with an equation, the easier it is to understand, remember, and use it at home or at school. As the actual equation is shown on your screen, you type in your numbers and let the Equator compute the right answer. Entering the numbers is easy; telling the Equator to compute the answer is even easier. Everything you need to know about using the Equator is on the screen at all times.

Graphiti

Graphiti is a mathematical function plotter; the display can be two-dimensional or three-dimensional, and can show polar functions, based on your input. Functions must be in one of the following forms:

x = f(y)	2-dimensional vertical angles specified in radians
y = f(x)	2-dimensional horizontal specified in radians
z = f(x,y)	3-dimensional specified in radians
r = f(p)	polar specified in degrees

Functions. Absolute value, arc cosine, arc sine, arc tangent, smallest integer not less than input value (rounds up), cosine, hyperbolic cosine, cosecant, cotangent, cube root, E to the power of input value, largest integer not greater than input value (rounds down), rounds down (same as FLR), natural logarithm, natural logarithm (same as LN), base 10 logarithm, SEcant, sign of input value (positive = 1, negative = –1, zero = 0), sine, hyperbolic sine, square root, tangent, hyperbolic tangent.

Laser Graph

Laser Graph is a group of programs useful to engineers for producing high-resolution data plots on an HP LaserJet II printer. The types of plots Laser Graph can produce are: single line, multiple line, three dimensional perspective, X-Y plots, and scatter diagrams. It prints with a resolution of up to 2000 × 3000 dots (pixels), depending on your printer's memory.

All of the elements of a really good data plotting system are here. It does require patience and diligence to become familiar with how it works, but Laser Graph produces stunning results on your laser printer. All plot commands can be entered on the DOS command line. This allows a string of plot commands to be put in a batch file for execution without requiring constant user intervention.

LSTSQR

This software carries out a least-squares fit of data to a variety of functional forms, including linear, logarithmic, polynomial, and exponential. It allows multiple fits of the same data using the same or different functions without exiting the program. LSTSQR87 supports the 8087 math coprocessor.

The program has many measures of quality-of-fit, including standard deviation and the 95% confidence interval. The program also graphs and prints the fit if desired. You can input data by keyboard or from data files.

Math and Stats 2, 3, and 4

This series of programs computes mean, median, variance, and standard deviation, has a statistical calculator for small data sets and has Stat-SAK, a program that includes Normal, t, chi-square, f, binomial, poisson studentized range distributions, etc.

Mathplot

Mathplot is a set of graphics-aided programs that helps scientists and engineers analyze data. Routines are included for least-squares fitting polynomials to sets of data points, solving simultaneous linear equations, numerically integrating a curve defined by a set of data points, calculating simple statistical parameters for data sets, and for calculating the coefficients of a Fourier series for a periodic function described by a set of data points.

Programs are also included for calculating and plotting Fourier spectra and Bessel functions, for plotting data sets, and for plotting and comparing user-specified functions with data sets. Screen graphics capabilities are used by most of the programs, and the host computer must be equipped with a graphics card and monitor (CGA, EGA, VGA, MCGA or Hercules). Mathplot detects which graphics hardware is present and provides the appropriate drivers.

Math Tool and Convert

MTool is an easy-to-use mathematical tool used to define and analyze functions of a single variable. The speed, power, and ease of use make it ideal for science, engineering, and math students.

This program evaluates a function for a range of points or for a single point of the independent variable. It numerically evaluates derivatives and integrals. It finds solutions for the function and plots the function using CGA, EGA, VGA, or Hercules graphics.

Convert is a conversion calculator that accepts arguments from the command line and shows a list of equivalent quantities of different units (English versus metric).

This program is very useful because it converts a variety of units concerning distance, area, volume, time, rate, heat, energy, and force.

Matrix Calculator

This software operates on 20 matrix areas, numbered from 1 through 20. Each matrix can store 150 elements, the product of the number of rows with the number of columns. Because the matrices are stored in contiguous blocks, if only every other matrix (say numbers 1, 3, 5, etc.) is used, then up to 300 elements per matrix can be stored. Storage of 450 elements per matrix is possible if only areas 1, 4, 7, etc., are used; 600 elements can be stored if only 1, 5, etc., are used, and so on.

Because the program makes no boundary checks, the 151st element of a matrix is placed in the first position of the next matrix. Only one operation at a time is performed; i.e., only one multiplication, one addition, etc. Results can be displayed at any time using the print command. The storage limit of 20 matrices can be extended by using the disk for additional storage and retrieval.

The idea behind Interactive Regression is that the researcher knows his or her data best. Rather than provide canned variable selection routines, the researcher is prompted for the next step each time the model changes due to the addition or deletion of a variable. At any particular step, variables that are in the equation can be deleted, or variables that are not in the equation can be added.

Mercury

This program solves equations similar to Borland's Eureka. Easy-to-use, interactive, and powerful, Mercury evaluates mathematical expressions, solves for the roots of an equation, solves a system of equations, and maximizes or minimizes a function, with or without constraints. You can also evaluate derivatives and definite integrals, plot one or more functions, and print a report or a graph. Mercury has a built-in editor, pull-down menus, online help, and all the conveniences necessary to make it accessible to computer novices.

Metric-X

Metric-X is an extremely easy to use, very accurate program that converts metric units to inch/pound units, and vice versa. This utility program is DOS-based and has been engineered using ANSI/IEEE Std 268-1982 and FED Std 376B as a conversion factor base (for mathematical equations); the program also conforms to ASTM E-390's specification, except for differences in spelling, i.e.: "ter versus tre," as in meter versus metre, liter versus litre, etc.

This program uses drop-down menus to display/select any of 12 unit categories, which include: length, area, volume (dry/liquid), force, mass, pressure, temperature, energy, power, velocity, acceleration, and wavelength/frequency. The total number of "units of measure" that are supported (including all prefixed base-units) is 179. Several "unique features" set Metric-X apart from the other metric conversion programs. All conversions are processed using the powerful technique of "double-precision calculation," which accounts for the program's high degree of accuracy.

Metric-X has been designated as an "accepted metric conversion tool" by the U.S. Interagency Council on Metric Policy's Metric Operating Committee (ICMP/MOC), and has been submitted for review by: the Institute of Electrical and Electronic Engineers, Inc. (IEEE); American National Metric Council (ANMC); United States Metric Association (USMA); National Institute of Building Sciences (NIBS).

NLin

This is a computer program for nonlinear curve fitting. It requires a color graphics adapter (CGA) and a math coprocessor chip. Fitting functions to data is a frequent task in science and engineering. Implementing a linear fit is a simple task; however, powerful nonlinear fitting programs are not common, even on mainframe computers. NLin uses the Simplex algorithm, coupled with an adaptive Runge-Kutta integration routine, to fit complex functions efficiently on a microcomputer.

NLin fits multiple nonlinear algebraic or differential equations simultaneously to experimental data. Several repeated experimental runs can also be fitted simultaneously to take into account the natural variances involved in the experimentation.

In particular, the capability of NLin to fit models consisting of simultaneous differential equations has profound usefulness in the field of chemical and biochemical kinetics. In contrast to traditional programs based mainly on the Marquardt algorithm, NLin does not require either analytical or numerical derivatives. This greatly increases the number and complexity of models that can be easily fitted.

Graphical output is provided by NLin for visual assessment of fit. In addition, various statistical measures are computed to help assess the suitability of the model and the quality of fit.

The program is entirely menu-driven, and context-sensitive help screens are instantly accessible from any part of the program. A number of examples and tutorials are supplied so that within an hour a user should be able to fit complex functions using NLin.

Polymath

Polymath is a language developed for number processing. Much like Forth, Polymath is highly modulated and based on function words. Most words are defined with very simple tasks. These simple words are then used and integrated with other words to build more powerful word functions. An entire programming dictionary of function words can then be developed, which is actually custom created by the author. Specifically, Polymath has a variety of applications and special uses. It has a strong online documentation application that takes the first-time user through a menu-driven tutorial. It also has explain files that explain the syntax and use of any word currently in its main dictionary.

Polymath as a function processor works much like the Hewlett Packard programmable calculators. Although a language, Polymath can be used just as a calculator. Those familiar with the HP reverse Polish notation (RPN) know how simply complex formulas can be entered into the calculator. Polymath uses this same scheme, but with defined words or functions. Polymath has a number stack and an alpha stack that revolves numbers through it as new computations are created and old values are stored. Included on this diskette are a few dictionaries and application programs. It is best to view these application programs with a text editor; they all end in ".pm" and are very informative when learning the language. Polymath also has full graphic features to well round out this language package.

SPlot

This program plots two- and three-dimensional data to the screen of a PC with CGA, EGA, VGA or Hercules graphics. The plot can be sent to an IBM/Epson-compatible printer or a DHPGL-compatible plotter.

SPlot produces these types of plots: line, smooth curve, scatter, step, or histogram of up to 16,000 data points. Up to 200 sets of data can be plotted on the same two-dimensional plot; multiple data sets can also be plotted as a three-dimensional. Axes can be linear, log, or normal probability, with auto or manual scaling. Data can be fit by a spline or a least squares polynomial or exponential.

The program creates the plot by executing a set of plot commands that can be typed at the keyboard, selected from a menu, or read from a command file. On most PCs, the program detects the type of graphics card and uses the highest resolution available.

Universal Converter

No instruction is needed to use this program. If you get in trouble, the [Esc] key gets you out of almost anything. Experiment with all the options on the menu to see what it can do. If you use a mouse, generally the left mouse button selects an action (same as pressing Enter) or answers a Yes/No question affirmatively. The right button (same as Esc) abandons an operation and answers questions negatively.

The program is also Windows friendly. Because it is relatively small compared to many other programs used from within Windows, the Universal Converter requires little RAM space and swaps very fast. Even though it is a large program relative to the DOS 640K RAM limit, the Universal Converter can go TSR in some situations. There are two limits: You cannot pop it up from within a graphics-based program, and you cannot pop it up from the DOS prompt. If you have two megabytes of RAM, the converter uses it for faster service, assuming your extended RAM is not already occupied with SMARTDRIVE.SYS or other programs.

Astronomy Software

Astromeeus

This is a Windows-based astronomy ephemeris program. In addition to producing the traditional sun, moon, and planets phenomena charts, Astromeeus produces a monthly calendar of astronomical events with sunrise/set, moonrise/set for each day, the phases of the moon, and the time of any equinoxes, solstices, or eclipses. The calendar uses Windows vector fonts and can produce some very attractive output that's suitable for hanging on the wall.

Astromeeus is based on the book *Astronomical Formulae for Calculators* by the renowned Belgian astronomer, Jean Meeus. The calculations in his book yield very good results and apply for a wide range of dates, not just for the twentieth century.

The easy-to-use Windows interface provides: hypertext help, list boxes with more than 150 North American cities to set the longitude/latitude configuration, scrolling output, colorful icons, dialog boxes for converting Julian dates to calendar dates, and many other functions. Calculations and charts displayed on the scrolling window can be saved to an ASCII file.

The following calculations are available (Julian dates must be after 4713 B.C.):

- Position of the sun (apparent or geometric).
- Times of sunrise, transit, sunset, and twilight (Civil, nautical, or astronomical).
- Equation of time.
- Equinox and solstice times.
- Position of the moon.
- Times of moonrise, transit, moonset, and phase.
- Graphic display and readout of the position angle of the moon's bright limb.
- Times of full moon, first quarter, and last quarter.
- Eclipse times.
- Positions of the planets (heliocentric and geocentric).
- Rise, transit, and set times for the planets.

The phase or illuminated fraction for each planet will also produce a monthly calendar of astronomical events with sunrise/set, moonrise/set for each day, the phases of the moon, and the time of any equinoxes, solstices, or eclipses. The calendar uses Windows vector fonts and can produce some very attractive output, suitable for hanging on the wall.

Astronomy

These computer programs were created to help explain fundamental concepts of astronomy and to help amateur astronomers and casual observers locate the sun, moon, planets, comets and other celestial objects in the skies.

KEPLER is a tutorial program that solves Kepler's equation for elliptic, parabolic and hyperbolic orbits (public domain).

SIDEREAL is a tutorial program that introduces the user to the relationships between Julian and calendar dates and solar and sidereal times (public domain).

J2000 is a utility program that converts stellar positions, proper motion, parallax, and radial velocity from the standard epoch B1950 (FK4) to epoch J2000 (FK5). This method is described on page X of the 1985 edition of the Astronomical Almanac (public domain).

GALILEAN is a tutorial program that determines the position of the Galilean satellites relative to Jupiter. This program explains how to construct position diagrams like those found in the Astronomical Almanac and astronomy magazines (public domain).

The next set of astronomy programs requires the following inputs from the user: latitude and west longitude at the observer's location, calendar date and local time during the observation, and time zone and daylight saving time. Each astronomy program provides the following information for observation sites anywhere in the world: Julian Date and local sidereal time, right ascension and declination, azimuth and elevation (altitude), heliocentric and geocentric distance of objects within the solar system.

SOLAR determines the position of the sun. The elevation and declination angles computed by this program are corrected for the effect of atmospheric refraction.

LUNAR determines the position, semidiameter, and phase of the moon. This program compensates for the effect of flattening or "oblateness" of the earth at the observer's location.

PLANET determines the position, semidiameter, and phase of any planet of our solar system. This program also determines the heliocentric and geocentric distance of a planet.

COMET determines the position of comets. This program also determines the angle between the sun and the comet as seen by an earth observer. The user can also input orbit data for any periodic comet, and the program determines its location.

STAR determines the location of stars and other stellar objects. This program predicts the apparent right ascension, declination, azimuth, and elevation (altitude) based on either a 1950 (FK4) or 2000 (FK5) reference. The observer's local time of rising, setting, and meridian crossing of a stellar object can also be computed by STAR. The stellar position is corrected for the effects of precession, nutation, annual aberration, and atmospheric refraction. The user can also compute the right ascension and declination of an object from the azimuth and elevation of an observation. This observed position is corrected for the effect of atmospheric refraction.

ECLIPSE determines the characteristics of lunar eclipses. The user specifies a month and year, and the software determines if an eclipse will occur. This program provides the user with the type of eclipse, the observer's local time of maximum eclipse, and the eclipse magnitude. This program also provides the observer's local time at the beginning and end of each eclipse phase.

Astronomy and Aeronautics

This disk contains several useful programs relating to astronomy, earth satellites, and aeronautics. KEPLER is a tutorial program that solves Kepler's equation for elliptic, parabolic, and hyperbolic orbits. SIDEREAL is a tutorial program that introduces the user to the relationships between Julian calendar dates and solar and sidereal times. J2000 is a utility program that converts stellar positions, proper motion, parallax, and radial velocity from the standard epoch B1950 (FK4) to epoch J2000 (FK5). GALILEAN is a tutorial program that determines the position of the Galilean satellites relative to Jupiter.

SYNCSAT and TNODE let you determine the location of several earth-orbiting satellites. This information can be used for radio tracking, communications, and visual observations of earth satellites. ATMOS is a utility program that determines properties of the standard atmosphere. ROCKET1 determines the flight performance of a single-stage model rocket. GLIDER determines the maximum range and maximum endurance flight conditions of boost and rocket gliders. This program also applies to sailplanes, hang gliders, and radio-control gliders.

Astronomy Catalog

Star Catalogue lists the named and designated stars of the 88 constellations, generally of the sixth magnitude or brighter. Others of interest are included along with dimmer stars within ten light years of earth. About 1,700 stars are included in the data files. The statistics on each star include name, magnitude, spectrum type, distance, and other information. It also includes cross-referenced stars, whether the star is a binary, variable, or erratic variable, and the period of variation.

Constellation 88 presents a diagram of each of the 88 constellations, with their general coordinates, right ascension, and decline. Solar System gives essential facts and figures on the planets, moons, asteroids, and satellites within our solar system. Information includes distance, diameter, rotation, revolution, orbit, eccentricity, and much more.

Astronomy Notes discusses general information concerning stellar astronomy such as spectral classes, Hubble's Constant, Kepler's Harmonic Law, and Hertsprung-Russell diagrams. Constellation Names lists (surprise!) the names of the constellations. Astronomy Conversions contains information about different systems of distance, temperature, and time.

Deep Space 3-D

Deep Space 3-D is a powerful star-mapping program that uses a database of more than 19,000 stars (expandable to nearly 250,000 stars and 1100 comets); this program is used by a number of serious astronomy publications to create their star charts.

It produces all-sky views or horizon views for any day and time anywhere on earth. Selected portions of the sky can be chosen by constellation or coordinates, then zoomed in for more detailed views. Comet paths can be computed and added to the star maps to create finder charts or to visualize the orbits of comets from any point in space. A "What's Up" feature tells at a glance the times of sunrise, sunset, twilight, moon phases, and positions of the sun, moon, planets, and current comets for a given day and time. Once computed, the information is stored and can be added to the star maps.

The most unique feature of Deep Space 3-D is depth. It produces stereo 3-D printouts that show the stars or the orbits of comets in their true three-dimensional relationships when viewed with a stereo viewer.

This program is an ideal starting point for anyone who is looking for a reliable guide to observational astronomy. For centuries people have gazed at the nighttime sky and wondered what was out there. Deep Space 3-D helps answer some of those questions.

Earthwatch

Earthwatch graphically displays in color the majestic march through time of the earth-moon system, showing the 24-hour night/day cycle due to the earth's rotation, the monthly phases of the moon, and the earth's annual trip around the sun, with the passing of the four seasons. A map of the world (Mercator projection) slowly scrolls eastward across the screen as the continents in turn move daily from predawn darkness through sunrise and daylight, then through sunset back to night.

Watch the shape of the daylight zone change ever so slowly from day to day as the seasons progress, with the days shorter in winter and longer in summer for the northern hemisphere, and vice versa in the southern hemisphere. Sunrise and sunset times, along with percentage of daylight for your city or anywhere you designate, are calculated and displayed for the present date, or a date that you choose. The phase and position of the moon will also be displayed numerically and graphically.

A special almanac screen can be called up to display additional information such as a comparison of sunrise and sunset times with those of the previous day, and the number of days remaining until the beginning of the next season (solstices and equinoxes).

Follow the sun's progress along its annual trek around the Analemma, the figure-eight pattern found frequently on globes that shows the sun's apparent north-south excursions in the sky and the "equation of time." See at a glance the local time anywhere on the populated earth and observe whether that location is experiencing night, day, dawn, dusk, or anywhere between. Up to ten locations around the world can be highlighted on the display to help you monitor the status of various cities, etc. All of this is done accurately and in real time while it is actually happening.

If you wish, you can reset the date and/or time to monitor the earth in the past or future. Replace your ordinary calendar or menu graphic program with Earthwatch when your PC is in "standby" mode for an informative and fascinating earth-monitoring clock.

ECU

Earth Centered Universe (ECU) is a sky visualization program capable of simulating many of the phenomena of the country night sky. This includes the stars, planets, sun, moon, comets, and "deep sky" objects. ECU is an observing tool for the observing amateur astronomer, but it is equally useful to the "armchair" astronomer or other interested person.

ECU operates on a PC from within the Windows environment. This environment provides all of the "tools" necessary to support a highly graphical program like ECU. This manual assumes the user is familiar with the operation of typical Windows programs.

ECU was written because of the obvious lack of good, reasonably priced astronomy programs. Most "commercial" software products (and I have purchased three of the high-profile programs) suffer from very awkward (not intuitive at all) user interfaces and slow operation (especially without a math processor). ECU fills an obvious need.

SkyMap

SkyMap is a high-precision celestial mapping program for generating star fields on an EGA/VGA monitor and/or an HP laser plotter (or any similar hardcopy device that accepts the Hewlett Packard graphics language). Polar and rectangular map projections are supported, which can be plotted at any scale and centered on any coordinate. Stars can be plotted at any epoch, with defaults for Besselian 1950.0 (B1950) and Julian 2000.0 (J2000). Proper motion is included with precession. SkyMap's star data files are based on the Smithsonian Astrophysical Observatory Star Catalogue (SAO), epoch B1950.

The SAO contains some 259,000 stars, but only those stars with visual magnitude brighter than 8.1 are presently used by SkyMap. This 51,000+ star subset was converted from 150-byte ASCII format to a 49-byte binary format by Rob Matson for use with SkyMap and other astronomical software.

Although this software can be used simply to plot near-publication-quality star maps, its greatest feature is its ability to superimpose a time-tagged satellite trajectory. Using a standard NASA two-line element set, SkyMap can plot the trajectory of any low-earth orbit (LEO) satellite against the stars as it would be seen from any location on earth. (SkyMap uses Space Command's SGP and SGP4 algorithms, which are designed to propagate the element sets of satellites with periods less than 225 minutes.)

Similar to the way it handles satellites, SkyMap also supports trajectory plotting of rocket boosters. In this case, the two-line ephemeris file is replaced by a rocket trajectory file that contains a tabular listing of a rocket's position as a function of time. The format for each line of the trajectory file is: mission elapsed time (MET) in seconds, latitude in degrees, longitude in degrees, and altitude in kilometers. Negative values are entered for south latitudes and west longitudes. The MET values need not be at regular intervals, but they must always increase.

TrakSat

TrakSat is a general-purpose satellite tracking program; any satellite that has a NORAD, NASA two-line element set can be used. There are some limitations in the program, along with some assumptions. See the program's section on limits and assumptions for further study. The solution to the satellite motion that is used by TrakSat is completely analytic and therefore requires no numerical integration. This makes the program fast, even faster when a math coprocessor is used, since the solutions can be evaluated at arbitrarily large or small time intervals.

This program provides the user with a means of propagating NORAD element sets in time to obtain tracking information of the space object. Important note: The NORAD satellite data set included with TrakSat contains the elements for more than 600 satellites. However, this data will be quite old by the time most people obtain this program. The elements included can be used to run the program through its paces, but updated elements are required to make accurate predictions.

STSOrbit Plus

STSOrbit Plus is an enhanced version of STSOrbit. As a general rule, a 286 or better computer (AT-class IBM compatible) is recommended, and a math coprocessor chip significantly improves performance. Some users report acceptable performance on an XT-class machine with a math coprocessor.

STSOrbit Plus (which will be referred to as STSPlus from here on) displays the position and ground track of an orbiting satellite on a selection of maps ranging from a full map of the world to zoom maps showing considerable detail.

The program has special features implemented at the request of NASA astronauts and others for use during a NASA space shuttle mission. When used with NASA/NORAD two-line elements, the position and ground track of a variety of satellites (such as the space shuttle, the Hubble Space Telescope, the Gamma Ray Observatory, or the Soviet MIR Space Station) can be displayed.

The Special Location and Tracking Station displays show concentric *isocontours*, circles of equal satellite altitude. These special maps can be especially valuable for visual or amateur radio sightings. The initial premise for STSORBIT was to duplicate the wall map in NASA's Mission Control Center in Houston, Texas.

Chemistry

Chemical

Chemical is a molecular modeling program to help form three-dimensional pictures of chemicals. Atoms are selected from a periodic table, and electron orbital information is retrieved. The atoms are then bonded. The chemical is displayed as it is being constructed. The chemical can be viewed from different directions by using the up and down cursor keys and the V command. If desired, the hybrid and ionize commands can be used to alter the orbitals before bonding. Atoms can be bonded into groups, then the groups bonded to other groups to make large chemicals. The current version includes support for Microsoft mouse drivers.

ChemView is a companion program that shows a three-dimensional animation of the models generated with Chemical. ChemView requires an EGA board and monitor. ChemView is written in Turbo Prolog with the graphics routines written in Turbo C.

BioChem is a companion program that combines files made with Chemical into larger molecules. Chemical groups can be bonded together using covalent, ionic, hydrogen, or peptide bonds. This program provides much more capability than the group command that was included in older versions of Chemical. BioChem works with a CGA, EGA, or VGA monitor.

Crystal is a companion program that places atoms on a lattice framework to form crystal structures. The structure is defined by a text file. A text editor is built into the program. The text file is converted to an image that is shown rotating in three dimensions. An EGA or VGA monitor is required.

The Integral Scientist

The Integral Scientist is a hypertext information utility for scientists, science educators, and college or high-school students. This program contains the following:

Computerized periodic table and element information database. More than 50 items of information on each of the elements.

Flash-card learning mode. Helps the student study the periodic table.

Customizable text viewer. Allows the user to include additional information for easy access.

Units conversion calculator. More than 450 units of measure in 30 categories. Enter a value and instantly convert it to any of the corresponding units.

Chemical solutions calculator. A tool with which to manipulate the parameters of a chemical solution and perform "what if" planning prior to solution preparation.

Molecular weight calculator. Enter any chemical formula and calculate its molecular weight without having to look up the information in printed tables.

Solvent selector. Provides important data for choosing the correct solvent for solution preparation.

Dilution calculator. Eliminates mistakes when calculating and performing dilutions.

For the laboratory, The Integral Scientist provides the chemist with a tool with which to manipulate the parameters of a chemical solution and perform "what if" planning, prior to solution preparation. A solvent selector provides important data for choosing the correct solution solvent; a dilution calculator eliminates mistakes when performing dilutions.

The user can also enter any chemical formula and calculate its molecular weight without having to look up the information in printed tables. Included in The Integral Scientist is a units conversion calculator with more than 450 units of measure in 30 categories. A unit value can be easily converted to any of the corresponding units. Input and output values can be transposed, and the current input value can be scaled. The user can determine if the conversion is exact or approximate and can view the complete list of conversion factors at the press of a key. Category definitions are displayed when selecting the units category, and the category of measure can be found easily by looking up an unfamiliar unit in the index. Science has never been so much fun, but the information is serious.

Labcoat

This laboratory data and cost analysis program helps you manage a clinical lab. You can calculate item depreciation and evaluate extensive-test cost data for your equipment. Routine lab statistics and Q.C. charting are included. Labcoat handles data entry and calculations for lipoprotein data. You can calculate numerical data, mean and standard deviation, and range. Also included is an evaluation program for prospective lab assistants, based on such criteria as job knowledge, judgment decisions, patient awareness, organizational skills, and more.

PC-Calib

Chemists, researchers, and users of analytical instruments will find PC-Calib a multifaceted aid to the sometimes tedious task of instrument calibration or verification. PC-Calib's curves can have up to 8 points, with 3 replicates each, and can be measured in any of 12 different standard units. They can be log-log equations (log base 10) and either a first- or second-order polynomial least-squares fit. Experimental values or controls can be entered and checked against a generated curve. Data can either be viewed on screen, printed, or formatted as ASCII files for import into other software.

Mandelbrot Zoom Lens and Sci-Calc

Mandelbrot Zoom Lens lets you create, view, and magnify a Mandelbrot fractal and demonstrate the unique nature of these complex mathematical patterns that are found in nature. As a fractal image is magnified, the smaller details still look like the same fractal. The more a fractal is magnified, the more hours it takes to fill in all the details on the screen.

Sci-Calc is a scientific calculator with a simple, direct approach. Various standard formulas are shown on the screen relating x and y. You simply enter the numbers you want, and the program then calculates the results.

Physics

Nuclear Magnetic Resonance

This unique program displays the calculated spectrum, including adjustable peak broadening. Simulated plotting of very complex molecules, such as strong couple proton spectra, is supported. Parameters for up to seven spins (nuclei) can be calculated.

Physics Tutorial

This program contains a physics tutorial that presents the physical properties (mass, charge, time, etc.) in a graphic chart analogous to the chemistry periodic table. That is, the properties are arranged so that the relationships between the properties can be easily seen. Relativistically correct formulas can be obtained directly from the chart. This program provides users with a clear overview of physics in a very short time. It provides technicians, engineers, chemists and anyone interested in science with a far better understanding of physics.

Chapter 7

Databases

Data, data, and more data. Engineers and scientists have to contend with growing mountains of data. During much of their professional lives, they are inundated with data: experimental results, drawing lists, research notes, parts lists, lab supplies, periodical articles, new ideas, phone numbers, vendor part numbers, records, lab reports, bills of materials . . . the list goes on and on.

To prevent this flood of data from inundating you, it's necessary to categorize, filter, scale, adjust, sort, group, organize, and save your data so it can be easily recalled. A solution to this problem of exploding data is to use computerized databases. Modern databases can store virtually any type of information you can input, and the databases provide access to any or all of this stored information in a fraction of a second. Once accessed, this data can be sorted, grouped, alphabetized, edited, viewed, and printed out. This chapter discusses the basics of databases and lists a variety of free and low-cost databases that can solve virtually all of your data problems.

Why Use Databases?

Databases are computerized lab notebooks, reports, phone directories, memos, parts lists, notes, statistical data, and filing cabinets that have been converted into an electronic media storage that is possessed with a perfect memory and instant recall. Databases provide:

- Reusable data. Any data you enter (text, graphics, numbers) is stored and can be later recalled, sorted, edited, grouped, or combined with existing data to create new databases, charts, graphs, reports.
- Fast search. Even though you might enter the data in a random order, you can search and locate that information quickly, even though it is in the midst of thousands of other records.

- Sorting. The data you have entered can be sorted in a number of ways: e.g., alphabetical, all in the same zip code, people with the last name of Johnson, all drawings relating to an engine piston . . . in virtually any combination of groupings.
- Reporting. The data you searched for and found can then be displayed on your screen or printed out in a report.

Having your information in a database is just like having an enormous file cabinet full of randomly filed and accumulated information that you can access and find in a few seconds, all from the comfortable chair in front of your computer.

Anatomy of a Database?

A database is basically an organized collection of related information that can be recorded, then later automatically sorted, updated, and easily searched using key names, words, titles, etc.; the information can then be viewed and/or printed.

The basic elements of a database are illustrated in Figure 7.1. A record is one complete set of data, such as the data required to record all the information about a drawing. Each record is divided into a number of fields. Each field, in turn, stores a single set of information, such as a drawing title or a drawing number. When you need to recall this data, all you have to do is to input the information contained in one of these fields (such as the drawing number), then all of the pertinent information about that drawing appears on your screen, or it can be printed out on your printer.

A database usually has an index associated with it that allows you to retrieve stored records. You can sort these stored records in ascending (A to Z) or descending (Z to A) alphabetical order, or in many other types of sorts such as by date, zip code, city, company name, etc. You can usually update databases with individual entries, or with global entries throughout the entire database. The basic steps in creating a database are:

- Setup. This initial step consists of designing the form or format to be used for recording the data, e.g., establishing the field names, the length of the fields, etc.

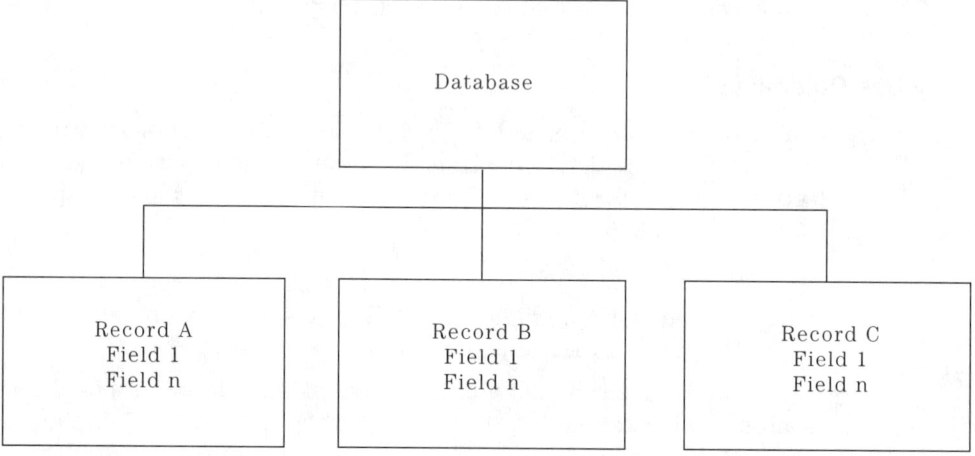

Figure 7.1 Anatomy of a database.

- Inputting data. This can be accomplished either manually by inputting data from a keyboard, or by importing data from a compatible file.
- Organizing data. This is one of the main reasons for entering data. Now you can index, search, sort, select, and/or group this data.
- Outputting. The final step is to output the data you have input, edited, massaged, and manipulated into the format you request. This can be presented on your screen for review, or it can be printed out on your printer for a permanent record.

Database Basics

The reason why a rather severe limitation is imposed by restricting the database fields to a maximum size is simply speed. If your computer can't assume that all of your records are the same size and that all of the information is in standard slots, the software has no choice but to initiate a time-consuming and tedious character-by-character search through each individual record until it locates what it is looking for.

For this reason, designing the forms or templates to be used in your databases is a very important step because you have to accurately anticipate the type of information you're going to need to record. You have to analyze and decide the approximate length of each field and the type of information that will be entered (alphabetical and/or numerical).

You have to create a clear, succinct name for each of your fields. The database software uses field names to identify information, so be sure to use concise but easily understandable field names. If you want to perform an alphabetical last-name sort, for example, the software knows precisely which field this information is located in and uses the data stored in this field for the sort.

The longer the names and the longer the length of the fields, the more memory is used, and the longer it takes to access that information. And you have to make sure that the fields you create are exactly how you want them to be, since you will have little or no opportunity to change them later after you have entered considerable data.

It's usually a good idea to assign a unique number to each record for purposes of reference, for use in sorting and reporting, and for general record-keeping purposes. Another suggestion is to use all capital letters in the field names so they stand out and also to show you where the field name ends and the field entry begins.

On the form you design, you might also have to specify how the entry is to be justified (left, right, or centered) when the records are recalled and displayed or printed. A data field might require a specific input format such as mm/dd/yy or dd/mm/yy. A phone number format might be specified as 218/692-3412 or (218) 692-3412, or the software might be programmed to accept both types of input, then output the data in a previously established format. Another item to consider in designing the fields is that you might need to specify that each new drawing number or purchase order, or whatever is entered, must be unique so the software has to check to make sure the number has not already been used. Some types of databases offer ready-made templates that you can use for data entry. However, these are usually very limited in their applications.

Once you have loaded a database, you can't be expected to remember the precise spelling or correct number of every piece of information you input, so another desirable capability of database software is its ability to use wildcards in making searches.

For a single unknown letter or number, the question mark (?) is often used. For example, searching for ?ex could locate sex, hex, rex, tex, and a host of others. For a wildcard to replace a group of letters or numbers, the ellipsis (. . .) is often used. If you input . . . sen as your search criteria, it becomes a suffix and you would locate Jensen, Adolphsen, Paulsen, plus others, but not senior, senator, or similar data. And, of course, a database must have growth potential so you can edit and delete entries, and continue to add new entries.

Two Major Categories

Databases can be divided into two major categories as diagrammed in Figure 7.2. Structured databases are a category in which each maximum field length, data content, and location has to be specified before data is entered. Because of the rigid structure in which data are always stored in the same location and in the same format, structured databases are much easier and faster for searching and zeroing in on the needed data.

Free-form databases, however, are the type in which the format of the data need not be specified ahead of time. This type operates much like a random set of notes that you have created over a period of time. The free-form database is made up of individual records with field entries, but without fields. Thus, you can input any type of information you want to. Because of their unstructured format, free-form databases are more versatile in their applications. However, they are much more difficult to search because the data are stored in a random manner. The entire contents of every record in the database must be searched to locate specific data.

Structured Databases

Structured databases can be further broken down into the subdivisions illustrated in Figure 7.3. The flat-file database covers a specific type of data. Each individual record contains all of the information pertaining to that particular topic. It does not have the relational capability of the relational type discussed later. If data is to be entered into a number of different flat-file databases (e.g., parts-list database, drawing-list database, spec-list database, etc.), it must be entered into each one individually if flat-file databases are used.

Relational databases, however, can link different sets of data together. Data in one database is "related" to data in another database. Then you can enter one key field

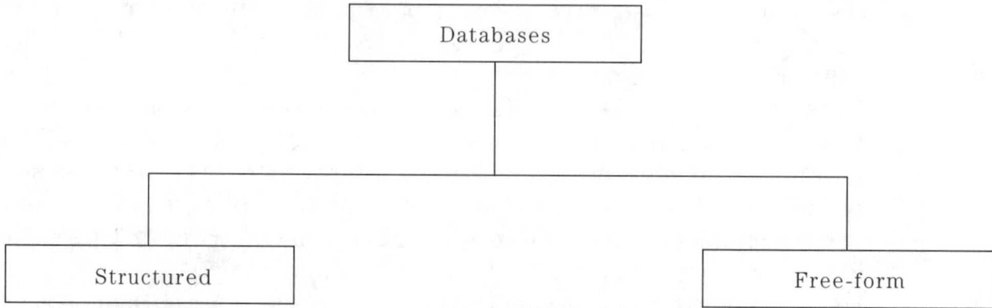

Figure 7.2 Database catagories.

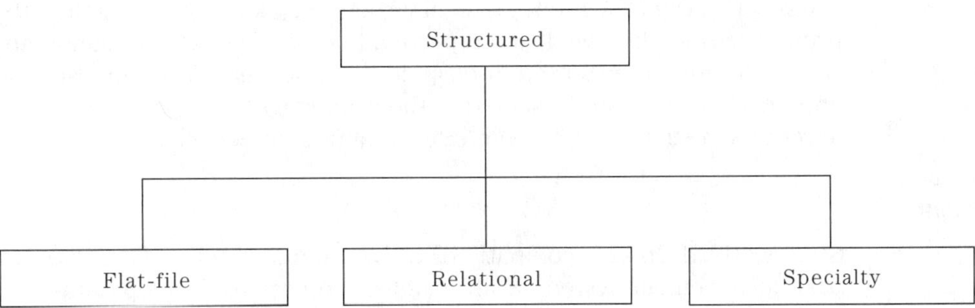

Figure 7.3 Structure databases.

in your master file, and your reports can show all data available from all relational databases that match that key. This not only gives you access to a variety of data, but it also saves typing the data into each individual record. For example, if one database contains an employee's personnel record, and another database contains the job title and pay, when the employee marries, the name change and related information needs to be input only once, and all the relational databases will be updated automatically.

Some of the more complex databases provide a programming language that lets you perform sophisticated linking of databases, such as taking data from several databases and performing a calculation based on the data. For example, assume an employee's data exists in several databases: one containing the employee's personal information, a second the employee's salary, a third the hours worked, and a fourth the employee's payroll deductions, etc. The database software can then link all of these databases, calculate and print the employee's paycheck, then print out the mailing label for sending the check. Some programs also offer compilers that turn command files into separate programs to make them run faster.

Specialty databases are the types that are limited to one specific application, such as a listing of books and magazines, sports records, a listing of computer software, a church membership list, etc.

Flat-File Databases

File Express

File Express is very popular among both first-time database users and nontechnical users. Easy to use, it makes extensive use of menus and prompts. Setup is straightforward: you define the field names, type, length, and format. Creating custom data-entry screens is also easy. File Express has an auto-dialer and a simple text editor for mail merge.

Although it's a flat-file database, File Express can look up data in other files by using programming commands called formulas. This fast, powerful DBMS has a 16 million record-per-file, 120 fields-per-record, 250 characters-per-field capacity. It has unique formula-based fields, very fast sorting, a report writer, duplicate file-record deleting, search on multiple fields, selective indexing, and much more. It can import and export data to and from dBASE III and Lotus .WK1.

Using File Express, information of virtually any kind can be put into the computer, printed, sorted, displayed, and reported. For example, a list of names and addresses can be entered, then sorted into zip code sequence or last name sequence, printed onto mailing labels, and updated as the list changes. The same data can be included in reports or simply displayed on the computer screen when needed.

Staff

Staff is dBASE III Plus compatible but has an independent user interface and database management system. Staff enables the user to develop data-intensive applications without traditional programming effort. The system is pull-down-menu driven, and the windowing technology used offers a very powerful and user-friendly database management system, adequate for both the novice computer user and the expert dBASE user and developer. Staff accommodates the following user categories:

- Personal computer users who deal with large amounts of information and are not specialized enough to develop their own customized database applications.
- Users who are interested in independently resolving their data management problems without depending on external collaborators or on a centralized EDP department.
- Developers of specialized database applications directed to a large number of users without detailed knowledge of PCs.
- Users of the specialized databases created and distributed by the application developer described in the second user category.
- The dBASE advanced users, or the users of other application software developed under a dBASE-compatible developing environment such as Clipper, Foxbase, Quicksilver, etc., who want to manipulate DBF files efficiently without using dBASE or other compatible products.
- Corporations with a high degree of MS-DOS microcomputer implementation who are using the industry-standard dBASE and who are in the process of training the personnel for database management tasks.

A very intensive use of database management applications can be achieved by means of Staff without the need of huge investments for the highly priced dBASE software, and the extensive training requirements of this program. The main features of Staff are:

- The interactive dBASE III concepts and functions are maintained.
- Many dBASE III Plus functions can be performed by means of clearly defined pull-down menus. Examples are: creation of new database files, appending new records and deleting unwanted ones, appending and inserting new fields, and deleting unwanted fields.
- Powerful commands for editing the information stored in the database.

- Packing a database file and zapping (deleting at once) the whole content of a database file is possible.
- Extremely fast search of very large amounts of indexed information. Plus much more.

Relational Databases

Data Master

This powerful, easy-to-use program allows you to create, modify, and work with dBASE III databases. Many users find Data Master easier to use; it includes the ability to convert Lotus 1-2-3 worksheets into dBASE database files and to convert dBASE database files into Lotus 1-2-3 worksheets. While the primary emphasis of Data Master is on the use of database files, you can also use it to browse and do simple editing of Lotus 1-2-3 worksheets. Symphony worksheets, .WKS, and .WK1 files are all supported. Data Master works in a multiuser local area network environment and on stand-alone PCs. Features:

- Uses existing dBASE files apart from their normal application.
- Creates brand new databases for any kind of list you want.
- Changes the structure of databases to meet changing needs.
- Creates Lotus 1-2-3 worksheets from dBASE databases.
- Creates dBASE databases from Lotus 1-2-3 worksheets.
- Browses and edits data in Lotus 1-2-3 worksheets.
- Enables you to design your own custom reports with Data Master's report maker. Plus much more.

Dream

Dream (data retrieval, entry, and management) is a menu-driven database management system with relational capabilities, and it possesses extreme power and flexibility. Much like dBASE, it can custom design database applications, which include reports, sorting abilities, query abilities, and data entry and retrieval abilities without writing code. The program comes complete with more than 200K worth of online and manual documentation that leads the user, step-by-step, through the program. Dream handles more than 32,000 records per data file with more than 1,500 characters per fixed record length.

One of the strong features of this software is that it accommodates both technical users and end users by providing two modes of operations. The technical user (or "application designer") is supplied with powerful, yet easy-to-understand functions with five levels of complexity that are based on user expertise with the system.

The Designer creates files, designs reports, sorts, queries, etc. With this power, the entire database application can be built and made menu-driven. Later in the nontechnical or end user environment, someone without programming knowledge, such as a data entry person, can use Dream in a very structured, limited, and easy manner to perform file manipulations from the finely constructed system. In short, a technical user can build a powerful system for someone else to use by employing menu-driven options that are nontechnical. Model-S lets end users produce "ready-to-use" screens without having to use the screen painter of Dream. Features include:

- Two modes of operation: technical and end user.
- Three levels of data protection.
- Multiple views of data.
- Powerful query and selection facility.
- Multiple record updates (batch processing).
- Edit, format, and range tests on input.
- Computations on input/output, including computations with dates, auto-generation, and duplication of data fields and records.
- Single data file reports that are post or concurrent with data entry.
- Scrolling and searching within generated reports.
- Relational operations for multiple files.
- Multifile reports that are post data entry.
- Procedures for multiple report generation.
- Data restructuring.
- Copying/moving data from one file to another.
- Sort on fields, subfields, and special formats.
- Screen painting for data entry and reports.
- Full-screen operations.
- Windowing technique for multiple data files.
- Unique report layout features.
- Summary reports with totals and subtotals.
- Four character sizes for reports and labels.
- Multiple look-up and data transfer from outside files. Plus much more.

Fastfile

Fastfile is RAM-based and programmed in assembler language. It's fast and easy on laptop batteries. Typical features include:

- Add records.
- Global edit fields.
- Find and edit records.

- Math calculations.
- Clone new files.
- Import export files.
- Add, delete, edit fields.
- Print/display reports.
- Create new DB designs.
- Mailing labels/merge.
- Write sorted reports.
- Load and save files.

Unique features are:

- RAM based (both program and data).
- Compressed files save disk space and laptop battery life.
- Uses single floppy disk (without disk swapping).
- Eye-blink speed—programmed in assembly language.
- The same screen is used to add, edit, and search for data.
- Screen reference to global calculations like BAL DUE.
- Instant indexing when a record is added or edited.
- Table and relative fields saves entry effort.
- Automatic report edit when fields are added or deleted.
- Automatic record protection.
- Automatic "balance forward" updates for billing.
- Grade book with automatic record replication. (Very useful for student and sports statistics).
- Subtotal reports with averages.
- Nonzero averages, totals.
- Keyboard field calculator.
- Makes matching and multiple choice tests.
- Automatic searching and reporting for similar files.
- Comes with ready-to-use programs.

Capacities are 32 files per directory, 2,000 records per file, and 32 fields per record. The program comes with predefined database applications:

- Billing.
- Checkbook.

- General ledger.
- Invoices.
- Client list.
- Inventory.
- Teacher's grade book.
- FAST quiz.

FreeFile

In its simplest form, FreeFile does your filing for you. You define what pieces of information you want to store. FreeFile stores the information you enter, then lets you print it in various forms. You can think of your database as being made up of rows and columns, just like a spreadsheet. Each row is a record, and each column represents each field (or piece of information). Your file (or database) is made up of a number of "records." A record is a collection of pieces of information. These pieces of information are called "fields."

A field might be a person's name, or social security number. A special feature of FreeFile is its ability to use "computed fields." These fields are a lot like formulas in a spreadsheet. You can also "import" data from other programs. This allows data from another program to be put into a FreeFile database. You can import data from a BASIC program if you like. You can also "export" data from your database to other programs (e.g., to your spreadsheet program). Data to be exported can be selected (e.g., select only cases where the Due Date > 09/01/99) just as you would when you have FreeFile write out a report for you.

Probably the most significant design feature of FreeFile is that although it is loaded with many advanced features, these features stay in the background until they are needed. For example, to print data from your database, select the print option and select to print. You need not specifically set any print options. You might later decide to change the order of the items on the page, or select only certain items to be printed, but that is always optional. Once you are comfortable with the program, you might want to learn how to do many of these more advanced features. It is not a program you are soon to outgrow.

The maximum number of databases opened at any one time is one. The maximum number of indexes for any one database is 10. The maximum number of records in any one database is 2 billion. The maximum record length is 1,000 characters. The maximum number of fields in a record is 100. The maximum field length is 65 characters. The maximum number of sorts for the print or export features is 100. The maximum number of compares for print and export features is 25. The maximum number of computed fields (formulas) is 100. The maximum number of calculated fields on a report is 100 (numeric and computed fields will be totaled at the end of the report). Comparison for date fields will be sorted correctly (e.g., 01/01/86 will follow 12/01/85), and can be used in computed fields (e.g., Start date–End date). Computed fields are similar to formulas in spreadsheets.

NewBase

NewBase doesn't require you to learn special syntax or complicated commands. The program is easy to learn and use; even beginners can become productive immediately. Yet it is powerful and sophisticated enough to satisfy the needs of the most demanding power users.

Advanced users can program NewBase to create customized menu-driven applications. Among its many features are:

- A capacity of 32,000 records.
- Online context-sensitive help.
- Keyboard macro keys.
- Multilevel password protection.
- Mail merge for word processors.
- Global search and replace.
- Calculated fields.
- Note fields.
- Lookup table.
- Recover damaged files.
- Customized data entry screens.
- Import and export files.
- Memory-resident version.
- User-defined search filters.
- Programmable features.
- International date support.
- Screen blanker.

AutoPilot lets you create custom menu-driven applications. AutoPilot simplifies repetitive tasks and greatly reduces training and support when using NewBase in a multiuser environment. Search filters let you permanently store your search preferences in libraries for instant recall. With search filters, it is not necessary to type in the search criteria every time you want NewBase to search a database. The software makes no special demands on you to enter data or commands in their proper case. For example, to pop up a note field when a record is displayed, pressing either capital N or lowercase n gets the job done.

PC-File

This popular database manager has pull-down menus, 175 context-sensitive help screens, easy macros, relational capabilities, searches that are faster than dBASE, a sophisticated report writer, and it can produce pie and bar charts directly from the database. It has a built-in, rudimentary word processor. It allows memo fields up to 5,000 characters. PC-File creates and reads dBASE III/IV. Postnet prints zip+4 and bar codes.

PC-File has two methods of setting up new databases. You can choose Fast and specify the names and types of fields you want, and the program sets up a form. Alternatively, the Paint option lets you place the fields anywhere on the screen and create up to five screen-length pages. Just type a label, place the cursor where you want the field to start, and click on the Data button to specify field format. You can also set masks, which are codes that specify the acceptable values in a field, default values, and formulas. The software also allows lookups between files.

Other features include:

- Fast deletion of duplicate records.
- Multiple input screens.
- The maximum of 128 fields.
- Automatic phone dialing and call logging.
- Up to nine different report subtotal levels.
- Multifield indexes.
- Global find.
- PostScript output option for graphs.
- Support for WordPerfect merge files.
- Very easy to use database.
- dBASE compatible.
- Analyze data with relational retrieve and posting.
- Search/query bar codes.
- GUI-mouse or keyboard.
- Charts and graphs.
- Custom reports, letters, labels. Plus much more.

Wampum

Wampum is a menu-driven, pop-up (TSR), dBASE-compatible database management system. It permits even novices to develop powerful applications in minutes instead of months. Virtually every dBASE command and function is supported by easy-to-use menus. Versions are available for stand-alone PCs or NetBIOS-compatible networks. In addition, users can choose between Clipper, dBASE III+, and FoxBASE+-compatible indexes. The FoxBASE version also supports a simulated dot prompt that permits a user to enter virtually any dBASE command. Simple programs up to 500 lines also can be written and executed using the FoxBASE version.

MenuMaker

MenuMaker is a data entry screen generator for use with Wampum or dLITE. The developer draws the data entry screen as it should appear to end users. Then MenuMaker writes the necessary code to support either Wampum or dLITE applications. Wampum Form Letter Editor provides an easy-to-use editor for developing form letter applications with FoxBASE versions. Badge-O-Magic is a name tag and badge generation program that uses dBASE III-compatible databases to produce printshop quality name tags on any HP LaserJet II-compatible laser printer. The software includes documentation on disk and 5 soft fonts with point sizes ranging from 15 to 40 points.

Wyndfields

Wyndfields is a general-purpose database program that's useful for keeping track of any type of structured information. Fields are displayed as column headings across the top of the screen, with each record on a separate line down the screen. If the combined width of the fields is wider than the screen, the screen scrolls horizontally, a column at a time in a circular manner. Because Wyndfields supports a 43-line mode, you can see several records simultaneously, similar to what you would see looking at a large, paper, columnar pad.

Data can be added, changed, removed, found, displayed, analyzed, and printed easily using the friendly pull-down menu system. Creating and printing mailing labels or detailed reports can be accomplished painlessly from within the program. Multiple indices (ways of ordering your data) are maintained automatically for you, and several methods are available for extracting exactly the information you need from your files.

The program can have 70 different fields, 2 billion records, and 10 different indices, each with 10 key fields in either ascending or descending order. For more advanced users, Wyndfields also offers extensive string and math functions for use in searches and calculated fields, and its date and time tracking abilities are unsurpassed.

In addition to the procedures outlined previously, Wyndfields offers many advanced abilities such as:

- Creating and printing detailed tabular reports.
- Creating form letters using the built-in text editor, or creating data files for use with mail merges using popular word processors.
- Redesigning the appearance of the table display and form layout renaming, deleting, and copying databases and other files.
- Using formulae (and numerous calculation functions) to create calculated fields and to do extremely detailed searches.
- Updating or deleting multiple records based on search conditions.
- Transferring data between different databases, and translating data between the Wyndfields format and other common file formats.

Wyndfields for Windows

This relational database for Windows allows up to 20 different databases to be open at one time. Databases can be linked together for easy transfer of data between them. Information can be viewed in (configurable) table or form format and can be globally updated. There is user-defined indexing for each database. The program also prints custom labels and reports and has many powerful search features (quick searches, conditional searches, etc.).

Sports Databases

Bowl101

Bowl101 is a user friendly, totally configurable, yet powerful bowling-league management program. This program handles most of the scoring systems available. The program is designed to do all the work. About one minute per team is all the time it takes to enter the scores. A 16-team league can be input in 16 minutes, including subs after the initial setup. All the user has to do is to enter the 1, 2, 3, or 4 games bowled. Features:

- From 2 to 36 teams.
- 1 to 10 bowlers per team.
- 100 subs.
- 1, 2, 3, or 4 games bowled.
- Swap subs and/or bowlers.
- Powerful edit features allow you to scroll forward and backward through teams' bowlers and subs while editing.
- Proper case function aids in data entry.
- Split season.
- Position rounds.
- Match points.
- Peterson points.
- Difference between team averages.
- Team averages forfeits.
- Auto lane assignments.
- Mixed leagues.
- Bantam.
- All male or all female leagues.
- Separate handicaps for male and female.
- Minimum and maximum handicaps for male and female.
- Individual points for game and/or series.

- Keep entering averages for up to 21 games or have the computer calculate them from the start.
- Zero high games weekly or have the computer keep them for the entire year.
- Add or remove teams or bowlers at any time.

The Bowling League Record System

The Bowling League Record System (BLRL) is a set of programs to facilitate accurate record keeping of individual and team bowling scores following the guidelines of the American Bowling Congress (ABC) and the Women's International Bowling Congress (WIBC). This evaluation copy of BLRL provides the following features:

- Easy installation program for floppy or hard disk system.
- Female, male, or mixed leagues bowling; either scratch or handicap scoring is supported.
- Bowling scores input keyed off the league bowling schedule and team lineup.
- List of league substitutes displayed using PFK 1 during input of scores.
- Weekly scores input can be stopped and restarted at a later time using PFK 2.
- Add/replace bowlers in a team lineup during scores input.
- Automatic archiving and update of bowler and team records, scoring statistics, and win/loss points.
- Separate handicap base, handicap percentage, and maximum handicap for female and male members in mixed handicap leagues.
- Season-to-date league standing report containing bowler and team score and team win/loss statistics.
- Bowler and team series more than game scores for league standing report.
- Edit/update capability of bowler and team records, initialization data, and league bowling schedule. Plus much more.

Diamond Data

Diamond Data keeps track of the members of a team throughout the season, an individual player across several seasons, etc. The built-in merge facility allows multiple databases to be "added" together to form a composite database. This gives the ability to have each season as an individual database, combining them to provide lifetime statistics, or have each game be a separate database and merge them to form season totals.

Exactly how you organize your databases is up to you. Diamond Data covers the entire hitting, pitching, fielding, and base running stats. Even games can be entered to keep track of how the season is going. Diamond Data tracks the following offensive statistics:

- At bats.
- Batting average.

- Doubles.
- Game winning RBIs.
- Games.
- Hit by pitch.
- Hits.
- Home runs.
- On-base average.
- RBIs.
- Runs.
- Sacrifices.
- Slugging percentage.
- Stolen bases.
- Strikeouts.
- Triples. Plus much more.

Reports can be sent to the printer, screen, or a disk file.

Football Pool

This program makes running a football pool both easy and fun for a beginner or expert on football. First, set up both the team schedule and people who will take part in the fun. Print the 16-week schedule and make copies to form a package to hand out to each person. They mark a W or X for the team they want to win. The left side is theirs to keep, and the right side is handed in with their name on it. Notice that at the right-hand side of each team name is a letter. Use this letter to code in their picks for the week. A word of advice: Wait until you have most of the picks before entering a week's picks.

Once all the picks have been entered into the system, the "F4" function prints a list of everyone in the pool and their choices for the week on an 8½-×-11 sheet of paper. This is great for anyone who wants to know on Sunday how they're doing against everyone else.

After the Monday night game, enter the team codes that won on this screen. For games that end in a tie, use a lowercase "t." After you press Enter, the system prints out a standings sheet to your printer. Again you can make copies of this and pass them out to your players. Try out "F6." It's used if you need additional picks or status sheets; up to 99 can be printed. A feature here prints a check-off list. It's for you to keep track of which picks have come in for the week.

GolfHCap

Providing handicaps for golfers per USGA procedures has always been an odious, error-prone job. Now, with the addition of slope-ratings and calculating a handicap index, it is even more so. Many courses have turned to the computer services being offered to relieve themselves of this time-consuming responsibility. The program Golf-HCap makes it possible to handle this work by the normal course administrative staff.

This program not only calculates the golfer's handicap index in accordance with procedures set forth in the USGA publication "USGA HANDICAP SYSTEM," it also prints membership lists, mailing labels, score input forms, handicap index cards, as well as a monthly report to be posted in the clubhouse showing scores, handicap index data, and both old and new indexes for each golfer. The program also provides, on request, a printout of all the handicapping data for any single golfer to help resolve any problems.

This form shows the last twenty scores, course ratings, and slope ratings and also flags the handicap differentials that were used in calculating the players' handicap index. The program handles 12 groups of golfers, with each group having up to 500 members. If more are required, the program can be run in a different directory that allows an additional 12 groups, and so on.

PC Flt Plan

The program makes operational flight plans for visual flight rules (VFR) or instrument flight rules (IFR)—fixed wings or helicopters. PC Flt Plan is not made with airline operations in mind. Most such carriers use big mainframe systems, fed with wind prognosis for large parts of the globe from centers several times a day. PC Flt Plan should be quite useful for pilots who fly smaller airplanes or helicopters, and who have access to an IBM-compatible PC.

PC Flt Plan is easy to use. If you take a close look at the sample routes, you'll understand them. A flight can be defined to the destination and to one or two alternates. In addition, a list of other diversion fields can be created. The program has a nav database with positions, magnetic variations, and frequencies for airfields, navigational aids, intersections, etc. The data included are examples only (they are correct, however), but the users can enter their own waypoints into the database as they create or edit routes, and in that way build up a database. The program also provides a small register where some key data for the different aircraft used can be stored. The routes created can be saved on disk so that a collection of frequently used routes can be compiled. There are four different flight plan printouts that can be selected. One of these is a kneeboard size flight plan.

Personal Golf Tracker

This program provides an easy way to keep track of your golf scores. It has the following features:

- Keeps track of your scores.
- Displays your personal best on each course you play.
- Displays average scores on par 3s, 4s and 5s.
- Compares your scores to par, bogey, or double bogey.

- Graphically shows the percentage of time you shot par, bogey, or double bogey.
- Shows percentage distribution of eagle, birdie, par, etc.
- Graphically displays your high, low, and average hole by hole.
- Tracks six special aspects of your game of your choosing.
- Graphically displays your trends of six aspects of your game of your choosing.
- Prints a hard copy of the rounds you've entered.
- Prints summaries of your performance on all courses or all golfers on a single course.
- Enters up to 99 rounds on 64 courses of 96 golfers.
- Analyzes rounds before and after a given date.
- Calculates handicaps.
- Checks your consistency on each hole or on rounds played.
- Keeps summaries of your performance across all courses played.
- Ranks golfers according to summary information you select.

Softball Coach

Softball Coach maintains the statistics for the coach of a softball team, which might explain some of its emphasis on offense rather than defense. Softball Coach aids a softball coach in many areas of managing the statistics of a team. It is particularly good at maintaining individual statistics during a softball season. The software can also be used by an individual player to keep track of statistics across multiple seasons. This is easily accomplished by treating each season as if it were a player. In this way, a player's "career" can be examined at a glance by looking at team statistics.

The following lists the statistics this version of Softball Coach maintains. Some are calculated from other statistics entered and can't be changed except by changing the statistics on which they are based:

- At bats.
- Batting average.
- Doubles.
- Game-winning RBIs.
- Games.
- Hits.
- Hit by pitch.
- Home runs.
- On-base average.
- Runs.

- Runs batted in.
- Sacrifices.
- Slugging percentage.
- Stolen bases.
- Strikeouts.
- Triples.
- Walks.

Softball Coach has mouse support built in, but it has no support at this time for a floating point coprocessor.

Ultimate Bike Log

The choices on the main menu are:

- Personal information.
- Bicycle information.
- Ride information.
- Gearing calculator.
- Registration printout.
- Target heart rate.
- Print empty logbook.
- Calorie consumption.
- End program.
- Log book printout.
- Component printout.
- Names of riders printout.

Once you get to the Personal Information section you will be presented with the following information to enter:

- Name.
- Address.
- City.
- State.
- Zip.
- Birthdate.
- Phone.
- Previous miles.

No field is mandatory. If you want, you can fill in only the fields you desire. Use this to keep track of all your cycling friends.

Once you get to the bicycle information section, you will be presented with the following information to enter:

- Name of the bike.
- Headset.
- Previous miles.
- Hubs.
- Serial number.
- Quick release.
- Brakes.
- Rims.
- Brake levers.
- Spokes.
- Chain.
- Tires.
- Color.
- Freewheel.
- Crankset.
- Pedals.
- Front derailleurs.
- Toeclips.
- Rear derailleurs.
- Forks.
- Bar.
- Stem.

Once you get to the ride information section, you are presented with the following information to enter:

- Date.
- Miles.
- Hour.
- Min.
- Sec.

Average will be figured for you if you enter the miles-hours-min-sec fields.

Group Databases

Merit Badge Tracker

Merit Badge Tracker (or MBTrack) is a menu-driven program that stores valuable information on Merit Badge Counselors and presents that information in various reports of use to the Council, District, or Troop during its operation. Before any reports can be produced from the program, the user must give MBTrack certain data to process.

This input process starts by adding records to a counselor roster. It is continued by updating individual counselor records with the specific merit badges and other skills which that specific counselor is registered to counsel. Depending on the amount of data provided to MBTrack, the program generates Counselor rosters as well as a roster of Merit Badges and the Counselors (with addresses and telephone numbers) that counsel that particular badge.

Additionally, MBTrack produces an annual verification letter to each counselor so that the continued excellence of the program can be monitored by the responsible individuals within BSA. The number of scouts/scouters that the system can track is limited only by the amount of storage medium; systemically, MBTrack is designed to accommodate up to 999 counselors, although the actual limitation is more a matter of the storage medium used.

Specialty

Book Librarian

Book Librarian, an easy-to-use book database management system, is ideal for managing a personal, professional, church, club, or business library. You can quickly search your library by title, author, subject, or book type. The program prints reports to the screen, printer, or disk file and holds 16 information fields about each book:

- Title.
- Author (first and last).
- Subject.
- Book type.
- Edition.
- Pages.
- Copyright year.
- Catalog #.
- ISBN.
- Purchase date.
- Price.
- Publisher.
- Location.

- Note1.
- Note2.
- A memo field that can hold up to 30 lines of information about each title.

Database management functions include:

- Prev.
- Find.
- Top.
- Last.
- Edit.
- Add.
- Del.
- Search.
- Report.
- Browse.
- Quit.

Reports. Eight fixed format reports, plus a self-design report. Reports might be printed to screen, printer, or disk text file. The program can print on continuous form 3-×-5 cards.

Context-sensitive help. Press F1 to bring up a relevant help screen at any point in the program. Mouse support.

Screen saver. A built-in screen saver activates when the keyboard is not used at a user-customizable interval.

Browse and find windows are movable, scrollable, and resizable. You can find any title, subject, author, or book type in approximately one second or less. You can search the database on any combination of fields. The program can hold more than 2 billion titles, approximately 900 on a 360K floppy if the memo field is not used. The memo field uses disk space only if the field is used.

Book Minder

This database system is specifically designed for collectors, researchers, and others who like books or magazines. The program is completely menu-driven, easy to use, and allows three levels of information cataloging: standard, research, and librarian. You can catalog books, newspapers, magazines, articles, and chapters by title, author, subject, or any of 10 other variables. Entries can be made for individual articles. You can determine the location of the printed material by any one of up to 21 characteristics. Searches can be easily performed using complex multiple cross-references. The program has a capacity of 10 million entries per data file, and reports can be printed.

Find-X

Find-X, designed with the computer novice in mind, is easy to install and use. Totally menu-driven, very little computer knowledge is required to use it. This is a "universal indexing and cataloging" system to keep a short identification, up to 9 items of data of up to 30 characters each, a data, and an item number.

With this powerful system, you can conduct generic searches, search with Boolean AND and NOT logical parameters, nest parameters within parentheses, and search for words or phrases. This database is particularly useful for items whose characteristics are difficult to identify by casual inspection such as photographic slides, photographs or documents in files, record collections, items kept in different places, research notes, books, etc.

Information Unlimited

This is a collection of information databases and includes Demand Research Corp Directory, which gives phone numbers and addresses for more than 40+ computer companies, 1,000 MIS and human resources executives, and 800 major corporations and financial institutions. It also includes the Bates Directory of U.S. Daily Newspapers (1,500+).

Membership List

This program maintains membership lists for a club or organization of up to 2,000 members. Each member record contains the name, address, home and work phone number, whether or not dues are paid, the date when the dues are up, the committee the member belongs to, and a 40-character comment line.

The software addresses envelopes, folders, labels, and prints file cards and most any other form you can get for a 9.5-inch tractor feed. Address records can be printed by restricting to any field of information, including a particular day fees are due, or you can sort members into various categories. Other features include the ability to import and export data in several standard formats, and support for two- and three-across labels.

Membership Tracking System

This program keeps track of both membership and financial records of clubs or organizations in a single program. Its simple and intuitive menu system makes it easy for both computer novices and experienced users. The program tracks cash flow, attendance, and activities and provides a comprehensive set of reports and mailing labels.

Organize Everything

This is a collection of programs; each is a specialized database for a specific field. The fields covered are:

- Art collection.
- Books and magazines.

- Classical music.
- Coins.
- Comic books.
- Fabrics.
- Guns.
- Jazz.
- Memorabilia.
- Model railroad.
- Photographs.
- Plants/garden.
- Records and CDs.
- Sports cards.
- Stamps.
- Videotapes.
- Wine.

Professional Librarian

This data retrieval system stores citations of articles, books, etc., so you can later find them quickly. It includes add, delete, edit, browse, sort, search, and print functions. Records include subject, title, author, date, citation number, brief description, and more.

Spybase

This research tool provides source listings on tens of thousands of groups and individuals, including political organizations, foundations, celebrities, scientists, politicians, world leaders, and more. You can search by name or group, and it provides you with a list of books and periodicals (with page numbers and publication dates) containing the desired information.

Free-Form Database Managers

3×5

3×5 is an information management program that offers speed, simplicity, and versatility in entering, indexing, and retrieving free-text and structured data. It's as simple to use as 3-×-5 index cards. You can use the integrated text editor or any other word processor, text editor, or computer program to enter data in any format: words, phrases, sentences, paragraphs, or variable-length files.

Searching files is fast because of 3×5's unique indexing method (signature screening). Just enter a Search key, and the software finds the first and all subsequent occurrences of the key in your file. Moreover, when the key is found, it is displayed in the context of the sentence or record in which it appears. As each occurrence is found, you can continue the search, edit the record, print the information, transfer records from one file to another, or end the search.

You can search for partial keys, i.e., an incomplete word or phrase. And you can search for combinations of key terms in the same sentence or record by using the conjunctions "AND" (symbol = &) or "OR" (symbol = ¦). You can limit a search to specific fields, or you can search and edit up to five files at a time. Standard ASCII, WordStar and WordPerfect document files can be searched directly or imported in 3-×-5 files for editing and indexing, and 3-×-5 records can be exported to ASCII, WordStar, and WordPerfect files.

Here are the features of the program:

- Up to 15 variable-length fields.
- Sort records by one or two fields when copying or printing.
- Underline and bold commands.
- Improved indexing for records greater than 512 bytes.
- Faster searches.
- Multiple print formats for mailing labels.
- 3-×-5 cards.

Here are some suggested applications for 3×5:

- Indexing personal literature collections.
- Quickly searching client records.
- Card catalogs for small libraries.
- Mailing lists.
- Cataloging and cross-referencing research notes.
- Writing notes and memos.
- Cataloging paper documents.
- Slide and video collections.
- Parts inventories.

Freebase

Freebase is a free-form database program that features:

- Free-form data entry using any ASCII text editor.
- Rapid search and unlimited narrowing of the search records by subsequent searches using case-insensitive search strings.
- Ability to print search records.
- Ability to write search records to a disk file.
- Keywords to link records.

Getit

Getit can search for any word or string within your file. It is not case-sensitive, so you can use any combination of upper and lowercase characters. Your searches can be for a single word or string two separate words or strings related by AND or OR. If the AND relationship is used, both search "keys" must be present in one record to be successful. If the OR relationship is used, only one of the keys needs to be present to have a successful search. Getit also indexes your records. Indexing is the ability to make your records appear in a predefined order.

As a reminder program, Getit keeps track of appointments, birthdays, anniversaries, or any other information that is tied to a date. When starting Getit, only the GETIT.DBF database is searched for "remind" dates corresponding to the current system month and day (year is disregarded). If a match is found, it is displayed on the screen. You are then asked if you want to search for further reminders for the day. With this in mind, you might want to have Getit started by your AUTOEXEC.BAT file so you can see your reminders first thing every day. Searches for a particular reminder date can be initiated if you want to check for something in the future or the past. Output from Getit can be directed to a printer or to a file.

Information Please

This text storage and retrieval program stores and retrieves full-text information such as memos, catalogues, briefs, letters, invoices, articles, product descriptions, scientific papers, and so on. Enter information from a keyboard or import information from an ASCII file. Search, display, and print entries using the keyword or full-text searches that allow multiple AND and OR criteria. Relate text to a PCX graphic or another program.

The user interface is an easy-to-learn pull-down menu (Windows-like) interface. Example "textbases" databases could store information such as: how to respond to PC-DOS error messages, information on toxic chemicals, directory of national parks, poison control information, franchise opportunities, and patent infringement case studies.

Instant Recall

Instant Recall is ideal for the busy person who has many ideas and doesn't know where to store them: the business executive, the writer, the personnel administrator. This program replaces awkward filing cabinets and restrictive database systems with an instant cross-referencing system that lets you store bits and pieces and retrieve them together so you can turn them into a coherent whole. And it lets you cut and paste information from Instant Recall's records right into another document without exiting from either program. It is a memory-resident, free-form database program with its own word processor. Data can be entered in whatever form you like, without worrying about field sizes, and the date can be recalled without worrying about keywords. Because it runs at the same time as your other programs, the process is highly interactive; you can quickly get the data you need, or enter new data and go on working.

Here are some applications: You can use it as an index to where you've put notes, articles, etc., in your file cabinets, desk drawers, or computer files. You'll never forget where that important data is again. You can use it to save bits and pieces of information that you'll need later but you didn't have a convenient place to put it—things like airline flight information you often use, people's birthdays, long-off meeting dates, ideas you've had that need time to percolate, to-do lists, notes on how to use other programs, notes on documents and spreadsheets you're working on, etc.

You can use it to gather together information that was ostensibly unrelated, or maybe very loosely related, when you typed it in (typed at different times). When you need to go back and find all of the information you have about a particular project or subject, it's easy . . . and you didn't have to worry about where to put the data when you typed it in, or how you would get it back. You can use Instant Recall for all of these things, and more. Best of all, you can do all of these at the same time, without clutter.

Memo Master

Memo Master is a free-form data manager that lets you enter information in your own words—totally unstructured—and retrieve it the same way. You don't have to designate a keyword or remember where you filed something. Whenever you need the information again, just pick any word in the record, and that's all there is to it. Even people who keep accurate notes don't always remember where they put them. Many important notes, phone numbers, etc., end up scribbled on crumpled envelopes or on matchbook covers that end up in the twilight zone.

Memo Master organizes all your random information for you. You can even have the program remind you of certain records on a given date. Forget those little yellow sticky notes that end up sticking where they don't belong most of the time. This software puts an end to digging through the trash or tearing your desk apart trying to find important notes. Just go to your PC, load the program, and everything is right at your fingertips. If you need a hard copy, you can easily locate a record and print it out. You can even import existing ASCII text files into the program. Features:

- Add new records to the database. You can import an ASCII text file, but the maximum file read is 32,000 bytes.
- Browse through each record.
- Change databases.
- Delete database data.
- Floppy data backup.
- Help file. You can search by REMIND DATE, CREATED or UPDATED DATE, and by TEXT.
- Create a new database.
- Sort database records.

ProFile

ProFile (professional filing system) is a practical database manager. It helps manage customer lists, inventory, and just about anything. It prints reports, invoices, labels, and form letters with its mail merge file feature. It also helps prepare bulk mailings with its advanced bulk mail sort. ProFile shares information with other software programs such as word processors, spreadsheets, or database programs.

The program gets you started quickly and easily. After you are comfortable with the system, you can expand it to use its more powerful features as you need them. No computer experience is required. Help is only a keystroke away, any time you need it.

ProFile supports linking multiple databases to one data entry form through the use of lookup fields. For instance, you could create a customer invoice form linked to a customer database and an inventory database. By entering a customer name, the software could find the customer address in the customer database and place it in the invoice form. Next, when you entered a part number, ProFile could retrieve a part name and cost for you. Lookup fields can either be stored in the current database or only referenced (stored in an external database). Lookup fields are very flexible and help you create very complex forms that are easy to use and speed data entry. Another feature is the ability to create a calculated field in a form.

Calculated fields allow you to perform math on the numerical fields in a database. You enter a formula that might reference other fields in the current database (or other databases via lookup fields). The formula might contain multiply, divide, add, subtract, exponent, plus nine functions (ABS, SQRT, SQR, SIN, COS, ARCTAN, LN, LOG, EXP). Using the preceding example for an invoice form, the form might include fields that extend the price of an item (price, quantity) and then total all of the items purchased.

Bibliographical Databases

Biblogic

Biblogic, a special type of database, can be used as a research/writing tool. This program lets you store, retrieve, and print bibliographic information and related remarks. Biblogic creates selected reference lists in many styles to satisfy the varied requirements of different publishers (e.g., Johnson, HL and Taylor, WV or H.L. Johnson, W.V. Taylor). It also lets you organize your own notes on references. The comments relating to a particular topic can then be selectively combined.

Articles are categorized by a hierarchical numbering system that you define. So retrieval is simpler than using keywords. You keep your comments on the articles in a different area so you can enter multiple reviews for a single reference and call up reviews on different aspects of an article independently. Output from the program is text files that you can input directly into most word processors. You can develop several databases on one computer. A database can contain several thousand references. Information from telelibraries can be read into a Biblogic database. A number of search methods let you locate a reference, even when it is not completely categorized. For example, you can search for a specific word in the title to locate it.

Jeprs

Jeprs (pronounced Jeepers) is the Journal and Editorial Personal Reference System. This menu-driven system keeps track of reference citations to the (scientific) literature. Jeprs lets you:

- Add references to your database (either directly, or by importing a MEDLINE, BRS/Colleague, ERIC, Psyclit, or comma-delimited formatted files such as Paper-Chase or Current Choice.
- Search these references with an easy-to-use but powerful menu system.
- Format these references for output to a text file, printer, or word processor in most formats and not lose boldface, italics, underlining, subscripts, and superscripts.

Online, context-sensitive help is available. A mouse can be used. Up to 64,000 references can be accommodated. References can be exported into MEDLINE or comma-delimited formats. Up to 30 formats of your own choosing can be used for exporting. Output can be sent to any combination of screen, printer, text-file, or word processor file. References can be sorted by author, year, or journal. Searches can be made not only for a single author and journal, but also for several authors in several journals using Boolean AND/OR/NOT logic, such as Hansen and Barstow or Mikey and Tennyson or Medicine and Nature and Psychomatic and English.

Miscellaneous Databases

Names and Dates

This is a combination address book, mailing list manager, and date reminder system. You can group names into categories to keep separate lists of clients, club members, vendors, etc. This program stores any type of address and allows you to attach up to eight lines of notes. There is an on-screen calendar with date event reminders. The program prints address books, Rolodex cards, envelopes, variable-size labels and wall calendars.

Software Inventory System

This database program is for keeping a detailed and informative inventory of all your software programs. With the Software Inventory System, you can keep track of how much you've spent on software compared to retail cost, the backup status of your program diskettes, ratings for each program, average ratings for groups of software, and much more.

The Software Inventory System is designed for maximum ease-of-use, quick and intuitive data entry, and concise output in the form of a variety of informative reports. Online help is within reach from almost anywhere in the program simply by pressing the F1 key.

Software Suzie

This program keeps track of all the computer software a person acquires. It keeps track of: software name/title; publisher (address, telephone number, technical support line); cost, place acquired; date acquired; vendor, description; category; version; serial number; date registered (for shareware); comments about the program, etc. The program prints records to disk or file.

TerraTime

This program displays a map of the world and shows the local time all over the globe so you know when you can call foreign contacts during the appropriate hours.

Vendor List

This list contains commercial, shareware, and miscellaneous bulletin boards, vendor sales support, technical support, and fax phone numbers that the author has found useful as the Senior Analyst/Programmer and Operations Manager for a small DP Service Bureau, Computer Consultant, and later as a Senior Network Support Engineer for a major national distributor of computer hardware and software. Much of the list has been compiled from industry publications such as *Byte*, *PC Magazine*, *PC Week*, *PC World*, *InfoWorld*, *Service News*, *Lan Times*, *Lan Technology*, *Lan*, *Lan Computing*, *UNIS World*, *UNIX Today*, *Lotus*, *Data Base Advisor*, *Computer Language*, *Dr Dobbs Journal*, *Computer Shopper*, *Communications Week*, etc. Every effort has been made to ensure the accuracy of this list.

World

In case you have to locate a specific city or country, this program displays a map of the world in five different sizes (down to about the size of a state). It includes city finder, distance calculator, and state/nation capitols locator.

Mail Lists

Ultimate MailList Manager

Ultimate MailList Manager (UMM) is an exceptionally easy-to-use software program that allows you to store detailed mail list information by category. You can maintain a separate category for customers, prospects, friends, members of your organization, and much more. The ability to maintain unlimited detailed notes by date, time, and subject for each entry makes this an excellent sales prospect system. There are even several user-definable fields. Organize a special mailing that spans categories (i.e., some of your associates, a few of your prospects associates, and many of your customers). Ultimate Mail List Manager does this and much more.

With this program you can also import and export data in three different formats, print mailing labels in one across or two across formats, use the optional Report Writer to define your own formats, or export data into one of several graphical label programs. It is completely menu-driven. Pop-up help screens are always available. There are scrolling lookup tables throughout the system. Menu options include:

- Maintain mail lists by category.
- Maintain mail lists by name.
- Mail list categories.
- Import data.
- Export data.
- “Special mailing” utility.

- User-definable parameters.
- Mail list category descriptions.
- Telephone directory.
- Mail list file listing.
- Notepad subject summary.
- Label test pattern (one up).
- By mail list category (one up).
- Special mailing (one up).
- Label test pattern (two up).
- By mail list category (two up).
- Special mailing (two up).

zipMail

This program handles up to eight separate databases online, unlimited offline. Labels are printed one-up to four-up (Cheshire).

- A unique "rapid review" mode allows edit, print, delete, recall, transform, insert, and selective fields on edit or print.
- Auto dialer and many more features.
- Import WordStar or ASCII letters for MailMerge purposes.
- Up to 36 "search codes" (customer unique) per database, with individual or combination code selection for search/print functions.
- Appends "mailing codes" for automated follow-ups.
- Automated nonlabel lists.
- "Code count" to predetermine needed mailing pieces ahead of time.
- Optional "repeat addresses" in add and rapid review modes.
- Function keys in add mode can be personalized.
- Automatic appending of city with entry of zip code (must have Zip.dbf in \zipmail directory—download zip_CITY.zip).
- Online checking for possible duplicates in add mode.
- Automatic, selectable salutations (record unique).
- Management and deletion of duplicate entries handled in three ways: automatic alert during the add function.
- Interactive screen display of duplicates.
- Print report of all potential duplicates. Context-sensitive help <F1> at most data entry points (GETS). Download shareware Ezeditor by John E. Bean.
- P.E. (ZMAILSUP.zip). It interacts with the Zipmail mail/merge program.

Database Utilities

DBFEdit

You can use DBFEdit to:

- View several database files at the same time and edit any one you like.
- Edit the fields of a record.
- Cut and paste between different database files and different fields.
- View the file structure.
- Send the file structure to a printer.
- Mark a block for delete and recall.
- Delete and recall a block of records.
- Copy a block of records in the same file.
- Print a block of records.
- Draw a line graph on a block of records.
- Draw a bar graph on a block of records.
- Draw a stack graph on a block of records.
- Save the block of record as an ASCII text file.
- Search a file for certain string.
- Change default color.
- Use a pop-up calculator.
- Calculate the sum.
- Calculate the mean.
- Calculate the maximum and minimum.
- Calculate the standard deviation.
- Calculate the variance.
- View an index expression of an index file.
- Copy file structure to another file.
- Filter a record.
- View the filter expression.
- Append a record.
- Go to anywhere in the database through the use of simple key combinations.
- Delete a record.
- Recall a record.
- Exit to DOS shell.

dBASE Utilities

This program uses and manages dBASE files, giving you instant access to the files. A report writer compiles information for any type of report that prints to disk, file, or printer. The program also includes utilities to create merge files for WordPerfect, and it includes a search utility and tutorial for dBASE programmers.

PDT Tool Kit

The PDT Tool Kit includes several things to help you manage, edit, convert, and repair data files—dBASE files or fixed-length files created by other programs. Uses include:

- Quickly create dBASE files with any field structure.
- Create "dictionaries" of field names to serve as a permanent reminder of what fields in dBASE files are for.
- Create "subsets" of (or modify) dBASE files; you specify which fields to include.
- Delete fields, add fields, and change field lengths, types, number of decimal places, etc.
- Quickly convert dBASE files to comma-delimited ASCII files.
- Or, do the opposite: read ASCII files and save the data to dBASE files. These small, fast programs simplify sharing data between different programs.
- Review, understand, or repair dBASE files.

Odd

Odd is an easy-to-use program that provides several utilities often not included in many dBASE-compatible database programs and applications. These utilities include: optimal database size analysis, duplicate and nonduplicate record checking, and distribution analysis. All can be executed with a filter (selected search criteria) applied.

Odd does not alter database information in any way. All database modifications must be made in the original application or database program.

Optimal size analysis. The optimal database size analysis feature simply scans the database and reports the maximum amount of field usage for character and numeric fields. It helps generate reports and modify database structure.

Duplicate and nonduplicate record reporting. The duplicate and nonduplicate record reporting feature helps you find duplicate records and nonduplicate records through a menu-created "key." The duplicate key can be all or part of a field(s), and can be issued phonetically (sounds like). For convenience, the "key" can be saved, then recalled with a single keystroke.

Distribution analysis is an indispensable tool for many users, but, unfortunately, it is not included in many packages. The distribution analysis has two modes of operation: manual and auto. In the auto mode, the frequency distribution of numeric, date, and character fields can be generated with minimal keystrokes. The manual mode has the same functionality as the auto mode, but it also allows the user to customize the partition sizes and use dBASE expressions.

Each mode reports basic statistics such as total, average, min, max, variance, and standard deviation. Some nonstandard (but interesting) statistics are also reported. Distribution analysis keys can be saved and then recalled with a single keystroke.

Chapter

8

Scheduling, Estimating, and Finances

For many, cost estimating, scheduling and creating milestones, updating cost estimates and schedules, and attending and participating in meetings where these topics are covered are tasks that are demanding, dreaded, and disliked. Yet these activities are indispensable to the success of engineering and science projects.

Although the dart-board approach for cost estimating and scheduling has much to recommend it (easy and quick) and is still being used occasionally (especially for estimating government projects), the advent of the PC, along with outstanding "what-if" software, has convinced more and more scientific and engineering professionals to abandon the multicircled prognosticator and rely on these excellent modern tools for predictions and updates. In the pre-PC age (actually that wasn't too long ago), using "what-if" speculations was laborious and virtually impossible for large programs. But now modern software accomplishes these tasks almost instantly and effortlessly.

Early project management software was programmed to run on huge computer mainframes and track gigantic projects. Organizations such as NASA and the Navy were instrumental in their initial development. If you had an atomic submarine to build, or an orbiting satellite to schedule a launch for, you used an enormously complex project management package. However, with the advent of small computers, everything has changed. Now, just as you can use a PC spreadsheet package to perform once impossibly complex analyses, you can also use PC project management software to obtain the same benefits once reserved for large scientific and engineering organizations.

What If?

For effective scheduling, you usually begin with start and finish dates and select milestones (which hopefully do not become millstones) to monitor progress along the way; you then input this data into your computer program. Invariably, however, the schedule bulges out because there are too many milestones, too little time, and too small a budget to accomplish what you would like. So you must seek compromises—alternative methods of accomplishing tasks. Do not be discouraged. You simply ask yourself several what-ifs, such as, what if we:

- Assign more personnel to the project?
- Extend the completion date?
- Change the specs?
- Go on an overtime schedule?

With modern scheduling and budgeting software, these variables can be input, their impact immediately evaluated, and a revised schedule and cost reflecting these changes can be created instantly by the software. These what-if speculations can be used not only during the initial creation of the schedule and budget, but also continuously for troubleshooting during the performance or nonperformance of a project.

The ongoing data records you accumulate, the history you amass as the program progresses, is another vital benefit of computerized cost estimating and scheduling. These records are invaluable assets and serve as "20-20 hindsight" data for analysis and reference in estimating future projects. This data lets you identify the optimists, the pessimists, and the realists in the engineering and scientific staff.

Estimating and scheduling are pretty much learn-as-you-go processes and grow-as-you-learn activities. Software stores actual versus estimated data and recalls this past data for reference as you undertake new programs.

The software covered in this chapter provides the basic tools for estimating cost and schedules, for what-if inputs, for continuous updating of costs and schedules, and for recording your estimates versus actuals for the duration of the entire project.

Personal Information Managers (PIMs)

Alarm

Alarm provides a memory-resident, on-screen clock in the upper- or lower-right corner. On the hour and half-hour, a short tune is played (the tunes are programmable by you). The program stores up to 10 alarm times per day, which are indicated by another tune and a flashing reminder message at the bottom of the screen, all of which are programmable. The clock display, chimes, and alarms can be toggled on or off individually. Alarms can be specified for day of the week, daily, or single-time activation.

Organize!

Organize! is an innovative, cost-effective PIM for Windows. No matter how disorganized you are, this program tracks all of your important projects without the complex and rigid control systems most PIMs use. Cluttered with unused functions, other systems steepen the learning curve unnecessarily. With Organize!, you add new functions only as you need them. Learn as you go.

One of the easiest-to-use systems ever, Organize! looks and feels just like a diary. You flip pages, like the real thing, with binders that click open to insert more pages. The Word-in-Context system manages all random and structured personal information that you enter into your diary. You never lose your way with this software. If you're looking for a program to help you record contacts, schedule appointments, track projects, or control random notes and ideas, this might be your solution.

RAMDisk

RAMDisk is a handy program for the typical small-business user. RAMDisk will:

- Schedule upcoming events and allow you to set up to 20 alarms per day.
- Pass messages among workstations on a LAN.
- Act as an electronic "in-out" board to keep track of where office workers have gone.
- Hold phone numbers, addresses, and notes about your acquaintances.
- Maintain a record of your expenses and the time you spend working on various projects.
- Allow you to establish numerous minidatabases, such as to-do lists, general ledger account numbers, and client number cross-references.
- Provide a programmable calculator with a long "tape," the ability to enter corrections into a long sequence of calculations, and the capability to transfer results to other programs.
- Coexist and exchange data with TickleX (the ultimate scheduler) and BillPower (integrated Time/Billing/Bookkeeping/GL).

Remember

Remember is a computerized address book, mailing list manager, and date reminder system. It stores an unlimited number of names, depending on the amount of disk storage you have available. You can have an unlimited number of notes (comments) about each name, or you can group your names into categories to keep track of clients, club members, pizza places, or any other grouping that is important. Each name can be associated with any number of "date reminders" such as birthdays, appointments, etc. You can also have an unlimited number of general "date reminders" that are not associated with names (for holidays, etc.).

Remember has unusually large areas for name-and-address information. The flexible address format is designed for U.S., Australian, and Canadian addresses, but it works quite well for all countries. The mail merge facility has optional salutations that can have default or custom greeting options. The screen calendar function displays any month you choose, with "date reminders" highlighted and date details available. Printed calendar pages, similar to a monthly wall calendar, are personalized with your date reminders. Features include:

- Prints wall calendar with your date reminder messages.
- Prints mailing labels (10 formats, including laser).
- Prints Rolodex cards.
- Sequence by name, company, or zip.
- Six report formats.
- Addresses envelopes (including laser).
- Five-line address, thirty-three characters/line.
- Supports foreign address formats.
- Has repetitious date add.
- Holds four phone numbers (home, business with extension, fax, and other).
- Unlimited notes area that uses disk space only if filled in.
- Mail merge with several salutation options.
- Export, import, group names/dates by categories.
- Print individual labels or envelopes.
- Several mass name/date/note processing options.
- Holds an unlimited number of date reminders.
- Context-sensitive help.
- Documentation file on disk.
- Text scan.
- Resume print at any point (label jams, etc.).
- Print individual roster pages when information changes.
- Date alarms.

Reminders

Reminders is a perpetual calendar, coupled with a tickler file, for important activities and dates. All date reminders are organized on the month and day of date. After you enter a date for a birthday, anniversary, or accounting deadline once, that date reminder is then shown on the correct day of the week, year after year.

Display reminders are by date, code, date range, code and date range, or in summary for any one month in calendar format. You can list to printer by date range or code and date range. This product is ideal for newsletters, clubs, groups, or your personal special date book. Features:

- Menu-driven.
- Mono or color monitor.

- Help on all input fields.
- Memory space check.
- Self-installing.
- Can be personalized.

Reminder for Windows

Windows Reminder is a new time-management program written exclusively for Windows. The program is simple to use, yet powerful enough to do your work for you without you having to be there.

Windows Reminder lets you set up and view your appointments as easily as possible. Items or tasks are sorted by date and time, which places all items needing attention at the top so you can "see" what is next on your list. A time is not required, so you can still enter a task that needs to be done that day. Enter in a time, and Windows Reminder automatically sets the alarm to display your message or run other programs. Windows Reminder displays the time when running as an icon so you can always keep track of what time it is.

Set up "to-do" or "task" lists for yourself and others to keep track of dates, times, and milestones. Set alarms to remind you of items that have to be done or places you have to be. You can set the reminder to chime on the hour, half-hour, and quarter-hour. If you have Multimedia installed, you will hear the chimes from Control Panel. Set alarms to start other tasks without your being there.

Generate status reports at any time based on the items you have done and the items that are still on your list of items that need to be done or need your attention. Collect the time spent on each item for time management reports. Windows Reminder keeps track of when items are completed and creates a log file when you delete your items to a file.

Schedules

Easy Project (Simple Project)

Easy Project (AKA Simple Project) is a project management tool to assist computer professionals in the following areas:

- Evaluating projects to determine risk, financial attractiveness, and cash flow.
- Planning all aspects of a project, including task decomposition (work breakdown structure), labor and financial requirements, task interrelationships (dependencies), risk, and the actual schedule.
- Tracking expenditures, including both hours and dollars, as they are incurred.
- Predicting overruns or underruns as early as possible.

Some of the more innovative Easy Project features in this area are:

- Use of a hierarchical approach to entering and viewing project information.
- Project windowing support.
- Screen-oriented data entry with field-specific prompts displayed.

Easy Project

Easy project is a tool to assist anyone who manages projects. The program encourages the manager to define a project into a series of phases and tasks. Easy Project facilitates the planning, scheduling, and tracking of all types of projects. The manager plans the activities or tasks in a top/down outline fashion.

The GANTT chart displays the project graphically by using time bars that span the time period of each task. The extensive reporting module helps document the status of the project. Easy Project features:

- The ability to create projects with 2000 tasks.
- GANTT charts detailing planned versus actual progress.
- Automatic scheduling.
- Extensive reporting.
- Context-sensitive help.
- dBASE file compatibility.

A sample project is provided to assist in using Easy Project.

Employee Scheduling Assistant

This program helps department managers keep track of vacations and other activities such as sick leave, education, personal days off, work time and wages, shift assignments etc. Help is available by pressing the F1 key. "Pop-up windows" and "bounce-bar" menu selections make the Employee Scheduling Assistant very user-friendly. Two editors are provided.

Editors are screens displaying calendars that show weekends and special event days such as company holidays that you can specify. Editors also allow you to make changes to current schedules. A schedule change in one editor changes the data in both editors.

Editors can be viewed for an entire year for one employee; the department schedule can be shown for each month or for a 24-hour period. Work start and stop times are graphically displayed and can be sorted by start and stop time. A window can be opened for every date, and each employee can leave a 4-line, 30-character-per-line note. Seven types of reports are available:

- An employee schedule report (one year).
- A department report (by requested month).
- A seven-day report for posting.
- A department statistical report.
- Selected employee information printout.
- Messages for employees between two dates.
- All messages for the department for one date.

EasyTrend

EasyTrend is for forecasting future projections, based on historical data, and is especially useful for managers who do not have extensive knowledge of statistics.

At the main menu, four available options are displayed:

- Forecast for a single specific future time. Use this option for any forecast where a specific figure is needed, either a given level of activity at some unknown time, or a unknown activity level at a specified future time.
- Forecast for a time range (trend analysis). This option helps you to forecast a trend in future values based on historical data.

The remaining two options perform the same basic function. With EasyTrend you can store your data used in calculations to a file, which can later be used for other calculations.

GANTT

GANTT is a charting program that can be used in a variety of ways. The charts can be created from a variety of sources. GANTT displays are commonly used throughout industry for management presentations of project status and resource use. Time is displayed on the horizontal or x-axis, and a list of items are displayed on the vertical axis. A horizontal line or bar is drawn from the starting time to the ending time for each item.

This program has four primary applications:

- Produces GANTT chart transparencies for meeting displays.
- Directly drives a video projector from a PC for meeting displays.
- Creates GANTT charts to be inserted into the text files of word processors.
- Is used directly to view project status.

To get you going, a DEMO, a TUTOR, and a "sister" program, SCHEDULE, a limited working model of another sophisticated project management tool, are provided.

GANTT charts can be printed directly on an IBM matrix or graphics printer. GANTT allows you to direct the printout to a disk file rather than to the printer; this creates a file that can be embedded into documents through the use of a word-processing program.

GANTT Charting

GANTT Charting produces GANTT charts lengthwise on paper. Up to 20 task names of 30 characters or less are input, and the maximum time span ranges from 36 months to 24 months, depending on the length of the longest task. Time span increments are in weeks, space permitting.

PC-Project

PC-Project lets you organize and control complex projects. It uses the critical path method to produce a project schedule displayed in the form of a GANTT chart. It helps you manage complex jobs so you can finish them on time and under budget. PC-Project views a project as a group of tasks or activities. When you supply the estimated duration of each task, and the relationships between the tasks (the predecessor tasks and successor tasks), it automatically calculates a schedule for you and displays it as a GANTT chart. With this chart, you can see when each task should begin and end, the current status of each task, and whether or not the task is on the critical path.

Like a spreadsheet, PC-Project immediately shows you how any change alters the project's finish date and cost. This lets you try repeated "what-if" scenarios until you attain the optimal project schedule.

PC-Project also has a resource worksheet to track payroll, supplies, and equipment costs. The worksheet lets you list your employees and equipment and assign them to tasks, so you can coordinate your resources efficiently. The resource worksheet tracks costs to date, anticipated costs, final cost, and cash flow.

Project Management

This program contains a complete PC Project Management system. It handles 1,000 tasks, subcontractors, variable start dates, and holidays.

Project Management uses CPM (critical path method). It calculates the critical path, cash flow, and cost reports, and it produces GANTT charts and precedence networks. The system handles all necessary functions internally or allows sophisticated users to build and modify files making use of system editing software. PCPM is almost completely menu-driven and can handle up to 500 tasks, subcontractors, variable start dates, and holidays. It is a surprisingly sophisticated tool for those in need of a good planning utility.

TaskMan

TaskMan is a task-tracking program created to help keep track of tasks or activities. Anyone needing to keep tasks or activities organized can make good use of this program. It is an inexpensive alternative to the project management programs that are commercially available. Many commercial programs require more information to be entered than is needed to manage many projects.

This program is self-documented through the use of an integrated help file. The tasks, resource, and status databases, along with their related key files, are created the first time TaskMan is run. The program has a very flexible printing routine. You have the option of either printing all of the tasks or selecting a subset of tasks based on a variety of selection criteria.

TaskMaster

TaskMaster, an all-purpose, menu-driven project management system, helps you manage all types of projects, assignments, tasks, schedules, correspondence, etc. This schedule and task-tracking program can create a database with customized forms/records that describe your activities, such as individual tasks and schedules for a particular project. And you can print custom reports and list all past-due assignments, all assignments for a particular project, which assignments are due for completion on a specific date, etc.

It supports specialized forms and records; custom reports track the progress of multiple projects. Menu-driven, it uses pull-down menus and is easy to use. Also provided are a pop-up calendar and a calculator for constructing and editing your records and forms. The software is compatible with standard dBASE III+ database files and supports an unlimited number of records.

TimeTrak

TimeTrak provides a proportional display of time-dependent events. It can be used at a single workstation or simultaneously at many terminals on a computer network. Up to four schedules (with up to 999 resources in each schedule) can be viewed simultaneously so that normally isolated events can be compared in the most convenient way. In addition, each schedule can be split into multiple views with different display attributes—to a limit of eight splits on screen at one time.

Individual events can be located by time or by matching the contents of a data field within the event record. Up to 20 user-defined fields can be included within each event data record. Detailed information for each event is available simply by locating the cursor over the event and pressing the Enter key. A user-defined window is displayed that contains all the details that were stored with the selected event.

Typical applications of TimeTrak are:

- Real-time event management with one-second resolution.
- Personal time management.
- Staff scheduling and leave planning.
- Time management for skills, resources, and facilities.
- Work schedules by hour/day/week.

Estimating

Business Analysis

Business Analysis is a collection of 16 financial and statistical business programs. Some of the topics covered include accounting, asset management, labor, inventory, and stock holders. The programs help both large and small companies obtain a better picture of how their company is being run. By using these programs, the user can make decisions about the day-to-day operations of their company.

Budget Analysis is a financial program that helps you get a picture of your spending patterns in relation to your income. The program gives a summary of your income and expenditures, as well as the percentage of surplus or deficit in relation to your income.

Business Financial Analysis produces a combination income and cash flow analysis projection for a company for a five-year period, based on projections supplied by the user. It can be used for either a service or product-oriented company. By changing the projections, the program can also be used to do what-if calculations.

E.I.M.S.

E.I.M.S. (Equipment Inventory and Maintenance Scheduler) maintains an equipment inventory, user-defined maintenance costs for each item, and scheduled maintenance for each item. Maintenance can be scheduled for almost anything needed. Although based on time schedules, this can still be used for at least checking other maintenance-based systems.

The program offers virtually unlimited record capacities. It also uses auto lookup tables and scrolling record windows for ease of use. Depending on your particular needs, a quick check can be made periodically to see what maintenance is due. You'll probably want to print scheduled reports on a timely basis for this purpose. A date prompt is provided so as to enter a later date if desired. This lets you plan a schedule for perhaps a week or month ahead of time. A display option is also available for fast checks and maintenance updates.

Although the program is primarily designed for scheduled maintenance and costs, repairs can be maintained just as well. Scheduled maintenance days have to be provided, however. To avoid listing repair records in the scheduled maintenance reports and displays, create separate records for this purpose. You can enter a large number of days such as 9999, which will create a future date that will be meaningless for all practical purposes.

Job Cost

This Job Cost system provides a simple and straightforward method of entering budgets, capturing costs, capturing estimated percent completions, and projecting the status of complex projects. This capability of projection of status of any single item of budget prior to completion, along with the capability to summarize all items of cost to the job level and its state of completion, allow for the correction of trends that will cause a loss of dollars long before those trends have a major impact on the final status of any one job.

Reports produced include:

- Edit reports for Expenses Actually Incurred.
- Summary of all Jobs. Prints a job status summary report in job ID order sequence of all jobs currently on file.
- Job Status Report can be done by job and cost center for all jobs or for a specific job.
- Job Recap. Expense posting recap for all jobs, a specific job, and with or without activity date framing.

PC-Estimator

PC-Estimator is a complete, stand-alone system that does not require the use or assistance of any other software. All you need is your PC. This program is high-productivity software for use by general and specialty contractors, engineers, architects, manufacturers, cost estimators, consultants, and other individuals who need to estimate costs for projects of all sizes and types, from bridges to electronic equipment.

PC-Estimator is easy to use, and it increases efficiency and productivity in cost estimating. It was developed by construction professionals who are also experts in computers and represents a powerful cost-estimating productivity tool. Some benefits to be gained are:

- A complete, stand-alone system.
- No other software necessary.
- Immediate on-screen access to data through lookup windows.
- Material price list and cost data accumulation.
- Advanced, pull-down, menu-driven structure.
- Powerful, time-saving work assembly feature.
- Built-in digitizer interface.
- Instant recap for "what-if" analyses.
- Unlimited size databases; they grow as your applications grow.
- Databases for markups.
- Labor and equipment rates.
- Standard cost codes.
- Vendor tracking, etc.
- Different markups applicable for each estimate.
- Multiple estimates carried out simultaneously.
- Built-in comprehensive reporting.
- Several different levels of detail in estimate reporting, including itemized and built-in mark.
- On-screen reporting for instant "bottom-line" evaluation.
- Bid item reporting for heavy contractors.
- Supports standard and wide-track printers.
- On-screen takeoff calculator with memory.
- Many other advanced features.

PC-Estimator is based on a user-definable cost code and unit cost structure, so it is suitable for a wide range of industries, trades, and private and public organizations, ranging in size from one-person to medium, to large-size contracting companies, engineers, and architects. Because of PC-Estimator's optional industry-standard cost code database and sample structured estimates, even people with little or no cost-estimating experience can create a detailed estimate of a project in a short amount of time.

Project Managers

PC Project

PC Project is a project management software package for organizing and controlling complex projects. It offers numerous features, some not found in products costing many times as much. It's easy to use and features Lotus-like menu-driven operation. Complete context-sensitive help screens are provided along with a complete display with constantly updated critical path. Easy-to-remember names, rather than numbers, are used to identify tasks and resources.

The number of tasks and resources are limited only by installed memory. The number of resources and prerequisites for each task are limited only by installed memory. Tasks can be assigned fractional resources, and fractional resources can be assigned to the project. You can view tasks by resource, as well as resources by task, and you can view prerequisites for each task, as well as successors, those tasks dependent on a particular task's completion. Cost remaining is always available by resource or project. You can view a bar chart for resource loading, and you can use the numeric keypad for all data entry. The software provides quick operation with speedy updates.

PC Project lets you break a project down into smaller tasks so you can estimate the required time for each. After you input the relationships among these tasks (i.e., which come first and which later), a complete schedule is automatically drawn up. Also, PC-Project allows you to assign employees or equipment to tasks, then the program uses this information to give you complete cost breakdowns.

After work has started, the software helps you keep a tight rein on these costs and your scheduled completion. You are always shown the crucial path, that sequence of tasks in which any delay will result in late delivery. Given any change in your schedule, you are instantly informed of changes in costs or deadlines. Because of its unique capabilities, project management software is destined to join the spreadsheet, database, and word processor as an essential microcomputer software package.

Quick GANTT

The GANTT chart is a popular way of displaying schedules. It is constructed with a calendar on the horizontal axis and a list of events or tasks on the vertical axis. The duration of the task is shown as the horizontal line. The GANTT chart tends to have a stepped or "waterfall" appearance as the tasks go down the side and the duration line moves further to the right.

Another tendency is for many of the tasks to be dependent on one another. The second task often must wait for the first task to be completed before it can start. This dependency can be useful in constructing the chart. For example, if you know when task one is to begin and its flow time (duration), that task two begins with the end of task one, and that task three begins with the end of task two, etc., all that is required to build dependencies in the chart is to know when task one starts, the flow time of each task, and the order of task execution. If task one is delayed, then the second task is also delayed.

As a result, the chart can be thought of as a series of dependencies. GANTT charts are popular for project control because they are easy to construct and graphically show the tasks, their dependencies, and their durations. As a task is worked or completed, it is often "colored" in to more easily display project status at any point in time.

Some scheduling programs account for resources, cost, and other items. But in most projects, the schedule is the overriding concern. The need then is for a simple tool that can depict tasks, their flow time, and their relationships, thereby providing an overall view of when each task of a project is predicted to occur.

The objective of Quick GANTT is to rapidly create, maintain, and print a project schedule in the form of a GANTT chart. Quick GANTT treats tasks in a schedule very much like a spreadsheet treats cells. Tasks can be set up as dependent on one another, such that if one changes, all dependent tasks are changed to agree.

Up to 500 tasks can be scheduled over any time span. Weekends can be included or excluded, along with specific user-defined holidays. The minimum increment of scale is one day. Tasks of less then one day appear as one day flow on the GANTT chart. The printed copy can be up to 255 characters wide. Most of the fields on the report are optional, so only necessary data needs to be presented. Quick GANTT creates and maintains a GANTT chart on the PC screen. Each task requires only three pieces of data to create a useful chart:

- A task number or name.
- The task duration or flow time in days.
- The identity of an absolute start date or the preceding task on which the task is dependent.

As a task is created or changed, the task duration is displayed as a bar in the chart portion of the screen, along with other tasks and their durations. In this way you can see the position of the task with respect to both calendar dates and the other tasks. Once the data has been entered, the chart can be printed or sent to a file, where it can be edited for a report. Once a report has been created, the report titles and specifications are saved with the file.

General Scheduling Aids

CalNotes

This program is an electronic calendar and database program that helps you organize important information. CalNotes is a flexible, powerful, easy-to-use program. You can set variable and fixed events (holidays), associate notes up to 150 lines long with each day, set alarms, and store telephone lists. You can make "to-do" lists and perform many other related functions. And you can use a mouse, keyboard, menu, or "hot keys" to activate the functions. (A mouse is not required.) A full-screen calendar with room for holidays or reminders every day is also available. Features:

- Automatic carryover of unfinished "to do" items.
- Free-form notes up to 150 lines long for each day.
- Ability to set up to 20 alarms during the day.

- Built-in telephone list.
- Simple, intuitive operation—no confusing commands to remember.

Contact Plus

Contact Plus is an excellent way to keep track of people in your life. It helps you organize your phone calls, correspondence, notes, and future appointments. You can keep multiple databases and set up your own user-defined fields. It is a great application for salespeople. The software logs the last 30 phone calls and the last 15 letters to each client in your database. It features an auto dialer and a way to keep track of incoming calls.

Contact Plus also interfaces directly with WordPerfect and WordStar. This is not an ASCII or text file interface; it generates instant follow-up letters in native word-processing format for your clients by simply pressing a key.

Corporate Information Manager

This LAN/multiuser version dramatically improves corporate communication and productivity by providing access to the information that personnel need to perform their job more effectively. Features include:

- Online lookup of personnel by name, phone number, location, employment status, or department.
- Special treatment of calls.
- Alternate phone number etc.
- Online creation, updating, viewing, and management of messages. The software generates reports based on various criteria, including phone directories, with the option of being sent to printer, screen, or disk file for later processing.

Pop-up color menus, scrolled lookup displays, and context-sensitive help, plus built-in multiuser/LAN support, are included. Also provided are a full security system with multiple access levels and encrypted files. To help build the database faster, the program also incorporates a separate system administrator program, online lookup of phone numbers, and copying (cloning) of records.

The Contact Manager database tracks clients and lets you store information about clients and friends, including company, address, phone, alternate phone, E-mail information, comments, free-form memo information, and more. It also generates reports and mailing labels. Also included are a private reference database for the system administrator and a "name/number/description" database with a comments field that can be used for whatever you want, including tracking company property/equipment, vendor account information, etc.

The program can be used on a single PC or network and can easily be used by telephone operators, clerks, receptionists, secretaries, sales personnel, managers, and anyone else who deals with the phone. The program uses "pop-up windows," and anyone can become proficient with its use in a matter of minutes. It is very intuitive.

With a little play, most features become readily apparent This cuts down on training time and reduces documentation size. Users hate to have to constantly look things up in manuals. The program is very user-friendly.

Daily Calendar

Daily Calendar is a daily diary and calendar, a name and address database, and a to-do list manager. It tracks projects, phone calls, letters, clients, and accounts. It was written to be simple, easy, and menu driven. The documentation gives you a quick tour of the program. This program provides a daily calendar, to-do list manager, name and address database, label, list, and form letter utilities, and an extensive ability to query data for output. It is menu-driven and keeps appointments, prints calendars, and allows entry of multiple appointments and unlimited text entry.

It has extensive record-keeping ability; any amount of text can be added to any calendar entry. This feature allows logging conversations, keeping records, and tracking projects and correspondence. Text can be imported or exported. Text search ability is provided; entries can be retrieved and assembled into a document by searching for keywords or phrases. The to-do list manages a list of items by category, priority, or date. The dialer dials home or work numbers from within the daily calendar.

Dayo

Dayo (a support-call management system) gives you the platform for managing all your support and/or service calls. It is, by default, a multiuser system, allowing almost unlimited numbers of users to simultaneously access the same data on a LAN/network, i.e., Novell and LANtastic. Options, features, and reports include:

- Printer setup.
- PIN list.
- PIN labels.
- PIN phone list.
- PIN expanded report.
- PIN usage/history report.
- Open call log list.
- Daily call log report.
- Incident/support call log report.
- Category #1/problem list.
- Category #2/problem list.
- Resolutions list.
- Customer list.
- Customer labels.

- Configuration status report.
- Calculator.
- View reports on screen.
- Company name.
- Company address.
- Company telephone.
- Last reference number.
- Allow customer add.
- Customer location.
- Phone log.
- Location.
- Current directory.
- Browse category #1.
- Browse category #2.
- Browse.
- Resolutions.
- Serial number codes.
- Note screen size.
- Assign PIN blocks.
- Browse phone log.
- Default excluded options.
- Use passwords.
- Passwords.
- Maintenance/browse.
- Global pack and re-index.
- Miscellaneous database routines.
- Screen fonts.
- System statistics.
- Report Printer Port.
- Auto daily indexing.
- Color or mono monitor.
- High-intensity color setting.

dbDial

This is a Rolodex-like database program for dialing voice and fax phone numbers if you have a Hayes-compatible modem. It can go resident in memory as a TSR, letting you work in other applications (like a spreadsheet or word processor) and call on db-Dial without having to exit from the other program. Important features:

- Menu-driven.
- Rolodex-like application with mouse support and context-sensitive help.
- Runs as a stand-alone program or as a fully functional 7K TSR.
- Stores up to four different phone numbers per record and dials them if you have a Hayes-compatible modem.
- Stores up to two billion variable-length records and instantly sorts databases six different ways.
- Writes more than 2K of notes for each record in the database using a built-in professional text editor.

There is speedy access to any record in the database using abstract or rigid search criteria covering every field (even the notes). An incremental search capability locates records the instant you start typing a name or company. It keeps a log of every call you make in a CDS-compatible file.

It also records output language and lets you merge letters, generate mailing labels, and export whole databases. A powerful filter function temporarily "masks" records if they don't meet your criteria. Utility functions let you move records to another database, delete the notes for every record, and so on. You can import and export data in different formats—with or without notes.

Fone

Have you ever wondered ahead of time what it would cost to make a telephone call? Fone provides you with the answer, as well as "real-time" phone/facsimile call pricing, by showing accumulated charges on screen during your call. This program is a super telephone-call pricing program for point-to-point voice, fax, E-mail, and BBS calls. The extensive database includes information for residential, business (WATS, etc.), and international calls, with data for several national and international vendors. A personal telephone directory capability is included.

You can even keep track of the price of an ongoing telephone conversation in real time while you exit to a DOS shell and do other work at the same time. A volume discount calculator is also available if you qualify for volume discounts from your phone vendor. Hot menus make navigation through the extensive features a breeze. This program is useful for anyone who uses a telephone.

Features: The program is extremely user-friendly for even the most novice of users. You can price residential interstate calls for many vendors, business (WATS) interstate calls for many vendors, and international calls to more than 130 countries. And you can use an attached modem to place calls. It also provides a personal phone directory, hot menus, an extensive help file, and it records frequently priced call sequences. The personal directory is limited to 100 records, and intrastate (inside a state) call pricing is currently available for within California, but not for inside other states.

Hotnet

Hotnet is a hot-line phone calling/tracking, single/multiuser help desk system. It is your help desk. It logs calls, but even more, Hotnet does it simply and quickly. When the phone is ringing off the wall, you have to struggle with your software. Hotnet queries your calls simply and quickly. You can recycle your solutions; if you've solved a problem once, use that solution. Don't look up the same obscure error messages again and again. Let Hotnet give you the answers.

Finally, Hotnet manages your information. It tracks your progress and proves your efforts to management with Hotnet's reporting functions. The program runs on a NetBIOS-compatible network and has been tested on Novell Advanced Netware, ELS, and on TopShare. It uses standard NetBIOS record lock in. Hotnet should work on LANtastic, Netware Lite, and other peer-to-peer networks. Hotnet supports both color and monochrome video monitors.

Managing People

Managing People is a series of programs covering this basic topic. One program helps a person improve his/her attitude, or of a subordinate. The purpose is to help make attitudes more positive than they are. Mostly, it is just making up one's mind to be more positive. If people don't want to be more positive, they won't. But if they want to, and they just need a little help achieving it, then this program is for them.

The communication program helps a person improve his or her communication effectiveness. We all have problems from time to time with communications gone sour, and we are sometimes not sure why. We might think we know, but we are not seeing the real problem. This program is part of a series for managers.

The next program helps managers (or anyone for that matter) become better decision makers. As you run this program on decision making, you are presented with 25 questions that you should answer with a "Y" or "N" input. This program might not be what you expect the first time you run it. However, it gives you a big "plus" in your decision-making abilities.

No program can actually make decisions for you. Each job is usually too complex, and only you know all the variables to make the proper decision. This program focuses on the universal areas of decision making to help you improve your "batting average."

The Planning program helps individuals improve their skills as planners. Everyone assumes that they plan and are fairly good at it. The truth is that most of us don't plan as often as we should, or we plan poorly and run into major problems right away. This program on planning makes us more aware of the steps in planning, which must begin with clear, well-defined results to be achieved (goals).

PC Tickle

This program helps you organize your schedule better, displays your daily agenda on your monitor, maintains a history log of tasks accomplished, prints a copy of your schedule to be used as a checklist, lets you view a schedule from a previous date and view a schedule for a future date, balance your checkbook, track and total your expenses, track and total your caloric intake and output, and edit your calendar file with PC-Tickle's own minieditor.

The program is flexible and easy to use. Best of all, it doesn't forget. The software reads any ASCII text file, or you can use the internal editor to create your calendar file. However, the program cannot process files created with word-processing programs that store special formatting characters in the file.

TakeNote

TakeNote is a Windows personal filer, a combination address book and auto dialer. It stores names, addresses, and multiple phone numbers on predefined cards, and manages and maintains multiple address files. It auto dials any phone number, creates a list of frequently used speed-dial numbers, and prints mailing labels, reports, and 3-×-5 cards. The software searches using names, numbers, and "sounds like."

You can put file names on TakeNote's predesigned file cards, build a card deck, then sort data any way you like—by name, address, any field or sequence. You can search for any bit of information anywhere on the card, or search an index line or specific fields. Data is just a click away. Point and click with the mouse, and you have an individual on the phone. Push a button, and you have the name on a label.

Here are some of the predesigned cards: business contacts, credit card, home inventory library, music inventory, personal contacts, personal recipes, rolodex cards, and software library. Technical specifications are:

- Maximum number of decks opened concurrently per session: 8.
- Maximum cards per deck: 32,000.
- Maximum number of indices per deck: 1.
- Maximum number of sort levels per index: 3.
- Maximum notes field capacity per card: 8000 characters.
- Maximum card size: 8474K (business card, international).

TimeTrac

TimeTrac keeps track of multiple (nested) interruptions. Users can be busy, be interrupted, have that interruption interrupted, and still keep track of how the time was used. Personnel who provide client support while working consider TimeTrac's interruption-handling feature essential. Because the software uses clock time, neither the program nor the computer have to be running to record time spent on an activity. This allows memory to be used for large programs and permits the use of portable computers for field work. If tasks are started or completed while the program is not running, they can be entered later in a manual mode. The software computes time in units from any number of whole minutes up to one hour. The user chooses if it rounds, truncates, or computes partial units, and will also compute a minimum amount of time. Features include:

- Flexible key structure and reporting options.
- Ability to track interrupted tasks.

- Optional TSR mode that allows you to "pop up" over other applications to log your activities using a "hot key." It provides work group support either through a LAN or on remote portable computers.
- Exports to spreadsheet, database, and accounting programs.

TimeTrac is different because it tracks and calculates time for an unlimited number of nested interruptions. It is not a billing system, which makes it appropriate for corporate staff charge-out systems, large professional firms who need to feed an existing billing system, and consultants who want to continue using word processing for billing.

The program uses menu-driven rules for the unit of time used, minimum amounts of time, and how time will be calculated. It integrates the time records of a group of people charging time to the same project, with or without a LAN. It exports its data for analysis and processing to spreadsheet, database, word-processing, and accounting programs.

TimeTrac includes its own screen blanker, so you can leave the program running for long periods without etching your screen. Complete documentation for TimeTrac is embedded in the program. Simply press the Help key twice, and the manual appears. The manual can be printed by pressing one additional key.

Wcron

Wcron is a periodical task scheduler. It was inspired by UNIX's Cron utility and is very commonly used on UNIX systems. In fact, most UNIX systems depend a lot on Cron.

Cron (and Wcron for that matter) are most useful for machines that run 24 hours a day, but they can be of a great help even for casual users. Wcron lets you specify the time of execution of specific tasks. It can be used for one-shot executions and for periodical executions. People use it to make regular backups, remind them of something, and to start background tasks such as file systems.

NPA

NPA is a comprehensive area code, state, city, and exchange locator. NPA stands for numbering plan area, which is "phone speak" for area code. The data files accessed by NPA were compressed from nearly three megabytes of raw data containing directory information from our country's telephone system. Other area code programs simply show you the state and a few larger cities. NPA goes way beyond that. For example, you can locate all area codes in a state, display the central office cities within those area codes, and even display all of the valid exchanges (sometimes called the prefix or NXX in telephone company lingo) belonging to the city.

Even Erskine, Minnesota, shows up if it has its own telephone exchange. You can search based on state or area code, narrow the search by specifying a city search mask, search for cities having a certain number of exchanges, or search for a range of exchanges. And once all this is located, you can display either with exchanges listed or not listed. You can specify all parameters on the command line, making NPA easy to use, or if you just say "NPA" with no parameters, a full-screen version appears with online help. Either way, NPA displays its output in a buffer that lets you browse forward and backward with your up or down arrow keys or <Page up> <Page down> keys.

DialM

This telephone dialing program times and logs calls. Each directory entry includes the name, phone number, and three address lines, plus one "note" line. Each line can contain up to 32 characters (a "Rolodex" type entry). Printing functions let you print out the phone directory in its entirety or print an individual directory entry. You can even address an envelope. The log file can also be viewed or printed. It requires a Hayes or Hayes-compatible modem on either port, COM1 or COM2. The program will run without a modem, but, of course, dialing is not possible. A mouse is useful but not required.

DialM uses a telephone number directory file stored on a disk that is created and updated from within the program. File size can be specified by the user. Maximum size depends on available memory and is about 600 entries. The default is 500 entries. Each entry contains a name, the number, and a three-line address and a one-line note or other descriptive string. Any number of directory files can be available on disk, and the one to be used can be selected from the command line or after the program is running.

Directories can be sorted by name, address, phone number or "note." Numbers to be dialed can be searched for from the main menu and can be selected from a dialing directory window, or numbers can be entered from the keyboard. Ten dialing prefixes (A through J) are stored by the program and are created by the user. For example, prefix A might be "1-800." The number A-555-1212 would then be dialed as 1-800-555-1212. Prefixes let you dial numbers longer than will fit in the space provided in the dialing directory, and prefixes make it easier to enter often-used area codes, etc.

WorldTime for Windows

WorldTime for Windows is an easy-to-use worldwide clock utility that runs within the Microsoft Windows environment. It displays the current local time in any of more than 150 metropolitan and geographic locations around the world. The program is flexible, too. Location selection is not limited to the standard location database. With just a few keystrokes, custom location names can be substituted for any of the standard locations. This feature lets you customize the WorldTime location/time displays as you choose. All configuration settings can be saved and will be in effect each time the program is run.

Calendars

ApCal

This program is a time management tool for use on PCs. Extremely easy to use, it contains flexible, customizable screens with online help. ApCal displays full-screen monthly calendars with day blocks in which you can type notes, just like a wall calendar, and print to any printer. The program has a customizable holiday file for setting holidays, birthdays, anniversaries, etc., which are automatically included in the proper calendars.

ApCal allows you to set a default appointment screen to match your schedule. It does not limit you to a certain time interval between appointments. Appointment features include: quick scan, search, move, copy, print, and more. An alarm can be set to remind you of an appointment a given number of minutes before the appointment, even if you are in another program.

The software also contains a unique feature called "boot alarms," which are displayed when you boot up your computer. These are great for reminding yourself of a particular event that occurs on a certain day but not at any particular time. All alarms can be set to occur over a given time period and will keep resounding if you "snooze" them when they are displayed.

Appointment Book

Appointment Book is a general-purpose appointment maker that can be used in a small business or home environment. It easily allows the user to keep track of appointments for up to 10 office members per appointment book (multiple appointment books are allowed). The Appointment Book application makes it easy for the user to enter, edit, and delete (single or selected range) appointments; it also enables users to search for an existing appointment, search for an open appointment for a designated office member, print appointments, and it has many other features that allow the office (or home) user to easily maintain an appointment schedule for every office member.

When an appointment book is opened, the user is given a display of the current day's appointments. Appointment displays are broken into two windows (morning view and afternoon view). The user can toggle back and forth from each view. Appointments are shown in 10-minute intervals. Each appointment is displayed in reverse video next to the appointment time for the duration of the appointment. If a time slot contains an appointment for more that one office member, an asterisk is displayed next to the appointment time slot, indicating to the user that there is more data to be seen. (The user can use the configuration function to view office member appointments or use the edit function to quickly bring up a list of today's appointments to view any hidden appointments.)

Around the House

Around the House contains a database that maintains names, addresses, phone numbers, birthdays, whether to send a Christmas card, and a selection box asking why a name is included. You can also print a mailing label from this database. It also has another module for a home inventory. This allows the user to track items. It has slots for model and serial numbers, value, and a selection box for location and type of item.

Another module allows the user to keep notes. It can be used for anything from a personal diary to a recipe keeper. An appointment calendar is included that provides a variety of printouts. A small "word processor" that writes text files in ASCII allows the user to write mail-merged letters or edit any files on the disk. An additional module shows the user's income, and another lists all his expenses. Another option lets the user see "what's left." A variety of printouts and predefined reports are provided. This program also allows the user to create user-defined reports from any database, and it permits changing of screen colors.

BackSoon

Do you work in an office? Are you part of a busy family? If the answer to either of those question is "yes," you undoubtedly have lots of little scraps of paper all around your home or workplace. Notes from co-workers, family members, neighbors, or secretaries manage to pop up everywhere, but these pieces of paper somehow never manage to get thrown away; they accumulate all over the place. Some we never see until it is too late; others become such a common sight that they completely lose their effectiveness as notes or reminders.

BackSoon replaces all those annoying little pieces of paper electronically. While electronic mail has begun to replace such things as the interoffice memo, PostIt notes still end up stuck to computer monitors with messages from visitors and callers. The program also serves as a screen saver—not the mindless, flying-toaster variety that always seems to pop up just when you don't want it. BackSoon performs the following functions:

- "Blanks" screen to prevent phosphor burn-in.
- Saves screen contents for later restoration.
- Displays a customizable scrolling message.
- Accepts messages from visitors on your PC.
- Stores messages for printing or later retrieval.

Calendar Mate

Calendar Mate is a full-featured program for creating personalized calendars. You define events to the program, and it displays them on the calendars when printed. Three types of calendars can be printed: a monthly block calendar, a one-page yearly calendar, and a daily appointment calendar/scheduler.

The monthly calendar displays up to three events per day, and any of the calendars can be printed in English, French, German, or Spanish. In addition to the attractive printed calendars, the program's ability to handle different types of repeating events is superior to any other calendar program around. It handles simple intervals such as "yearly" or "weekly," but it can also handle "third Monday of the month" or "Wednesday–Friday," for example. It can even delete or reschedule single occurrences of a repeating event.

There are several powerful features, including:

- Prints daily, monthly, and yearly calendars.
- Simple installation procedure.
- Automatic computation of Easter and lunar cycles.

- Mouse support.
- Calendars can be printed in English, French, German, or Spanish.
- Julian dates can optionally be included on monthly calendars.
- Events that repeat on a regular basis can have a single (or several) occurrences rescheduled without affecting the other occurrences of the event.
- Future dates for a repeating event can be displayed.

CallNote

CallNote is a small program (54K) that provides a pop-up window in DOS to write notes of up to 100 lines at 38 characters per line, and it keeps track of phone numbers in a Rolodex file. Phone numbers can be dialed through the modem; all you have to do is pick up the phone when you get an answer. You can use a modem set up as COM port 1, 2, 3, or 4.

Notes you write are stored as plain ASCII text, and phone numbers are also stored. You can add a note to a number when using PhoneBook. Commands are single letters and can be either uppercase or lowercase. Because word wrap is activated in places where it is useful, there is no need to enter a carriage return at the end of the line; just keep typing. The only exception to this rule is when responding to the "TOPIC" line in the NoteBook section.

DataBank

DataBank consists of the following independent modules: Reminders, Calendar, Notepad, Phone Book, and Timetable. Each module maintains its own database. Whenever you need help, press F1. Help is context sensitive, so you get different help screens at different points of the program.

Reminder lets you enter, delete, edit, display, and search information in a calendar. You can enter ten lines (plus one line for a header) of information per one day. You can delete from the calendar one day at a time. If you do not want to delete all the information for a given day, use a function modify instead of a delete.

Calendar displays one page of a calendar for a current month so you can check which days of a week correspond to specific dates. You can move through the calendar forward or backward, one month at a time. A current day is highlighted. The calendar can be displayed from the main menu or from reminder.

Notepad lets you store your notes or useful information. You can create many notepads, giving them meaningful names and then writing related information on the appropriate notepad. You do not have to remember the names of your notepads; a list of all notepads will be displayed on the screen, and you will be prompted to select one using arrow keys. When you select a notepad, you can start writing new information or editing the old notes. The notepad provides you with many text-editing capabilities, so you can think of it as a small text editor.

Phone Book lets you maintain names, telephone numbers, and addresses. You can display one page (ten records at a time), move forward and backward, enter new data, modify or delete existing records, and search the entire database for a given name.

Timetable consists of seven days from Sunday through Saturday. You can display one day at a time. If you have some fixed activities during the week, you can enter, for each day of the week, the hours and description of them.

Small Business Adviser

This unique expert program puts the tax, legal, and business knowledge of a Harvard lawyer and "Big Six" CPA at your fingertips. It is a real "no-brainer" to install and use—even for computer novices. It gives your small business instant and current tax, legal, and other business-related help on a vast range of tax, legal, and other subjects, federal and state (for whichever state you select).

The program covers IRS, OSHA, ADA, ERISA, EEO, labor laws, workers compensation, environmental laws, and hundreds of other easily accessible topics. The software creates multipage checklist of business startup requirements in two minutes or less, customized for your business and the laws of your state. It has built-in worksheets/outlines for creating business plans and other documents. It has easy-to-use income and estate tax calculation/planning modules and consulting sessions in which the program asks you detailed questions to help you make key business decisions like whether to incorporate, choose Sub S status, etc. There's nothing else like this program at any price.

To-Do Lists

Telemagic

This program's features are:

- Name/address/phone number database. (Allows instant retrieval and filing for important contact information.)
- Appointment calendar and reminder system.
- Auto dialer.
- Word processor.
- Tickler file.
- Printouts of complete information/reports/labels.
- Thought bank.
- Online help and user-definable help.
- Easy lists.
- Customizable program.
- Filters.
- Magic rebuild.

More than 50,000 companies, including IBM, Xerox, and Coca Cola, use the business version of this program to help their sales and customer service people, secretaries, and executives. It helps write letters and keep track of people to see, things to do, and places to go. It also has an easy-to-use menu system.

Accounting

DAYO Business Applications

DAYO is a collection of multiuser business applications that provide a system of inventory control, purchasing, invoicing, customers, vendors, sales, and more. All DAYO products are multiuser and designed to work on a PC-based local area network (LAN) like LANtastic or Novell. DAYO can, however, be used in a stand-alone mode, running on a single PC. DAYO is full-featured, with features being added frequently.

There are hundreds of reports for DAYO. Each module has many reports specific to itself, and its Smart Reports contains many other reports. DAYO uses databases, report forms, and configuration files that are dBASE III+/IV compatible. This means you can use the popular dBASE programs to edit, view, and report on data from the many DAYO databases.

A simple database management module is included with most of the modules to provide you with additional means to modify, view, and report your data. DAYO places no limitations on the number of records or users. Current modules include:

- Inventory.
- Back Orders.
- Smart Reports.
- Point of Sale.
- Accounts Receivable.
- Materials Requirement Planning II.
- Price Book.
- Accounts Payable.
- Menu.
- Quotes.
- General Ledger.
- Clock.
- Index.
- Look Up Databases.
- Payroll.
- Passwords.
- End of Period/Purge.
- Bar Codes 3 of 9.

- Time and Billing.
- Little POS.
- Update and Conversion.
- SC.
- SMS.
- Report Writer.
- RMA.
- Simplest POS.
- Simple POS.
- OEM POS.
- OEM Quotes.
- BRAIN.

Spreadsheets

A spreadsheet is a powerful, dynamic, computerized pad and pencil that lets you add, subtract, multiply, divide, and quickly accomplish all types of ongoing calculations with ease, accuracy, speed, and efficiency. A spreadsheet can become your "computerized crystal ball," allowing you to "guesstimate" the results of various combinations of scheduling and cost factors (time, money, material, etc.) by instantly performing "what-if" calculations to present the final results of a variety of potential future scenarios. The programs that follow are powerful, versatile, and designed to meet a variety of cost and scheduling situations.

ALite

ALite, an easy-to-use spreadsheet program with graphics, has only minimal hardware requirements. The program runs on any PC with 256K or more (that's right, only 256K) equipped with a monochrome, CGA, or EGA monitor. Graphics can be printed on any 9-pin Epson or graphics-compatible printer, or PIC files can be created for inclusion in other programs. Features:

- 2,048 rows by 128 columns.
- Mathematical, logical, statistical, financial, string, user-defined, and date functions.
- Macro command language.
- Five types of graphs with merge and split-screen features.
- File merge, file extract, and file linking.
- WKS and WK1 file compatibility.
- Data sort.
- Data regression, and much more.

AsEasyAs

This program is a powerful Lotus 1-2-3 clone. Its features are:

- 8,192 rows by 256 columns. Enough room for the most demanding user.
- Expanded memory and virtual (disk) memory models. Simulate EMS on your hard disk, or floppy, to build huge spreadsheets.
- Multiple views (up to 6) of your worksheet with the ability to layer, zoom, size, and move each view.
- Fully mouse-compatible. Quickly expand column widths, move around in the spreadsheet, highlight ranges, move views.
- Powerful graphics—bar, stacked bar, line, xy, pie, hilo, cumulative, area, radial, polar, wall, and strip graphs with options like log scales, split-screen graphics, merged graphs, etc.
- Prints combined text and graphics on the same page using your dot matrix or LaserJet printer. Great for report and document preparation.
- Print preview mode (text and graphics) complete with page breaks, headers, footers, 200% zoom, ruler lines, etc.
- Cell formats including box, engineering, and label.
- More than 100 {MACRO} programming commands including FOR LOOPS, file I/O, Comm Ports I/O, Pick Lists, etc.
- More than 100 @FUNCTIONS for math, statistics, finance, logic, strings, dates, and even user-definable functions.
- Spreadsheet linking. You can dynamically relate information in your current worksheet to worksheet(s) on disk.
- Built-in utilities.
- Text find/replace.
- Macro recording.
- Auditing goal seeking, and more. (Other spreadsheets sell these features as expensive add-ons.)

EZ-SpreadSheet

Think of EZ-SpreadSheet this way: The computer display is your paper, the cursor and keyboard is your pencil, and your PC is the calculator. Worksheets provided with EZ-SpreadSheet are made up of "mailboxes" or "cells" that hold formulas, text, or numbers. The cells are arranged in 64 columns by 512 rows for a total of 32,768 cells. These worksheets provide the needed information to tie all the elements together to get your job done quickly and accurately. If the worksheets provided are not adequate, EZ-SpreadSheet lets you generate worksheets specifically tailored to your unique needs.

EZ-SpreadSheet is a fast, efficient, and ideal tool to let you easily solve your math-related home or business problems. (However, it's a good idea not to let your kids see

EZ-SpreadSheet or it will soon be doing their math homework for them.) In most cases, you can just enter your own data into the worksheet, and EZ-SpreadSheet does all the calculations for you. Applications for the home include calculating budgets, loans, savings plans, etc. Business applications include everything from simple loan calculations to complete financial statements.

The EZ-SpreadSheet worksheet processes inputs much the same way a calculator does. (If you need to use a word processor, you need to load a different program.) The worksheet is divided into cells that can store text, numbers, and formulas. Text or numbers are displayed at the screen cell where they were originally entered. Results of a calculated formula are displayed in the cell where that formula is stored. The numbers or text displayed in this case depends on the contents of the cells referenced by the formula cell. Worksheets can be stored to disk and retrieved any number of times. Several worksheets are provided with the EZ-SpreadSheet package.

Why another spreadsheet? Well, most of the spreadsheets on the market require you to have a Master's degree in computer science and accounting just to get them going. However, EZ-SpreadSheet is an easy and friendly spreadsheet (only you can tell us if we have done our job right). Hopefully, you will feel right at home using EZ-SpreadSheet.

If you are an experienced spreadsheet user, we hope you find the quick simplicity refreshing and turn to EZ-SpreadSheet for all but your most difficult problems. In addition, EZ-SpreadSheet costs less than 10% of what the "big boys" do, but it provides more than 95% of their features (including compatibility with other spreadsheet packages via "DIF" files).

Pivot

Pivot uses a spreadsheetlike interface that allows you to view and print the contents of a *.WKS or *.WK1 worksheet file created by As-Easy-As or other spreadsheet programs. It supports spreadsheets up to 8,192 rows × 256 columns and has complete support for special string concatenation operations performed in the spreadsheet.

Pivot prints the contents of a spreadsheet in either portrait or landscape formats using a variety of print control options. Six high-definition character fonts are built right into the program (three portrait plus three landscape). They are Courier, Sanserif, and Tiny. Other special fonts are distributed on the diskette as binary *.TRP (portrait) or *.TRL (landscape) fonts.

A companion program, GCHAR, permits you to customize any character or create a new character set with a variety of editing commands. It even lets you preview the interrelationship of characters (when used in words) by viewing them in a special typographic mode displayed on a graphics screen. Characters are sent to the printer as binary graphics information, not as ASCII codes. The result is 18 different combinations of text quality that you control when printing a font. For example, a typical character is based on a matrix of 16 vertical dots by 24 horizontal dots.

Choosing a high-resolution output with four passes of the print head results in the highest-quality text that your printer is capable of producing. Of course, printing might be very, very slow. An alternate choice might be to select two passes of the print head, which doubles the speed but slightly reduces the quality of the character printed. Note: Some older printers might not be able to print with high resolution.

ProQube Lite

ProQube Lite is a very powerful, easy-to-learn, true 3-D spreadsheet. It has been a PC Magazine's Editor's Choice. Up to 512 possible pages in each 3-D worksheet can be rotated to enter, view, and manipulate data from any perspective. Other features include:

- File linking.
- Direct import and export of 1-2-3 and dBASE files to and from any page and any view.
- A powerful macro language with a macro recorder.
- Mouse support.
- Search and replace, and much more.

Rebel

The manner in which spreadsheets manipulate data has changed very little during the years. This is somewhat surprising, given the advances that have been made in almost every other area of the computer field. Check the oldest spreadsheet manual you can find, and notice how similar the formulas are to the ones you can write today. Rebel has addressed this problem by abandoning the *de facto* standard that has been adopted over the past decade and replaced it with a simpler, more powerful syntax.

A major effort has gone into making this one of the easiest (if not the easiest) spreadsheet there is to learn; it is also one of the most powerful "pure" spreadsheets you'll ever use. You'll find that the new syntax used by Rebel resembles a true programming language that builds on a few simple rules to perform more and more complex operations. The following are just a few examples:

Cell addresses. While most spreadsheets on the market today use letters to refer to the columns of other cells, Rebel uses numbers (e.g., AB12 would be equivalent to [12,28] in Rebel—row 12 column 28).

Indirect cell addressing. The row and/or column fields of a cell address can themselves be expressions, allowing cell addresses to be nested. This means that a cell reference can vary, depending on the values of other cells (very handy for table lookups).

Arrays. Up to 8,000 values (array elements) can be assigned to each cell. Each of these values can be accessed by including the optional third array element field in the cell address.

Multiple worksheets. Up to four separate worksheets can be loaded at one time. Cells from any of these worksheets can be individually addressed by including the optional fourth worksheet level field in cell address. Note: Both the "array element" and "worksheet" fields of a cell address are optional. Their values default to zero if not included.

Programmable range operations. A *range operation* can be defined that assigns a series of "formulas" to a group of cells, which returns a single value on completion. The general form of this range operation is similar to that of a standard cell address and can be used within other formulas in much the same way.

Over thirty (30+) operators. Unary, arithmetic, relational, logical, bitwise, ternary, and assignment.

Over fifty (50+). Math, trig, financial, string, date, and time functions.

Variable labels. These are rewritten each time the worksheet is recalculated.

Rebel does not stop there, however. Every effort has been made to make this the simplest spreadsheet you have ever used. Function keys are designed to act like built-in macros that perform complex editing tasks; you can shift cells around your worksheet by pressing a single key; undelete cells that were mistakenly lost; search and replace (with wildcards); cut and paste; split screens (unlimited).

Organizations

Church Membership Tracking System

This program (CMTS) is an exceptionally easy-to-use, full-featured, church management system. The software is completely menu-driven and provides context-sensitive help screens to guide the user through every step. The software is available in both single-user and network formats. The system consists of seven individual modules and offers extensive reporting capabilities.

There is also an optional report writer that allows the user to write additional reports or modify the many reports supplied with the system: membership, donations, fund accounting, attendance, visitation, setup, and additional reports. In all modules, reports prompt with an option to direct the report to either the screen or the printer. Reports directed to the screen can be scrolled up and down, etc. Reports can also be easily directed to disk.

Extensive membership information is maintained. Store vital information on each family and each individual within the church. Members' special dates, talents, and membership in church organizations can be entered based on user-defined tables. Reports provide family and membership details, a roster for any organization within the church, a list of members' special talents, and a report for any "special date category" and date range. Membership information can be exported into ASCII text format for use with the mail merge features common in many word processors.

Mailing labels can be printed for each family, each individual, or all members of a church organization. Donations can be entered by name or envelope number. A comprehensive audit trail of all giving can be accessed by member or by fund category.

Transaction detail information can be retained indefinitely if there is adequate disk space. You can elect to combine members of the same family for statement purposes or print statements for each individual. Store member pledge information by fund category. Both annual and "life-of-the-fund" pledge amount can be stored for each member per fund.

Excellent reports are provided to access transaction detail and current and prior-year quarterly totals. You can print a complete detail donation listing grouped either by member or by fund category of all giving for any date range. Donation information can also be exported into ASCII text format for use in spreadsheets.

Fund Accounting. This is a true double-entry fund-accounting general ledger system. An accountant should help with initial setup of the church's chart of accounts; however, once installed, automatic transaction types make the system extremely easy to use with little or no accounting knowledge.

The system provides budgeting capabilities for all sources of funds (revenue) and uses of funds (expenditures). Budget amounts can be entered on either a monthly or an annual basis. Prior-year budget information is also available. Several automatic transaction types are provided to assist the user with a limited accounting background. First, when entering bank deposits, simply indicate the appropriate source of fund distribution. The journal entries are then made by the system automatically.

Similarly, when cash disbursements (checks) are entered, indicate the use of fund distribution, and the entries are again made automatically. Complete records of all bank checking accounts are generated automatically. The system even allows you to print check registers and flag returned checks to assist in check reconciliation. There is also a transaction screen to enter fund equity transfers. A general journal entry screen is also provided to allow traditional debit and credit posting when necessary. Extensive reporting capabilities include a statement of revenue and expenditures and changes in fund equity, statement of sources and uses of funds, chart of accounts listing, trial balance, check register, and much more.

Other reports include a detailed audit trail listing of all transactions for a user-defined reporting period, budget report, and a current-year versus prior-year comparison. Information can also be exported into ASCII text format for import into spreadsheets, etc. Attendance can be maintained for up to three user-defined events (morning worship, evening worship, Sunday school, etc.) for the entire congregation. It is fast and easy to enter attendance information. You then have the information needed to recognize members with outstanding attendance and to call on those who have been absent. A "call list" based on last attendance date is especially useful.

Attendance can also be maintained for all organizational functions. Attendance can be entered for any frequency (daily, weekly, monthly, etc). The attendance module uses an "exception posting" approach. This means it is only necessary to post attendance records for members either present or absent, but not both. The Visitation module provides information on shut-ins, last visit dates, visitation routes, etc., and the module includes an appointment scheduler. Use the appointment scheduler to schedule visits and other appointments. There is even a memo pad to enter detailed information in a free-form text edition fashion. Memos can be associated with a member or a date.

The Additional Reports module permits additional reports written using the Clarion Report Writer to be added to the menu system. The menu is "soft," meaning that any reports you write can be added to the menu. In addition, custom reports written by Torbert Data Systems and consultants can be placed here without modifying the program. This version of CMTS is a single user version. It will operate on a network if only one user accesses the files at a time.

Chapter

9

Graphics, CAD, and Drafting

"A picture shows me at a glance what it takes dozens of pages of a book to expound."

TURGENEV

The emergence of CAD, drafting, and paint-and-draw software has been a vital innovation that has created talented illustrators (even artists) out of many of us who almost flunked drafting. All of the basic geometric figures can be easily and accurately drawn, shaded, colored, sized, moved, rotated, and replicated by anyone able to nudge a mouse. Humongous libraries of symbols, components, and clip art are available and eagerly awaiting to be imported and similarly manipulated to create mechanical, architectural, electrical, computer, and chemical masterpieces. Now all science and engineering professionals can create dazzling, quality graphics that can be used directly in proposals, reports, articles, and books.

Not only can you opine and create, but because your designs are now stored in memory, you can also easily access, recall, and modify your designs and creations with "what-if" speculations. (What if I increased the length of this member? Would it still fit in the mounting area? How much weight would it add? How much more would it cost to fabricate?) With the proper dynamic software, these speculations are instantly available. And if you don't like the results, most software programs have an "undo" capability to bring you back to where you started "what-iffing."

So the next time you're scheduled to write a report, a proposal, or perhaps even a memo, incorporate some graphics. You'll not only save thousands of words, time, and effort (try to explain how a gear train works without an illustration), but your document will also be much easier to write, more readable, easier to understand, and more succinct.

Graphics File Extensions

A variety of file extensions are used to designate specific types of graphics files. Table 9.1 lists a number of the file extensions you'll most likely encounter when dealing with graphics files. Fortunately, several easy-to-use software programs are available for conversion from one graphics format to another. They are listed later in this chapter.

TABLE 9.1 Graphics File Extensions

File ext.	Usually defines
BMP	Windows device-independent bitmap
CDR	Corel Draw graphics file
CGM	Computer graphics metafile
DHP	Dr. Halo PIC
DXF	AutoCAD
EPS	Encapulated PostScript
GEM	GEM Draw
GIF	Graphic interchange format file
HPG	Hewlett-Packard graphics plotter file
IMG	GEM Image paint format file
MSP	Microsoft Windows Paint
PCX	PC Paintbrush
PIC	Lotus 1-2-3 graphic file, object oriented
PPIC	PC Paint Plus
RLE	Windows bitmapped run length encoded
TFF	TIFF formatted graphics file
TIF	Tagged image format, scanned graphics
WMF	Windows MetaFile, vector format
WPG	WordPerfect graphics

General Graphics Files

3-D Viewer

3-DV displays 3-dimensional wire-frame objects or data with real-time rotations. It works on IBM PC/AT compatibles with VGA, EGA, CGA, or Hercules graphics. A Microsoft-compatible mouse is required. Options/commands are selected by clicking a mouse button on the appropriate selection boxes. Perspective or orthogonal projections can be specified. In display mode, the position of the view point on a sphere surrounding the object is controlled by moving the mouse; moving horizontally changes the longitude; moving vertically changes the latitude. Pressing a button takes you back to the options menu.

3-DV is written in Turbo C, with fast line-drawing routines in assembler. Speed is due largely to the use of 16-bit integer arithmetic, with careful scaling and arrangement of computations to maintain precision while avoiding overflow.

Polyray

Polyray is a ray tracer that renders a wide variety of mathematical shapes. The means of description range from standard primitives like box, sphere, etc., to a three-variable polynomial expression, and finally (and slowest of all) surfaces containing transcendental functions such as sin, cos, and log. The files associated with Polyray are distributed in three pieces: the executable, a collection of document files, and a collection of data files.

The display modes include the standard 320-×-200 VGA color mode, as well as the "mode13x" display mode for standard VGA boards, and 640 × 480 256-color modes for many SVGA boards. SVGA displays are only supported on 286 versions of the executable. The executable requires a PC compatible with at least a 286 and 287 to run. Real mode and protected mode versions are available. The protected mode version is recommended if you want to ray trace large data files or if you want to use the z-buffer for anything but small image sizes.

VesaView

VesaView views and prints GIF, PCX, and IMG/GEM images and HPGL plotter files. Arrays of up to 81 images per screen can be generated and saved as GIFs for easy cataloging. HPGL plot files can be saved as IMG files. Mouse support is provided. This software requires a VESA driver for your graphics card and adequate extended memory to hold the entire image. It supports VESA modes through 1,280 × 1,024 × 256; the size of the image is limited only by the amount of extended memory.

Quick panning of images larger than your screen resolution is provided. Color/brightness/contrast adjustments are available, as well as cropping of GIF images. There is excellent printing to an HP DeskJet or LaserJet. The program reads any size GIF (CompuServe), PCX (Zsoft 2 or 256 color only), IMG (GEM/Digital Research Inc., black and white) or an HPGL (Hewlett Packard plotter file). The program determines file type from the extension (.GIF, .PCX, or .IMG for images and .HPG or .PLT for plotter files). If the file is larger than the screen mode you selected, you can quickly pan around the image. The size of the image is limited only by the amount of extended memory you have. The program supports a mouse.

Screen Capture

Screen capture software can grab and copy a pictorial or text presentation from your PC screen and store it on disk for later recall and for import into other documents such as reports, proposals, etc. Using this software, you can select the area you want to capture and accomplish a number of manipulations on the captured image. Some of the screen capture software also converts the images to and from various formats such as PCX, GEM, WPG, etc.

GemCap

GemCap is a memory-resident utility that lets you capture screen images from programs, save them in separate disk files, and later load them into other programs. The disk files are compatible with any program that reads GEM.IMG files, regardless of whether the original screen was in text mode or graphics mode.

Grabber

This screen capture program saves the images on the screen to a disk file. Grabber does this in a way that makes it very easy for you to recall the images back to your screen at a later time, and exactly as they were originally displayed. The program is memory-resident, which means that after you start it, the program is always waiting in the background while you continue to run other programs.

You tell Grabber to capture the image currently being displayed on the screen by pressing a combination of keys that Grabber reserves for itself. The software "captures" the screen in the form of a file on your disk, called a Grabber "offspring" file. When you want to view the screen later, all you need to do is run this "offspring" file like an ordinary program.

Grabber can capture practically any size text image that your computer can display, up to 132 columns by 60 rows. The program saves all images, whether text or graphics, as executable program files with an ".EXE" file name extension.

Grabber Conversion Utilities

GR2PCX and GR2GIF are companion utilities to the Grabber screen capture program. You can use them to convert graphic images captured from your video display in the form of Grabber "offspring" files into either of the widely recognized generic image formats: PCX and GIF. In addition, if your computer has an EGA or VGA video graphics adapter and a color monitor, the GR2PCX utility can also convert 80 × 25 text images captured by Grabber to the PCX graphic format. "GIF" (pronounced like "Jif") stands for graphic interchange format.

Generally, there is no other widely used file format that can compress a graphic image into a smaller space than GIF can. The GIF format is supported extensively on the CompuServe Information Service and many other online BBS services that make graphic images available for downloading. In addition to these online services, however, there are several PC application programs that can employ the GIF format. These include word processing, desktop publishing, and "paint" programs, as well as numerous picture-viewing utilities. Therefore, you can use GR2GIF to convert almost any graphic image captured by Grabber into GIF format, so that the image can be imported into any word processing, desktop publishing, or paint program that "understands" the GIF format.

The PCX format uses a data compression scheme to reduce the disk space required to store graphic images. You can use GR2PCX to convert almost any graphic image captured by Grabber into PCX format so that the image can be imported into any word processing, desktop publishing, or paint program that "understands" the PCX format. These programs can convert almost any Grabber offspring file containing a graphic image into a separate PCX or GIF file. This includes any image originally captured in any CGA mode, any EGA mode, any VGA mode, and the Hercules monochrome graphics mode.

PCXDump

PCXDump saves the images on the screen of your computer to a disk file. The program is memory-resident; it is always waiting in the background while you continue to run other programs. You tell PCXDump to capture the image currently being displayed on the screen by pressing a combination of keys that PCXDump reserves for itself.

The images are saved on disk in the PCX format. A large variety of word processors, desktop publishers, and paint programs can import PCX files, e.g., WordPerfect, PageMaker, Ventura Publisher, PaintBrush, and many, many more. The PCX files are stored in the current directory, unless specified otherwise.

PCXDump uses less than 9,900 bytes of conventional memory (and no EMS or XMS). All pixel writes/read are direct to the video memory, hence really fast. It recognizes a large number of Super VGA chips and can use their resolutions as well. It's easy to use; just press the hot key and use the mouse or keyboard to control which part of the screen to save.

Some capture programs require the user to convert the output to a widely used graphic format such as GIF or PCX. PCXDump converts the images directly to the PCX format. You can change the hot keys to hot keys of your own choice.

PCXDump offers a couple of ways to "delay" the installation. This means that you can capture screens that you normally wouldn't be able to capture. PCXDump captures graphic images on systems equipped with any of the following video display adapters: Multicolor Graphics Adapter (MCGA), Enhanced Graphics Adapter (EGA), and Video Gate Array adapter (VGA). PCXDump is not designed to work with high-resolution four-color images.

VGACap

VGACap is a resident utility that captures pictures in MCGA 320 × 200 × 256, VGA 640 × 480 × 16, SVGA 640 × 480 × 256, SVGA 800 × 600 × 16, SVGA 800 × 600 × 256, SVGA 1KB × 768 × 16, and SVGA 1KB × 768 × 256 formats. VGACap captures and "Bsaves" 320 × 200 × 256 pictures; all others are saved to *.RAW files (i.e., "RAW" pixel dumps).

VGAFil

This is a conversion program for use in a batch file (*.BAT), or as an interactive stand-alone program. BLD/PLT files are 320 × 200 × 256 screens; all others are RAW/PLT files. VGAFil integrates directly with VGACAD and replaces BLD2GIF and RAW2GIF. Moreover, VGAFil converts and saves captured images to 32,768-color TGA (Targa 16), 256-color BMP (Windows), PCX, GIF files, or 16-color GIF files.

VGAFil is also a utility to customize and configure VGACap to take advantage of specific SVGA chip set hard coding for fast screen captures. VGACap and VGAFil can be viewed as a generic conversion utility system because almost any MCGA/VGA/SVGA image from 320 × 200 to 1,024 × 768 that can be viewed can be converted to/from GIF, PCX, BMP, or TGA format. If you can view files using less popular formats, you can capture and convert them.

TxT2PCX TSR

TxT2PCX is a unique, memory resident file utility that converts any text screen to a graphics screen and saves that graphics screen as a PCX file. The PCX file format is, perhaps, the single most used graphics file format in the DOS world. It is the native format of the PC Paintbrush series of programs and was developed by ZSoft Corporation. Most of the major graphics programs allow the importing and/or exporting of graphics screens in this format.

This is a specialized memory resident screen capture program. This software captures any text screen (25 rows by 80 columns) that you can display on your monitor and converts it to a PCX graphics file format.

Graphics Utilities

GDS

GDS is a graphics file viewer with a very intuitive user interface. GDS reads GIF87a, GIF89a (including comments), BMP, LBM, PCX, RIX, and MAC files. It writes GIF87a, **LBM, **BMP, **PCX, and **RIX files directly from the screen. The program queues up to 2,800 images to show in a repeated sequence.

GDS reads successive images ahead of time into EMS or system RAM to minimize draw time, increasing the appearance of the slide show. The software can lock a screen mode, reduce and enlarge images to fit, or scale and zoom to almost any dimensions. GDS quickly converts 256-color images to 16-color images with antialiasing and dithering to ensure spatial color consistency.

It scales images to almost any size, and reduces, enlarges, or corrects image scaling. GDS has an image cataloging section that reduces and labels arrays of images on the screen.

GDS creates spatially optimized palettes, uses two levels of dithering and two-dimensional antialiasing to reduce color distortion, and can reduce up to 2,304 images onto a single 1,024 × 768 graphics screen. Although almost everything can be done with the keyboard, a mouse is highly recommended. GDS takes great pains to support video hardware, and GDS supports most SVGA boards. It supports extended 24-bit DAC circuitry, which can provide 16,777,216 colors rather than 262,144.

GIF Utilities

Six GIF utilities are contained in this package. They are:

GDIR. Program to display information about all GIF files in a directory or GIF files specified on the command line. Information displayed includes size, number of colors used, and presence of a global color map.

CHK4GLTE. Program that checks all GIF files in the current directory on the presence of the GIFLITE header.

STRIPHDR. Program that strips the GIFLITE header of all specified files.

GIFPALE. Program that shows all colors that are used in the specified GIF file on screen.

GIFBGRND. Program that sets the background color of a GIF file. GIFPALET can be used to pick the appropriate color.

GIFSIZE. Program that changes the size of a GIF file to the nearest standard size, not changing the image in the GIF file but putting it in the middle of the screen (e.g., a 325 × 416 GIF file is sized to 640 × 480 and the image will be in the middle of the screen). It is also possible to size a GIF file to a user-specified size. And here's the good news: All utilities are free.

GIF/TIF/PCX Format Picture Viewer

The main features of this program are:

- Superfast display or print of GIF, PCX, and uncompressed bilevel/grayscale TIFF files.
- Compressed bilevel TIFF files (compression type CCITT G3-1D, CCITT group four, or packbits).
- Eight-bit palette uncompressed TIFF files.
- Optional image preloading to conventional or extended memory.
- Support for a wide variety of display types: CGA, Hercules, EGA, MCGA, VGA, 6300, Tandy, 8514/A, Super VGA.
- VESA Super VGA driver for all Super VGA cards that have VESA BIOS support.
- Print images.
- Prescaling, zoom, pan, and auto-scale features.
- Color control: adjust brightness, contrast, etc., after image is on screen.
- User-friendly, multiple interfaces.
- GIF89a support: Supports plain text, graphics control, comment extensions, restore to background, etc.
- Utilities: includes utility to convert interlaced GIFs to noninterlaced.
- Low memory requirements: can work with as little as 180K of free memory, even on very large images.
- GIF89a images that use "restore to previous" function might need more memory for complete restoration to previous. Also, the "image preload" feature uses additional conventional or extended memory.

GIFLite

High-resolution VGA and full-color scanners have produced a massive volume of high-quality digitized images. Most of these images available in bulletin board systems are stored in the graphics interchange format (GIF). Though the GIF format uses a compression algorithm to store an image, the size of the resulting file can still easily be more than 100K bytes or even more than 200K bytes.

If you are a GIF collector or a BBS sysop, you will find your GIF collection occupies a large portion of your hard disk. How would you like to save 25% of that expensive disk space? GIFLite is an easy-to-use GIF file compression program. It compresses GIF files so they require much less disk space yet still preserve the same quality and resolution.

With GIFLite, high-quality digitized GIF images can be further compressed by an average of 15% to 35%, depending on the complexity of the images. What makes GIFLite special is that the resulting GIF file is almost identical to the original file. Human eyes can hardly tell them apart. GIFLite conforms to both GIF87a and GIF89a standards. All nonimage information is preserved in the resulting GIF files. The compressed images can be read by standard GIF decoders such as VPIC v4+ and CSHOW v8.3+. Using GIFLite is much like using the other file compressors. If you have used any file compressor, you will find yourself already familiar with the command structure and options.

Graphic Workshop

Graphic Workshop is for working with computer graphic files. It handles most of the popular formats:

- MacPaint files.
- PC Paintbrush files with up to 256 colors.
- GEM/IMG files of the sort used by Ventura Publisher.
- GIF files of any size and up to 256 colors.
- EPS files.
- Encapsulated PostScript for some operations.

Graphic Workshop is a simple, menu-driven environment that lets you perform the following operations on the aforementioned files:

- View them.
- Convert between any two formats (with a few restrictions).
- Print them to any LaserJet Plus compatible or PostScript laser.
- Dither the colored ones to black and white.
- Reverse the monochrome ones black for white.

Using Graphic Workshop, you can have your image files in the formats that your software wants to use them in, all without keeping track of numerous funky utilities.

In addition, using the half-toning and dithering facilities of Graphic Workshop, you can convert full color digitized photographs (most notably GIF files) for use as really excellent black-and-white clip art, suitable for inclusion in your documents.

Graphic Workshop handles image files of any size. It uses extended or expanded memory if you have some, and disk space if you don't. It has a fast and easily understood user interface.

Image Gallery

Image Gallery helps you keep track of a large collection of bitmapped image files. It's a visual database that catalogs files, allowing you to add comments and keywords to each entry. Having created an Image Gallery database—a gallery—you can search it either visually, looking through the images by eye, or by keywords.

Each entry in a gallery also contains a small thumbnail of the original image. These thumbnails can be stored in either landscape or portrait orientation, and as monochrome or gray-scale images. Note that these thumbnail images are intended to be representations of the original images. They are, of necessity, quite a bit lower in resolution.

In essence, Image Gallery is an electronic photograph album. It allows you to attach notes to all your pictures and to find any picture or group of pictures you want quickly, something you can rarely do with a real photograph album. Invariably, everyone wants to look at the embarrassing ones, slowing your progress. In addition to searching through a gallery, you can print entries that correspond to a keyword search. You can also extract entries by keyword to another gallery and merge galleries together. A gallery can contain up to 65,534 entries; although at about five kilobytes per entry, such a gallery would be fairly enormous.

IMProcess

This program has many features:

- Image contrast enhancement.
- Histograms.
- Sharpening by convolution.
- Custom filters.
- Full control of the VGA palette.
- Clipboard.
- Five stroke fonts in five sizes and the ability to import bitmap fonts.
- FAT BIT editor.
- Many drawing tools, flips, mirror.
- Color cycling.
- Plasma and terrain fractals.
- SVGA modes.

- UNDO.
- XMS/EMS support.
- Edit up to five images at one time, etc.

This is a feature-packed program. Take the time to master it, and you'll produce images that you never dreamed of.

Memory. You need at least 512K RAM free after loading DOS and TSRs, etc. The more memory you have over 512K before starting the program, the better. The same goes for EMS and XMS.

JPG2GIF and GIF2JPG

JPG2GIF and GIF2JPG are a set of programs that convert GIF images to JPEG images and JPEG images back to GIF images. This conversion is needed because, typically, JPEG files are several times smaller than GIF files. So, by converting images from GIF to JPEG, you save time on the modem, long-distance charges, and disk storage requirements.

The disadvantage is that you can't view JPEG files directly; you have to convert them back to GIF files so that you can use your favorite GIF viewer. Two options are common to both programs:

- a: Automatically proceeds (overwriting any existing files and converting GIF89A files).
- k: Kill removes the original files after they are processed.

The GIF2JPG program has an additional option:

- q[n]: Specify the JPEG quality factor that will be used when compressing GIF files. The default is 55. A lower number results in a higher compression ratio but a lower-quality image. Specifying a very low quality factor (less than 15) will result in poor-quality images. Note that you can specify wildcards as part of the file name and that multiple file names can be specified.

PC ToolBox

PC Toolbox is a series of programs and utilities for manipulating and processing PCX file images. PCX-ToolBox contains numerous tools to view, scale, crop, dither, convert palettes, adjust, and print your PCX images. The program also includes several handy features such as copy, move, and delete files.

The utilities in ToolBox are all contained in easy-to-use pull-down menus and pop-up dialog boxes. Each menu and box is mouse aware, and selections can be made with the cursor (arrow) and Tab keys, or by clicking the left mouse button on your choice. ToolBox also contains help screens for each tool and menu.

The PCX file format is, perhaps, the single most used graphics file format in the DOS world. It is the native format of the PC-Paintbrush series of programs and was developed by Zsoft Corporation. Most of the major graphics programs allow the importing and/or exporting of graphics screens in this format.

ToolBox recognizes most of the chip sets and also recognizes VESA standard drivers. Choose the memory/video option from the file menu to see how ToolBox interprets your current system configuration. It also tells you if it recognizes your chip set. If the chip set is unknown, load the VESA driver for your card and restart PCX-ToolBox.

SqzGIF

SqzGIF is a multifunctional utility that resizes small or large GIF/PCX pics (up to 2KB × x2KB × 256) to variable sizes (i.e., 320 × 200 × 256 to 1,024 × 768 × 256) and adds the synergistic power and functionality of EGA2VGA (to create/add up to 256 new colors) with GIF images with less than 256 colors (i.e., 2 to 128 colors) for:

- Image processing/editing in VGACAD.
- Conversion to EGA in VGA2EGA printing or desktop publishing with GIFPUB (integrates GIFDOT).
- Reconversion to a smaller or larger 256-color GIF/PCX files, with VGAFIL (included).
- Bloading in BASIC or QuickBASIC (BLD and PLT files only).

With SqzGIF you can resize and smooth large 256 color pictures with correct aspect ratios, all sizes such as 320 × 240 (Jovian Logic), 320 × 400 (Amiga), 512 × 480 (TARGA), 640 × 400 (MACII), 640 × 480 and 800 × 600 (Super VGA) can be handled (up to 2KB × 2KB × 256). No graphics card is required.

Use SqzGIF to reduce a large GIF/PCX picture or expand a small GIF/PCX picture. VGA/MCGA users can pick any selected area and zoom it to full-screen size; you can treat large GIF/PCX pictures as "virtual screens." Interpolation is a method to add or remove pixels to "fit" an image to a particular size. No changes to the colors are implemented. You simply see more or less pixels in a defined image size. Because calculations are minimal, this algorithm is very fast.

VPIC

VPIC is a file viewer/converter for EGA, VGA, and Super VGA display adapters. It supports Ahead (A and B chip), ATI, Chips and Technology, Everex, Genoa, Headland (Video 7), Oak, Paradise, S3, Trident (8800 BR and CS and 8900), Tseng (ET3000 and ET4000 chip), Western Digital, and Zymos VGA chips in all VGA/EGA graphics modes. If your board supports the VESA standard, it is autodetected and used.

It is configurable for your display card using a plain text configuration file and the accompanying CONFIG and CVPIC configuration programs. VPIC displays, and converts to, the following graphics file formats:

- BIF Binary Image Format BandW Image Capture board files (display only).
- BMP Microsoft Windows 3.0 Bit-maps, uncompressed and compressed (change RLE to BMP extension).

- GIF (GIF87a and GIF89a) viewing, normal and interlaced (Fractint FRA viewing too).
- GIF (GIF87a) saves in normal (G), interlaced (I), inverted (U), mirrored (M), rotated CW (>) or CCW (<) 90 degree.
- LBM Deluxe Paint VGA format, regular and enhanced (IFF Electronic Arts files too). Save command is D.
- MAC file (display only).
- PIC Pictor/PC Paint and ViewPoint (ViewSonic) files.
- PCX, including the VGA version 1.61 of PC Paintbrush.
- CUT Dr. Halo machine-independent format with a palette file.
- SCx file formats for Colorix and EGA Paint.
- TGA Targa 8 through 32 bits/pixel normal and compressed file formats. Sierra DAC (32K colors) support, too.

Optiks

Optiks is a system for viewing, scanning, printing, and altering monochrome PC graphics files. Optiks reads and writes many different graphic file types and converts files from one type to another, sizing files and changing files in many ways. The program is easy to use and to learn. It is a very complicated program in some ways, but it has very easy paths to let you accomplish exactly what you need to do with a minimum of fuss.

Optiks is a program for manipulating black-and-white images. Optiks is a large area into which pictures can be brought. You bring in the picture; once the picture is in the work space, you can copy or alter it in various ways. You can save the picture by telling Optiks how much of the picture to save. You don't want to save the entire work space because you would quickly fill up the disk with pictures surrounded by a lot of empty black space.

The Optiks work space is considerably wider than your screen; you can move around in it. In this way you can accommodate pictures that are bigger than one screen. The width of the work space is called the *virtual width*. The length of the work space is called the *virtual length*.

Charts/Graphs

ExpressGraph

ExpressGraph is an easy-to-use business graphics program. Numbers can be analyzed much more easily and quickly when they are displayed in graphic form. ExpressGraph lets you display numbers in a variety of graphic formats. Then you can choose the display you like best and print your graph. Here are some examples of how people use ExpressGraph:

- Graphing sales dollars and gross profit on a weekly basis.
- Displaying financial data graphically.

- Analyzing budgets and expenses with a pie chart.
- Spotting trends amid a "sea of numbers."
- Providing summarized data in graphic form to accompany financial statements for quick analysis and trend spotting.

Graphs created with ExpressGraph can be displayed in one of ten graphic formats, altered in various ways, and printed out.

The Organization Chart

The Organization Chart reads in an organization file containing the titles and names of people or objects. The output from this program is a file (and optionally a printer output) that contains the titles and names enclosed by boxes and connected by lines that depict the relationship among the people or objects in the organization.

Besides business and government organizations, The Organization Chart can also be used to describe the relationship in family trees and directories that can be represented by tree structures.

Presentation Graphics

Collage

This program is a computerized show-and-tell software package to create colorful graphic pictures on your computer screen. You create and save these slides, combine created slides, and present the created and combined slides as a show on your computer screen. Collage also lets you create and maintain text files (upper ASCII included) with a number of good editing and block features.

The documentation is presented as text files with IMG extensions. These are chained together to present a slide show. Select the option "slide show" in the main menu and the "file slide projection" option by pressing F. Then enter "DOCU" and hit the Enter key. The entire documentation is then projected, frame-by-frame, on your screen. Hit any key to stop or continue, and hit "Esc" to quit.

CompuShow

With this software, you can display any of the popular graphic image formats by scrolling through your directory. Compushow is a graphics display program for CompuServe GIF and RLE, MacPaint, PC Paint, PC Paintbrush, Dr. Halo, EGA Paint, and ColoRix graphics images. It supports Hercules, CGA, PCjr/Tandy-1000, EGA, Super EGA, MCGA, VGA, Super VGA, the IBM Professional Graphics Controller, and 8514/A adapters. Super VGA support includes Ahead Systems, ATI, Chips and Technologies, Everex, Paradise, AST, Compaq, Dell, Genoa, Orchid, STB, Trident, Logix, ZyMOS, and Video7 adapters (and compatibles). CompuShow also provides 16-color graphics displays on a CGA and Tandy 1000SL/TL 640 × 200 × 16-color mode.

Files are selected for display from an on-screen file directory, which can be sorted by name, extension, size, or date, in ascending or descending sequence. You can log to a different disk drive, change subdirectories, copy, rename, and delete files. The program automatically selects the best display mode for any graphic, or you can display it in any mode available on your graphics adapter. Images larger than the screen can be panned vertically and horizontally. Colors can be adjusted on EGA, MCGA, and VGA systems, and the adjusted palette can be saved.

HyperSlide

HyperSlide creates interactive presentations based on graphic images and programmed control sequences. HyperSlide bases its graphic images on AutoCAD's slide file format. You must create slide (.SLD) files from AutoCAD using the MSLIDE command in order to user HyperSlide. HyperSlide and the HyperSlide Presentation Designer use the graphics mode of your computer.

Both support Hercules, CGA, EGA, and VGA displays. Both programs will work on SVGA cards, but only at standard VGA resolutions. A presentation created on one of the supported displays works on any of them. Using the presentation should prove simple. Use the mouse to move the arrow over one of the controls and press the left mouse button to activate the control. That's the essence of HyperSlide. Current control types are as follows:

- Link to another slide.
- Overlay another slide.
- Link to previous slide.
- Link to an external program.
- Link to an external program and pause.
- Link to DOS shell.
- Timed link.
- Timed overlay.
- Link to program exit.

PC-Foil

PC-Foil, and its companion program Foil-Edit, combine to provide a multipurpose tool for preparing a complete presentation or document on the PC. It allows you to make effective use of the printer to graphically emphasize your thoughts and ideas. Key points can be highlighted.

PC-Foil can be used to make overhead transparencies (FOILS) or to create attractive documents. When you combine the large print, the big bold print, and the boxes with your creativity, the effect can be dramatic. You simply develop the presentation or document using Foil-Edit, then print it using PC-Foil.

PC-Foil improves your productivity initially and when making changes. To make minor changes to the document or presentation, simply crank up the PC, make the minor change using the Foil-Edit full screen editor, then print out only the page(s) that contains the change.

Foil-Edit is an excellent general-purpose full-screen editor. You can use it for creating programs, as a word processor, or for just modifying any standard DOS file that you have (such as BAT files or database files). It is fast and easy to use. It offers both menus and commands, depending on your skill level. The online help facility allows you to get help at any time without losing your place.

PC-Foil can been used for any general-purpose documentation. It can and has been used for instruction manuals, user's guides, operational instructions, management reports, program documentation, mailing labels, name tags, stick-on labels, and many other documents. If you make presentations or prepare documents regularly, then PC-Foil can be of assistance to you.

PC-Talk

PC-Talk is a new product for business and personal applications. The library on this disk consists of more than 100 words that can be used in all sorts of possibilities. No special hardware is required to run the voice library. (It works through the PC speaker.) The voice commands can be placed in any batch file (*.bat) and added to your programs/menus. This software is very easy to use; you simply type PC-Talk at the DOS prompt, followed by the words you would like to say. (For example, to say happy birthday, type in "PC-TALK happy birthday.")

Also included in this package is a PCX display program (EGA/VGA, 640 × 350, 16 colors) that can be used to display PCX pictures and talk over them. Imagine creating a rolling PCX demo with real human speech describing exactly what the computer is displaying.

PCX-Show

PCX-Show is a highly stylized screen presentation program. It allows the user to create customized screen presentations using pictures/screens saved or captured in the PCX file format. PCX-Show features include multiple screen fades and segues, special effects, kaleidoscope screen effects, text display using customized and computer fonts, border and screen clearing effects, and more. Accompanying the package is a demo script to get the new user started.

sLide

sLide produces high-quality slides and overhead transparencies (both referred to as sLides). A sLide can be created, edited, saved, displayed, and printed using the sLide program. Overhead transparencies can be produced by copying the printed output onto transparent material. Photographic sLides can be produced by photographing the printed output.

Each sLide can contain up to 18 lines of up to 40 characters each. For photographic slides, the number of lines should be limited to 15 to fit a standard 35 mm slide. Experience shows that sLides can be read easily by an audience when the amount of material on a sLide is kept within these limits.

VuImage

If you work with .GIF/.TIFF/.PCX graphic files, you need VuImage. VuImage lets you display or print .GIF files and bilevel/gray-scale .TIFF files. It's fast and uses only

150K of free memory. VuImage automatically configures for your monitor with excellent color rendition on CGA, EGA, VGA, MCGA, and Super VGA. (Hercules monochrome is also supported.)

You can automatically scale a graphic to the size of your screen or manually scale it from 1% to 999%. Zoom in and out, pan to different parts of a large graphic, and easily adjust brightness while the picture is on screen. Print out any graphic (with six dithering options), and VuImage automatically sizes it for your printer. (Some .GIF images are provided.) In addition, you can specify a list of files to be automatically viewed for creating great slide shows and demos.

Paint

Desktop Paint

Desktop Paint is a bitmapped paint program designed especially for use by people who run desktop publishing software such as Ventura Publisher, PageMaker, Publish It, and so on. It is equally useful to users of high-end word processors such as WordPerfect. Desktop Paint allows you to quickly create and modify bitmapped graphics, either for use as stand-alone pictures, or to include in other documents. It features a wide variety of file formats, a powerful selection of drawing tools, an easy-to-operate user interface, and lots of room for customization.

Desktop Paint supports the following image file formats: MacPaint MAC (and PSF MAC files), Ventura IMG, PC Paintbrush PCX, TIFF, and WordPerfect WPG. Note that Desktop Paint only creates and works with monochrome files. Note also that it will only work with WPG files that contain bitmapped images.

Desktop Paint is a paint program rather than a drawing program. It does not support GEM, Corel Draw, Designer, DXF, or EPS files. Unlike many other paint programs, Desktop Paint features complete XMS and EMS support. It handles enormous images, pictures that unpack into several megabytes, if you have sufficient extended or expanded memory in your system. Desktop Paint requires a Microsoft-compatible mouse and driver. You can run Desktop Paint on any PC, even really old 8088-based systems. It will be a bit slow on very low end systems. If you do run it on a slower machine, it is a bit snappier if you use a Hercules-compatible monochrome graphics card rather than an EGA or VGA card. Most EGA- and VGA-compatible cards have Hercules emulation modes.

Desktop Paint (Monochrome)

This 256 Paint VGA Image Workshop is for use by anyone who has a desire to paint or edit images in 256-color VGA mode. The images created with 256 Paint are simple bitmap images, with no compression schemes involved. This product encompasses many diverse painting and graphics-related routines. This graphics program operates stand-alone. This means that you can install it somewhere on the path, then call it up by simply typing 256Paint at any time. Keep in mind that help is available most of the time. Probably the most comforting command available is the UNDO command. The following shape choices are available:

LINE. This allows you to paint a line. You must specify the endpoints.

RECTANGLE. This allows you to paint a rectangle. You must specify two opposing corners.

CIRCLE. This allows you to paint a circle. You will be asked to first specify the center, and then a point along the circumference.

ELLIPSE. This process lets you create odd-shaped ellipses. After supplying the center of the circle, you then provide the x and y radius, respectively. Unlike the circle routine, where the circumference point can be anywhere, the program isolates mouse movement so only appropriate points are selected for the x and y radiuses.

POLYGON. This is the most powerful of the shape processes. This feature allows the user to paint a shape that has up to 100 sides. This gives a tremendous amount of freedom to create shapes that are widely diverse. Because it is a nuisance to stop and figure out how many points the shape will have, the program allows you to just keep entering points until you decide it is time to either draw the shape or abort the process. To use this process, just start selecting points in the work area. Once you are done, the right mouse button interrupts the process. It is at this time that you inform the program whether you are done selecting points or have changed your mind and want to abort.

Desktop Paint 256

Desktop Paint 256 is a full-color bitmapped paint program that uses the extended resolution of Super VGA cards. It features a mouse-driven user interface, full-color manipulation and drawing tools, and a choice of popular image file formats. Desktop Paint 256 supports the following image file formats: PCX (as used by ZSoft's PC Paintbrush), GIF (as used by CompuServe), IFF/LBM (as used by Electronic Arts' Deluxe Paint), and TIFF. Note that the programs work with image files having between two and 256 colors.

However, if you're working with two-color monochrome graphics, you'll find that the Desktop Paint (monochrome) package is better for handling them. It's faster, requires much less memory, and has tools designed for monochrome applications. Desktop Paint 256 is a paint program rather than a drawing program. It does not support GEM, Corel Draw, Designer, DXF, or EPS files.

Unlike many other paint programs, Desktop Paint 256 features complete EMS and XMS support. It handles enormous images, pictures that unpack into several megabytes, as long as there is sufficient expanded or extended memory in your system. Note that you must have at least 300 kilobytes of EMS or XMS to edit a 640-by-480-pixel picture under Desktop Paint 256. The toolbox has the following tools:

SELECTOR (THE SCISSORS). Allows you to select areas of your drawing. Selected areas can be transformed using the tools in the Edit menu.

AIRBRUSH (OZONE-FRIENDLY SPRAY PAINT). Allows you to spray random areas of color pixels in your drawing window.

TEXT (THE BIG T TOOL). Allows you to draw text in your picture using the currently selected font, font size, and direction.

LINE. Allows you to draw lines.

ERASER (THE PENCIL END). Allows you to erase areas of your drawing.

RECTANGLE. Draws rectangles. Seems pretty obvious, doesn't it?

FILL (PAINT ROLLER). Performs a "flood" or "bucket" fill using the current foreground or background color based on the mouse button you use to set it off.

ELLIPSE. Draws ellipses.

UNDO. Undoes the last thing you did, then selects whatever tool you were using before you selected UNDO.

BRUSH (THE PAINTBRUSH). Allows you to draw free-form lines in the line color corresponding to the mouse button you're using and in the currently selected line width.

ZOOM (THE MAGNIFYING GLASS). Allows you to zoom in on a section of your drawing and make pixel-by-pixel changes to it.

GRADIENT. Draws rectangles that are filled with a graduated gray tone, the details of which are defined in the Set Gradient function of the Gadgets menu.

EYEDROPPER TOOL. Allows you to change the current foreground and background drawing colors by picking colors from within your picture. If you select the eyedropper tool and click in the work space with the left mouse button, the foreground color changes to the color of the pixel under the mouse cursor. If you click with the right button, the background color changes.

Paint Shop

Paint Shop is a Windows program that displays, converts, alters, and prints pictures using the file formats of TIF, GIF, WPG, BMP, PCX, MAC, IMG, PIC, and RLE. Altering includes stretch/shrink, trimming, dithering, palette manipulation, and more. Another feature of Paint Shop is screen capture. The entire screen, just a portion of the screen, or a specific window can be captured. Once captured, the user can treat the picture as any other picture, thus saving the picture to whatever format you want, instead of the format the program wants. Other features include:

- Printing pictures.
- Full clipboard support.

- An online help system.
- A user's manual. Viewing can be done in a window or full screen.

Ready, Set, Draw

Looking for a small paint program that accomplishes the basics without all the frills of a full-blown computer studio? Ready, Set, Draw provides the simple functions you want, and it works with or without a mouse. Ready, Set, Draw is a fun, easy drawing program with features like copy, paint, color, load and save pictures, and other basic operations.

Once loaded, you find a small X in the middle of the screen—that's your drawing pen. The top left-hand corner of the screen displays the pen status (up, down, erasing). When the pen is down, it draws; when the pen is up, it won't draw. When the pen is erasing, it erases each dot in the middle of the X. Eight direction keys (up, left, down, right, and diagonals) move the X in the corresponding direction.

From there, you can fill in enclosed areas, cut and paste portions of your drawing, automatically draw various-sized circles and rectangles, even change the color of your pen—all with simple key strokes.

TurboPaint

TurboPaint is an alternative to using PC Paintbrush and Deluxe Paint. One of the big drawbacks to using PC Paintbrush is that you have to reinstall the program every time you want to change video modes. Both PC Paintbrush and Deluxe Paint consume a great deal more memory than TurboPaint, too. So people with less memory can use TurboPaint, and those who like to keep their TSRs resident don't need to reboot with a different configuration before using it.

You also can run TurboPaint effectively without a mouse. Try that with PC Paintbrush and Deluxe Paint. And PC Paintbrush doesn't offer support for Tandy 16-color mode, which is about 25% of the market. The video modes that TurboPaint supports are Hercules 720 × 348, CGA 4-color 320 × 200, CGA 2-color 640 × 200, EGA 16-color 320 × 200, EGA 16-color 640 × 200, EGA 16-color 640 × 350, Tandy 16-color 320 × 200, MCGA 256-color 320 × 200, and VGA 640 × 480.

VGAPaint

VGAPaint is a graphics/drawing program that produces screen images in up to 248 colors at once. Fourteen brush shapes and four automatic shapes are provided. Image handling capabilities include:

- Sizing.
- Flipping.
- Moving.
- Copying.
- Saving images to disk.

Although the resolution of the screen is only moderate (320 × 200), the careful use of color mixing produces images with near-photographic realism.

Drawing/Drafting

Analyst Tool Kit

This tool suite enables a user to perform top-down system specification using the structured analysis methodology (data flow diagrams). Diagrams are entered and manipulated graphically, but they are stored in textual form for easy access by other tools.

Support is provided for data dictionary generation and analysis, as well as for analysis of the data flow diagrams. The structured analysis methodology allows a user to perform a top-down specification of a system from a data viewpoint. The methodology is not limited to software development, but it functions well at the requirements and system design levels also. The resulting set of diagrams and the data dictionary augment any documentation effort and provide a concise, clear picture that can be dynamically maintained along with the system it describes. The toolkit features:

- Eight levels of decomposition below the context diagram.
- Data dictionary evaluation limited only by available memory.
- Flat ASCII data file formats that are easily accessible to other tools.
- PostScript output from DFDs, suitable for printing or incorporation as EPS documents in publishing systems like Ventura Publisher.
- Full function on vanilla PC hardware configurations, with no mouse, math coprocessor, or expensive graphics cards required.

Charts Unlimited

This program integrates graphics and text processing into a single system to allow creation, editing, and printing of flow charts, organization charts, floor plans, electrical diagrams, GANTT charts, business forms, and many other documents. Charts Unlimited has an interface similar to 1-2-3. Charts are drawn on a large worksheet that is 256 columns by up to 1,000 rows. You can zoom in on the worksheet and view all of it at once.

A few keystrokes bring up various geometric shapes, bar graph designs, flow chart arrows, math symbols, and more. You can display, move, stretch, shrink, save, and replicate these symbols. Text can be entered anywhere as lowercase or uppercase. Most dot-matrix and laser printers are supported, and you can print your resulting chart horizontally or vertically. You can also magnify your chart, or a portion of it, up to 100 times.

EditSch

EditSch is a user-supported schematic capture package. It was developed to provide a low-cost, professional method of creating schematics and distributing these designs to others via floppies or BBSs. EditSch is compatible with EditPCB, a PCB layout design package.

EditSch supports a wide range of laser and dot-matrix printers, enabling almost anyone with an output device to generate schematics. Two programs are included in the program: EditSch and PrintDM. EditSch is used to create schematics, while PrintDM is used to print the schematic.

Two types of objects exist in EditSch: electrical and graphical. Electrical objects contain additional information necessary to construct net lists from schematic drawings. Wires, labels, analog gnds, and digital gnds are the only objects that contain electrical connection information. The dot and bus objects serve only to enhance the appearance of the schematic.

When an electrical connection is desired between two or more components, use the Draw Wire command. If you desire to name a net, use the Draw Label command. Draw Analog and digital gnds automatically tie nets connected to them directly to the ground net. The remaining objects (dot, line, arc, bus, circle, and text) contain no electrical information. Output can be printed on an Epson 24-pin printer or an Epson 9-pin printer.

EditPCB Intro

EditPCB Intro is a user-supported printed-circuit-board (PCB) design package. It was developed to provide a low-cost, professional method of designing PCBs and distributing these designs to others via floppies or BBSs. EditPCB supports a wide range of laser and dot-matrix printers, enabling almost anyone with an output device to generate PCB artwork.

Two programs are included in the EditPCB Intro package: EditPCB and PrintDM. EditPCB creates PCBs, while PrintDM generates PCB artwork. The program begins by opening a window to the PCB workspace. Windows can be resized or moved. A PCB is composed of up to eight layers. Layers can be turned on and off. The following layer assignments should be used to maintain compatibility between users and future products. Layer Description: 0—board outline, 1—pins and holes, 2—solder side traces, 3— component side traces, 4—silk screen, 5—solder mask, 6—reserved, 7—user-defined.

Edraw

Edraw (Easydraw) is easy to learn and produces professional results in a very short period of time. Edraw will revolutionize your drawing in the same manner that word processors revolutionized writing. The user does the thinking; Edraw performs the more repetitive and mundane aspects of executing a drawing.

Edraw allows you to design and edit a library of your own symbols. For schematic applications, you can design at a functional level, and you need not be committed to specific components too early in the design process.

With a few glue parts, engineers, teachers, students, and home hobbyists can design a circuit with free-form style. Because Edraw is easy to edit, the designer can experiment with drawings in much the same manner as word processors experiment with and fine-tune text.

This general-purpose drawing program was designed for technical people like engineers, teachers, students, and technicians. In addition to drawing lines, boxes, circles, symbols, etc., the user-supported program includes logic/electronic symbols and fonts that enable users to draw electronic circuits. Edraw loads/saves symbols, and it is not limited by the viewing screen. Edraw is fast; it features automatic panning and mouse support. Edraw requires a standard color graphic adapter (CGA) or an enhanced graphic adapter (EGA) and supports the Epson or compatible graphic printers.

EEDraw

EEDraw is a simple parametric drawing program specifically designed for electrical engineering (EE) types of drawings. The program fully allows new parametric types, through library mechanisms, and all the EE types are defined this way. This library mechanism is extremely useful in defining new types and can be used for other applications other than EE.

The following graphic devices are supported for display: Hercules, CGA, ATT, EGA, VGA and SVGA (through user-provided BGI driver). The following printing devices are supported (via the printing drivers): Epson compatible printers (8 pins), PostScript printers. The program comes with quite a common setup. If you have a mouse and a VGA/EGA/Hercules/CGA, then you can simply start EEDraw.

EZPlace

EZPlace allows anyone to enter parts, placements, and geometries of existing or custom wire wrap boards and printed circuit boards (PCBs). Leading Edge Assembly has coupled the current CAD technology with WireWrap, the ultimate in interconnect reliability, to provide you with a quick and convenient method for circuit board design and layout.

EZPlace, an interactive placement tool, features user-defined board templates, "keep-out" areas, design rule checking, and a component library that you can customize to meet your specific needs. The program also makes conversion to printed circuit board technology a snap.

EZPlace has several purposes: simplify entry, reduce errors inherent in placing parts on prepinned wire wrap boards, and allow users to place components where they want them. The data created by this program allows Leading Edge to deliver to the user a wire-wrapped assembly or PCB. The user should specify a board geometry or template, specify board outlines and keep-out areas, create special parts, place parts where desired, and then send the data to Leading Edge.

Flodraw

Although originally designed for computer systems diagrams, Flodraw accommodates free-format diagrams and allows as much flexibility as possible. This flexibility allows you to use Flodraw for other types of diagrams such as electrical schematics and organization charts.

Easy to use, Flodraw uses a grid of tab marks that are visible on the horizontal and vertical ruler lines to place your symbols on the diagram. The symbols are designed to merge at the tab marks, so your diagram is always aligned correctly. Cursor key commands let you jump easily from tab to tab, so you can build your diagram quickly.

Flodraw documents can be saved, printed, and edited again. All Flodraw document files are compressed to use minimum space on your disk. Flodraw supports the following three formats for printing documents: portrait 8.5 × 11 inches; landscape 11 × 8.5 inches (printed sideways on 8.5 × 11); large landscape 16 × 11 inches (printed on two 8.5 × 11). The large landscape format can also print 14.5-×-11-inch diagrams on wide carriage printers.

PC-Key-Draw

PC-Key-Draw is one of the most powerful graphics programs available for the PC and compatibles. It is a graphics package that can be used by anyone and everyone. PC-Key-Draw includes many features designed specifically for a mechanical engineer, such as calculate and measure. It also includes many features designed specifically for artists, such as pattern fill and area smear. Its ability to create symbol libraries is ideal for the electrical engineer or the architect.

Graphic game designers can use the drawing abilities, along with the ability to save an image as a BASICA subroutine, to create graphics that are ideal for animation. Teachers find the geometry aspects of the program useful in the math and science classrooms. Using such features as slide and text fonts, business people can create eye-catching slides for presentations.

Drawing with a computer can and should be very different from drawing with more conventional tools. PC-Key-Draw gives you all the features you need for free-hand drawing, but its power and the power of the PC lies in the many features that are not possible with conventional techniques. When you learn to use such commands as slide, rotate, revolve, mirror, spray paint, etc., then the real power of computer graphics will be in your repertoire.

PC-Draft

This is a high-resolution, pixel-oriented utility for drawing and drafting. Built-in functions let you draw circles, lines, and boxes; load and edit fonts; create bar, line, and pie graphs; create patterns to fill areas; cut and paste objects; and save objects to files for later use. You can record graphic keyboard macros saved in files for later playback and animation effects. Maximum resolution is 1,280 by 700 dots, which covers several screens and fills an 8½"-×-11" paper (at 150 dots-per-inch resolution).

CAPTURE.COM lets you capture screen images from other programs, save them in separate files, and load them into PC-Draft. You can export portions of the screen, full screens, and/or full-sized drawings in the GEM.IMG file format. This lets you import graphics created with, or captured by, PC-Draft directly into desktop publishing documents such as those created with Ventura Publisher.

Skem

Skem is a user-friendly schematic drawing program. It lets you easily and quickly draw and/or print professional-looking schematics, and it operates on any IBM or compatible computer with 640K of RAM and a 640-by-480 VGA video system. A 386SX-16 MHz or faster computer with a hard drive is highly recommended.

Skem is fully menu-driven for ease of use, and Skem is controllable via a Microsoft or compatible mouse. Symbols, horizontal/vertical lines, and text can be added to, or erased from, a schematic. Symbols can be rotated 90 degrees or flipped horizontally to provide 8 different orientations. Your schematics can be saved and later retrieved for additional editing or printing.

A schematic drawn by Skem can only be printed on an HP laser or compatible printers. Sorry, no dot-matrix printers. It has a built-in help function. Skem uses a four-dot-wide snap-to-grid (not visible) so the user can easily align symbols, text, or lines. The X and Y location of the drawing cursor or symbol is displayed on the screen to allow ease of placement/positioning. If needed, there is a 16.6-pitch landscape font for downloading.

Skem was developed to fill the need for a simple, easy-to-use schematic drawing and printing program. It is only intended to be used to draw and print schematics (nothing else). It is not a CAD program. A professional-looking schematic can be made in minutes. The drawing screen is 640 dots wide by 440 dots high. The maximum screen size of a symbol is limited to 56-by-56 dots.

All symbols drawn on the screen are constructed from drawing commands within files that have an SCR extension. Linked to each screen SCR symbol file is a .PTR (bitmap) file to allow printing of the symbol. The printer bitmap files are constructed in the SYMBOL EDITOR program at the time when the design of a symbol has been completed. The printer bitmap pattern is three times the dot resolution of its equivalent screen symbol. Included with Skem is a small 16.6-pitch landscape soft font, which can be downloaded to your laser printer if it does not have an internal 16.6-pitch landscape font.

TheDraw

TheDraw is a powerful screen image generator/editor for color and monochrome video users. TheDraw makes designing and figuring out a video screen display literally child's play. Its basic premise is: Be easy to use. The most novice computer users can make an ANSI text file screen with TheDraw in only minutes. From that point on, they can explore and use more of the horsepower available in TheDraw. The biggest problem facing most people is that they don't know what looks good until they see it. Manually editing a screen can literally take hours.

With this software, you can free-form until you really know what is perfect. With the picture in mind, you can quickly develop whatever is necessary or instruct the program to create a data file that can be used. Instead of hours, making a screen takes minutes.

It's great for making welcome messages or notes of importance. Another feature TheDraw fully supports is animation. With this facility you can develop quick and dirty presentations or simply make a generic screen appear with pizzazz. You can either let TheDraw make an animation sequence from a static image, or you can make one from scratch, thus letting you fully control the appearance of your work. TheDraw gives you a maximum of 16,000 separate animation "entries" to handle the most complex job. The final result can be saved either as an ANSI file (Ansi Animation) or in TheDraw format file.

TurboDraw

This entry-level CAD program combines an easy point-and-click user interface, plus the ability to generate user symbol libraries. TurboDraw produces professional-quality drawings for electrical, architectural, and mechanical engineers, as well as data flow, program flow, and structural charts for programmers, systems analysts, and database managers.

The menu access system and the spooling capability are two convenient features. You can redirect output to a file instead of to a device and print or plot output at a more convenient time or from a PC that is connected to the proper printer or plotter. Unique features:

- Symbols can be generated and incorporated into symbol libraries by the user.
- Pop-up menus.
- Automatic undo/delete, unlimited depth.
- Pattern fill.
- Supports all plotters, dot-matrix printers, and LaserJets with 512KB or 1.5MB.
- Supports Hercules mono, CGA, EGA, VGA, or MCGA.
- Mouse driver included.
- 300-dpi output, suitable for presentation artwork.
- .TIF, .PCX, and .PLT output for desktop publishing/word processing.

TurboDraw doesn't handle bitmapped images like some other painting/drawing programs, and it is best suited to creating technical drawings that consist primarily of lines and curves.

VersaMap

VersaMap, a menu-driven program, allows the user to draw outline maps on eight different map projections. VersaMap allows great flexibility in designing maps: The latitude/longitude boundaries and type of grid superimposed on the map can be specified. The software can plot maps using the digital map databases "The World Digitized" and "Micro World Data Bank II," which are available from shareware disk vendors. Individual geographic features, such as coastlines, rivers, islands, and political boundaries, can be plotted with the "Micro World Data Bank II" database. A "Micro World Data Bank II" map database is supplied with VersaMap.

The program also accepts user input in an ASCII file format. Maps drawn with the program can be output to the screen, to Epson-compatible dot-matrix or Hewlett-Packard compatible laser printers, as .PCX files for use in paint programs, as Lotus .PIC graphics files for use with word processors and presentation graphics programs, and as ASCII files suitable for import into technical graphing programs.

A dot-matrix or laser printer and a Microsoft-compatible mouse are optional; a math coprocessor is not required, but the software runs about four times faster if one is installed.

CAD/AutoCAD

Auto-Mate ACAD:CAD

Auto-Mate ACAD:CAD, a menu/file management system, allows users to select an AutoCAD drawing file from a menu and run AutoCAD with the selected drawing file automatically. Auto-Mate also works as a front end for several other CAD packages that use the .DWG drawing file extension, and Auto-Mate can be invoked with a drawing file piped at the command line. Also included are several file management functions and the ability to add up to two lines of text describing a drawing file or directory listed by Auto-Mate.

Drawing files listed by Auto-Mate can be sorted in either ascending or descending order, based on file name, date, or size. Auto-Mate generates a log of activity within Auto-Mate if a valid DOS file name is entered in the "LOGFILE" parameter in the Auto-Mate configuration file. Auto-Mate supports a Microsoft-compatible two-button mouse, and all Auto-Mate commands can be accessed using a mouse.

DanCam and DanPlot

DanCam and DanPlot are two utility programs that you can use to automatically manufacture parts. They run on your PC and use your personal computer's parallel port to send motion control signals to almost any TTL signal-compatible stepper motor translator/driver module. These programs use the parallel port to drive all the necessary control relays and to read all of the necessary home, limit, and pause switches.

DanCam and DanPlot read drawing files (made with DANCAD3-D.COM and control-stepper motors) in such a way as to follow the lines drawn with DANCAD3-D. You can attach stepper motors to almost any two- or three-axis machine tool or plotting device's hand-wheel (crank, knob) shafts to automate it.

Because DanCam and DanPlot are general-purpose computer-aided manufacturing programs, you can use them to automate most, if not all, of the types of machine tools you use.

Draft Choice

Draft Choice is an excellent, powerful, fast, object-based, floating-point CAD program for both the novice and expert user. The ease of pull-down menus is combined with the flexibility of vector-based elements to be scaled, modified, rotated, copied, or otherwise manipulated. There is a slide creation/replay option and, of course, printing to dot-matrix printers and plotting to HP-compatible pen plotters are supported.

Whatever the subject, whether it's your house plan, a drawing of an invention, or graphics for business presentations, you'll be glad to know that you can get the help you need from Draft Choice.

MicroCAD

MicroCAD is a superb but small drafting program that was created with the following objectives:

- Provide a diagramming facility that combines the ease of use of paint programs with the object manipulation capabilities of CAD programs.
- Provide an icon-based interface that allows the creation of complete diagrams without commands, menus, or keyboard entry.

- Provide for both the creation of two-dimensional diagrams, and the creation and manipulation of objects in three-dimensional perspective.
- Achieve fast display speeds without the use of graphics or math coprocessors.
- Minimize the program size.
- Allow fast and inexpensive distribution.

The program runs on any DOS or OS/2 system with a VGA or VGA-compatible video adapter and a mouse driver installed. The program runs on an 8086/88, 80286, or 80386/486 processor. Features only available on 80386/486 processors are noted. A high-quality monitor of at least 14 inches is recommended. Most of the program features are designed to be self-explanatory. The following capabilities are provided:

- Draw Line.
- Draw horizontal or vertical Line.
- Draw rectangle or square.
- Draw circle or ellipse.
- Draw cube.
- Create text.

Object manipulation includes:

- Find object (sequential).
- Grab object (on-screen find).
- Find object (sequential).
- Grab (on-screen find).
- Move object.
- Resize object.
- Clone object.
- Solid fill.
- Pattern fill.
- Change line types.
- Add an arrowhead to the end of a vertical or horizontal line.

Other capabilities include:

- Drawing semicircles and semiellipses.
- Tilting object.
- Rolling object.
- Spinning object.
- Erasing/unerasing object.

- Finding erased object.
- Grouping/ungrouping.

The current object statistics and related data are displayed at the bottom of the screen and include:

- The cursor position (X,Y).
- The diagram scale (⅛, ¼, ½, 1:1, 2X, 4X, 8X).
- The current object position (X,Y,Z).
- The current line slope (degrees from horizontal).
- The current square, rectangle, circle or ellipse area.
- The current object rotations (roll, pitch, yaw).
- The projection mode (PERSP, ORTHO, ISO).

NorthCAD-3-D

NorthCAD-3-D is a design and modeling program that enables you to create, view, and edit three-dimensional objects and images. The program is completely menu-driven, easy to learn and use, and features versatile commands with primary emphasis on conceptual design and illustration.

NorthCAD-3-D also features the ability to create drawings from any viewpoint around your drawing, and it uses a simultaneous command architecture that allows you to use several viewpoint commands, even during the drawing functions. Included with the program are a number of sample files that help illustrate the program's capabilities for both conceptual design and graphic arts applications.

PC-Draft-CAD

PC-Draft-CAD is an object-oriented CAD program for drawing or drafting on a PC. PC-Draft-CAD is different from most drawing packages because it stores drawings as basic elements (such as points and lines) in a database.

PC-Draft-CAD has all of the standard functions for drawing circles, lines, curves, ellipses, boxes, etc., with or without a mouse. It goes further with the ability to change the scale of drawings (great for logos!) and the ability to create keyboard macros. Pull-down menus make it easy to use.

Drawings can be exported in .GEM or .WPG file format to be used in desktop publishing programs like Ventura Publisher and WordPerfect. Several text fonts are available, as well as a database of commonly used objects.

PrintCADD

PrintCADD is a three-dimensional drafting program that offers the advantage of creating more than one view out of a given drawing. You are actually constructing a model, rather than a drawing, to view. As such, the process is somewhat more time-consuming than standard 2-D drafting, but it is not overly slow.

You are rewarded for your effort with an actual model that can be displayed and printed from almost any angle or distance. This also reduces errors because the process is similar to actual construction, and clearances and design flaws are much more apparent.

There is no doubt that in the long run 3-D drafting and modelling will save you time, money, and effort. Normal 2-D drafting requires the use of two axes, X and Y. Three-dimensional drafting requires a third axis, Z, considered the depth axis. Normally, when looking at a drawing, you consider the X axis to increase in value as it goes to the right and the Y axis to increase in value as it goes up. In this way, you can define any point on a sheet of paper.

Professional CAM CAD

Professional CAM CAD guides the novice or engineer through the complex calculations required to design *cams*. Cams are a rotating piece that imparts motion to a roller moving against its edge. Cams transform rotary motion into a translating or oscillating motion. The requirements imposed on cams vary from machine to machine due to the speed changes and the type of equipment it is to be used in. Professional CAM CAD allows you to select eight different motion contours from a table.

Various styles of cam contours are available in Professional CAM CAD. Harmonic, cycloidal, 3-4-5, polynomial, constant velocity, matching constant velocity, and dwells are available. You can input one motion contour for a sector and incorporate another motion for another sector in the same cam.

Professional CAM CAD displays the exact contour of the cam and allows you to view the cutter path. The display permits you to see the final shape using the cutter/cam follower of your choice. You can even replot the cam before actually manufacturing it, saving countless hours of development time.

Tommy's CAD/DRAW For Windows

This program is the number one shareware program in Europe. It is a full-featured program that lets you draw anything you can imagine. Drawing tools include:

- Line.
- Polyline.
- Triangle.
- Quadrangle.
- Rectangle.
- Trapezium.
- Parallelogram.
- Polder.
- Circle.
- Ellipse.

- Parabola.
- Zigzag line.
- Circle arc.
- Chain.
- Bezier chain.
- Spline.
- And more.

Tommy's CAD/DRAW For Windows combines the ease of use of Windows with an outstanding CAD program.

VGACAD

VGACAD is an all-purpose graphics editor, image processor, and paint program to create, enhance, or colorize digitized images. It provides 21 brush shapes with 5 different brush modes: normal, spray, air#1, air#2, and eraser.

There are 256 color fills, border tracing, and pattern/gradient fills. Three modes of zoom editing are also available. Air#1 (airbrush) adds a color mixture as you paint over your picture, just like a real air brush. Air#2 mixes colors in more gradually. Spray, when used with the mask feature, has the effect of placing a stencil over the painted area. You can colorize gray images in the same way old movies are reborn in color. Cut-and-paste functions are also fully supported.

VGACAD provides several ways of altering the size and orientation of an image. Rotate the entire screen 90 degrees or continue to 180 degrees to turn it upside-down. You can create a mirror image, or an upside-down mirror image, of any defined area.

VGACAD supports .BLD, .PLT, .GIF and .CLP file formats. Utilities are provided to convert pictures to VGA from CGA and EGA, and from VGA to EGA. Also available is a VGA print utility that prints any viewable color picture in seven user-selected gray shades. You can shrink large-sized pictures and edit large .GIF files (up to 32K × 32K × 256).

VGACAD has been used by such institutions as Lawrence Livermore Labs, the University of Pittsburgh, Warner Brothers, and the *New York Times*.

Home Design

Home Plan

Home Plan was written on an IBM compatible in Turbo Pascal to draw and print floor plan drawings on 9- and 24-pin dot-matrix and HP Laser compatible printers. To provide adequate resolution and aspect ratio, this program requires a color VGA monitor. Features include:

- User-defined scale.
- Adjustable cursor speed.

- Undo of the last action.
- Automatic dimensioning.
- Optional showing of framing (studs, joists etc.).
- Rectangles and circles.
- Deletion of user-defined screen portions.
- Selectable line styles.
- On-screen "odometer" to show horizontal and vertical measurements.
- Saving and loading plans to and from disk.
- Deleting plans from disk.
- Printing the current plan.
- Insertion of predrawn furniture and appliances into your drawing.
- Square-footage calculation.
- Ability to view and delete elements from the plan.

Chapter

10

Telecommunications

One of the least used and most valuable applications of the personal computer is telecommunications. With a modem and telecommunications software, you can "talk" to virtually anyone else with a modem, anywhere in the world. You can write letters and have them delivered instantly to their mailboxes and receive an answer in much less time than it takes the mail. You can go online via your modem and participate in a computer conference. You can access and explore some of the world's largest databases. These contain most of the current articles on all of the new developments in science and engineering, and they also contain what has been documented in books and periodicals in the past. Telecommunications opens the doors to the computerized libraries of the present and the future.

Introduction

Telephone communications software is the special software, the intellect, that instructs your computer (which "speaks" digital) to make the necessary "language translation" so your computer can "speak" analog over the telephone lines. But before I get into telecommunications, here are some important definitions:

Baud rate. The correct technical definition of *baud rate* is the number of discrete signal events per second occurring on a communications channel. Even though it's technically incorrect, baud rate in online systems is usually specified as bits per second (bps). Because it takes roughly ten bits to send a single character (a letter, a space, etc.), the character rate equals the baud rate divided by ten. For example, 2,400 baud is roughly equivalent to 240 characters per second.

Hayes compatible. Modems that are Hayes compatible use the Hayes command set, sometimes referred to as the "AT command set." It uses the prefix AT to grab your modem's attention. Originated by Hayes Microcomputer Products, the AT command set was so popular that it has become the *de facto* standard for modems.

File transfer protocols. Several different software protocols are used to transmit data via modems to ensure accuracy of the data. These special error-checking protocols are not as important for text, but they are vital for transmitting scientific and engineering information, chemical formulas, part numbers, technical data, etc.

Telecommunications Software

General requirements

To be optimum for online operation, modem software should have the following capabilities:

- Hayes compatibility.
- 2,400 or 9,600 baud, or up to 19,200 baud or higher if you are going to transmit and receive large amounts of data.
- Auto speed adjustment.
- Autodial, redial, manual dial, and busy notification.
- Handle various protocols.
- Receive and transmit text and graphics files.
- Emulate various terminals.
- Generate script files.
- Provide a running record of on-time.
- Auto answer.
- Text editor.
- Capable of performing data compression.

Most modem software has Hayes compatibility; be sure the one you purchase has this basic capability. For most purposes, 2,400 or 9,600 baud is usually adequate. However, if you're going to transmit and/or receive large amounts of data, you should be able to operate at 19,200.

Auto speed adjustment is needed so your software can shift down or up to automatically adjust to speed of the modem at the other end. Your modem software should be able to dial another system without your having to manually input the number each time. Your software should store all of the telephone numbers you need to dial so that all you have to do is switch to a dial menu and tap the proper key(s) to automatically dial into the online vendor system you selected. The ability to continue to redial a number until it's available is another positive feature.

Some modems also have a "number linking" or "chaining" capability that you can set up to automatically dial a series of numbers one after the other, skip to the next number if one number is busy, then return and redial the previously busy number(s).

Your software program should also permit you to dial numbers manually when you need to dial a number you use infrequently. And your software should also output a busy signal when your call doesn't go through when the line is not available.

Protocols

Phone lines are notoriously noisy. Noise scrambles the data transmitted to and from an online system. This scrambling might not be too important for normal text because the loss of a single letter usually does not alter the meaning of the text. However, when technical or scientific information, certain foreign words, formulas, or data are scrambled by noise and interference, the loss of a single character could be disastrous. To prevent this, error-correcting techniques are used. These techniques use various methods or protocols to check the validity of each part of a transmission. If an error is detected, that part of the message is discarded and a request is made for that part of the message to be retransmitted. Various protocols are used by different databases to handle errors in transmission. Your software should be able to handle a variety of these protocols, including XModem, YModem, and ZModem.

The ability to both capture and transmit files is necessary, not only for transmitting information to a database, but also so you can capture your two-way "conversation" with the database for later display and printout.

Some database vendors specify that your computer be able to emulate specific terminals such as TTY, DEC, IBM 3101, VT 100, ANSI, etc. This is usually specified in the software's setup menu. As a minimum, your software should be able to emulate TTY and ANSI.

The ability to create and use script files is also important. A script file program automates repetitive operations by setting up a miniprogram to control your telecommunications software. A typical script program automatically dials your vendor's phone number, waits five seconds, inputs your account number, waits another two seconds, then transmits your password, waits another second, transmits another command, etc., all without intervention on your part.

Most software programs also have a "learning" feature that records the stream of incoming data and your responses when you go online. The learning feature then generates a script file that copies and automatically repeats your responses at the appropriate times. Another script feature that is available, and that uses the computer clock, can be programmed to dial up a databank, query for a specific topic, then automatically send a message at a specific time that you set into it.

Your software should also have a running clock that informs you how long you have been online. Also desirable is a log that records the calls you have made, and who they were made to, for later review. It's surprising how quickly you can use up fifteen minutes in a database before you realize how long you've been online.

An auto-answer capability gives your personal computer the ability to automatically answer and accept incoming data when your computer is unattended. If you request information to be delivered to your computer during off-hours, or if you want to converse with other computer users, your software should have this ability.

An online text editor serves many functions. First of all, it allows you to compose and edit messages before you go online so you won't waste your online time. You can also use it to compose and edit your messages when you are online and working in the conference mode that requires real-time conversations. The requirements of an online text editor need not be too elaborate. It should have the basic editing functions, insert and delete modes, paragraphing, search, search and replace, etc.

Because time is money when connected to an online system, the longer it takes a document to be transmitted, the more the cost. One solution is to compress this information so that it requires less time to transmit it. A number of techniques are available to accomplish this; among them are the LHA, PKZIP, and ARJ programs covered in chapter 1. And your software should be able to capture and store your entire two-way, online conversation in case you want to print it out later for review or analysis.

General Telecommunication Aids

ATHelp

ATHelp is a program that lists all of the Modem AT commands. I put together a list of these commands and developed a program that displays the entire command set in windows on the screen. I have released this executable program to the public domain, so it can be freely distributed without restrictions.

JaxHost

JaxHost is a specialized host communication program primarily intended to support operations at the DOS level on the host. JaxHost is great for the office or home. It can be interfaced with other communications packages to implement more complex functions. Popular applications are home-to-office (and vice versa) communications, as well as operations between a company home office and its remote sites.

Software configuration and file maintenance on the JaxHost-equipped host are frequent uses. The program normally monitors the modem, and when it detects a ring, it answers the call while adjusting itself to record the caller's baud rate. Speeds of 19,200 bps down to 300 bps are supported. Upon successful connect, JaxHost greets the caller with a one-line message of your choosing and requests entry of a password, also of your choosing.

If the correct password is supplied and the callback mode is not in effect, JaxHost transfers control of the next part of the communications session to other software. Or a caller might next be presented with another communications application, such as a BBS-like package, or a menu of programs that can be run. If the correct password is not given to JaxHost in three attempts, the software hangs up and recycles.

For enhanced security against unauthorized access, Jaxhost's callback feature can be used. If the callback mode is in effect when the correct password is given, rather than give the caller immediate access, the program disconnects the caller and dials the telephone number specified in the configuration data. Access is granted only if there is a successful connect at that number. JaxHost makes up to three attempts to establish a connect.

If access is granted to the DOS level on the host machine (through CTTY or the equivalent), most DOS commands can be executed and so can other programs designed for "line-at-a-time" operation, usually capable of having console (keyboard and display) data redirected. File transfers to and from the host machine also are possible. A program called XFER supports Xmodem, Xmodem/CRC, and Ymodem (actually 1K-Xmodem) file transfers invoked from the DOS level on the host machine.

Communication Utilities

1stReader

This off-line mail reader lets you download messages from the BBS to your computer, then read and respond to them at your convenience. This way, you can enjoy "BBSing" without tying up your telephone line, running up excessive long-distance charges, or worrying about running out of time on your favorite BBS.

1stReader includes everything you need to manage your BBS messages and replies, from built-in telecommunications software to a filing system for messages you want to save. 1stReader works with any Qmail-compatible mail system. However, when you use 1stReader to call a BBS that uses Sparkware's Qmail Door for PCBoard, you can take advantage of extra features not found on other systems. 1stReader is easy to use, even for people who have limited experience with offline mail readers. All the major functions and features are available from a button or a menu list.

Brain

This utility program works on a bulletin board that is busy, without a person needing to be present. However, there are many applications for which this program could be used, such as electronic mail or transferring files at any time of day or night. This program can hold up to 500 commands. When the program is run, it reads the file and makes sure everything works (checks syntax, makes sure files to upload are available, etc.), then it executes the commands.

Except in the Message and Send commands, spaces are completely ignored by the compiler, and so is the case of the command. However, each command must be on its own line. There need not be commands on comment lines. The function keys can be used to put the keyboard online with the modem, to skip the next command in the command file, to edit the command file, or to terminate the program prematurely and return to DOS. While editing the command file you can cause the program to jump to another point in the command file, you can save the revised command file, or you can restart the program. These functions are displayed at the bottom of the screen.

Modem Doctor

If you have ever tried to install a serial card or a modem only to find out that you have the IRQ lines set incorrectly, the modem cable wired incorrectly, or a jumper set incorrectly, then Modem Doctor has the cure for these ills. If you have had trouble connecting to a BBS, was it caused by the phone line, your setup, or your modem? Maybe it could be something wrong with the UART, the settings on your serial card, or an incorrect assignment of the base port address. This program helps you solve all of these problems and describes some of the pitfalls involved in personal computer communications.

Modem Doctor checks every serial port chip (UART) register, checks cables and modems for proper handshaking signals, and informs you if it runs across some parameter that isn't set properly. Modems are also rigorously tested, including handshaking tests and modem self-connect tests, which simulate an online connection. IRQ assignment errors, base port addressing errors, and a host of other settings are checked. The program also informs you of the type of UART and modem you have installed and prints a copy of the test results to a printer or a file.

RCOMM—Resident Communications

Did you ever put off writing a letter, balancing your checkbook, or doing a spreadsheet because you wanted to download a file from your favorite BBS? With this program, you can do both. With RCOMM you can do memory-resident uploads and downloads while you do whatever you ought to be doing instead in the foreground. Features include:

- Hayes or Novation Autocat modem dial commands.
- Memory-resident XMODEM (checksum or crc) and YMODEM batch upload.
- Memory-resident XMODEM (checksum or crc) and YMODEM batch download.
- Memory-resident XMODEM (checksum) autodownload.
- Pop-up terminal mode.
- PC-Talk-compatible dialing directory.

Bulletin Boards

Free Speech BBS

Free Speech is a streamlined BBS designed for high message throughput. It can be installed in seconds and supports up to three languages (English and French text provided), with full foreign-character translation. Free Speech can be run as a standalone BBS, as front end, or as a door. Like all Pinnacle BBS products, Free Speech is a zero-maintenance system.

Sapphire

Each brand of BBS suits a different kind of sysop. Sapphire is specifically designed for stores, computer consultants, customer support lines, people who have never run a BBS before, and people who have run a BBS before and found it too much work. Sapphire is very easy to set up and maintain. A moderately experienced computer user can set it up in less than an hour and spend approximately one hour per week performing the various duties required of a sysop.

Sapphire's design is quite flexible, so you can modify it to look the way you want. Multiline systems are not supported by Sapphire. This is in keeping with the "install-and-forget" design of the system, but it does mean that Sapphire won't suit everybody.

SearchLight BBS

If you are looking for a bulletin board system that's both powerful and easy to use, try out the SearchLight BBS. Unlike other bulletin board systems, SearchLight BBS is user-friendly to both you (the system operator) and to those who call in to your system. It has an easy-to-understand menu command system, so users who call in to your system need no documentation or help from you to use the bulletin board system. If help is needed, the program has an online help system. Moreover, it has some significant features such as optional ANSI graphics and color selections for the screen, customized menus, public message boards, private mail, a full-screen text editor (compatible with WordStar commands), and a file transfer system with both Xmodem and Ymodem transfer protocols.

There is a complete electronic mail system with up to 24 message subsections, each with up to 300 messages. SearchLight BBS has enough system operator utilities so you can easily manage the system from a remote location and also shell to DOS to run a program or DOS command.

Wildcat BBS

Wildcat is a full-featured BBS with the easiest setup operation in the industry. A complete indexed documentation file is included and covers all program features. The message system includes private mail, forwarding, carbon copies, and return receipts. File transfers make use of Xmodem, Ymodem (batch), Zmodem (batch), Kermit, and any others as external protocols.

A branching questionnaire allows formatted answers for order taking or polling callers. Wildcat! features specific internal support for multitasking operation. Baud rates up to 19,200 are supported. Wildcat! lets you install an operational BBS and get a real feel for the power of the commercial multiline/LAN release.

Communication Programs

1-RingyDingy

1-RingyDingy is a general purpose, asynchronous communications program. It was designed using the "KISS" ("keep it simple, stupid") principle. Although it employs a simple and straightforward approach to communications, its capabilities surpass those of some of the more expensive communications programs on the market today. With 1-RingyDingy you can:

- Communicate at speeds up to 9,600 baud (limited by the speed of your serial port and modem).
- Exchange (send or receive) text files with other computers. While sending text files, you can dynamically send selected records from the file and skip past other records not to be sent.
- Exchange (send or receive) nontext (exe or com) files with other computers that support the "X-Modem" protocol.
- Use modems that don't support autodial.

- Use modems that do support autodial. This includes not only the Hayes SmartModem, but also a variety of other modems that use other commands to autodial a phone number.

You can also:

- Capture incoming data in a disk file.
- Set up your own "filters" to remove or change undesirable characters in the incoming data.
- Dynamically slow down the speed at which you transmit, so as not to overrun the computer at the other end.
- Dynamically alter your communications parameters such as parity, speed, and number of data bits.
- Build automatic log-on sequences of any length to enable you to log on to a variety of remote computers without having to remember phone numbers, passwords, and access commands.
- Change the colors on your display (if you have a color display) to suit your personal taste.
- Move back and forth between different directories on your hard disk for retrieving and sending data.
- Communicate properly with computers that use "XON/XOFF" characters.

In short, 1-RingyDingy is easy to use, easy to understand, powerful, and small in size.

Boyan

Boyan supports interrupt-driven communications with all serial ports and modems at speeds from 300 to 38,400 baud. Hardware and software flow control are supported for high-speed modems and networking. It emulates most popular terminals, including VT-100, VT-52, TVI-925, IBM-3101, ADM-3A, and ANSI-BBS. Terminal emulations use their own "secondary" key sets, which can easily be customized. Further emulations, as they are developed, can be easily installed.

Speedy, reliable file transfers use a variety of common error-checking protocols, including Xmodem, CRC-Xmodem, Ymodem, 1K-Xmodem, Ymodem-G, 1K-Xmodem-G, ASCII, Kermit, Ymodem-batch, and Zmodem. Additional protocols can easily be integrated. Hands-free "auto-downloading" is provided for any protocol that supports it, including Zmodem, Bimodem, and MPt.

Boyan script files can run communications sessions completely unattended, supporting handshaking, timing, conditional execution, block nesting, macro nesting, and much more. Script Learn feature remembers what you type at all times. After a session, the software can walk you through what you typed and build an automatic script file for you. You can edit the script before it is saved, so your next session on that system is hands-free. Sample script files help streamline your use of CompuServe and common bulletin board systems. A simple "trace mode" helps you follow even the most sophisticated macros and scripts.

The host mode allows you to dial in to your computer from a remote system, read and write messages, send and receive files, and run programs remotely. A built-in database holds the name, password, and user level for up to 70 users. Because the host mode is written as a Boyan script, it is fully customizable. With Boyan's automatic-file-name-entry feature, you never again have to type a file name twice before downloading a file. Boyan saves you time and money by allowing you to upload messages you have prepared offline.

Exclusive features. The program determines the line prompt character and handles awkward "word wraps." A usage log records the length of all modem connections and the efficiency of all file transfers for later reference. This log can verify your phone records, for example. It works with DOS at all times. Issuing any DOS command, running any program, and entering DOS temporarily are possible during a communication session. The program uses your favorite file lister and word processor/editor. A full, built-in file manager lets you scan disk directories, copying and viewing files at the touch of a key.

Commo

Commo is a very powerful, full-featured, telecommunications program. Small in size but large in function (written in .ASM), Commo's comprehensive macro programming language can automate any process. Commo handles high-speed modems with ease, is fully DESQview aware, and includes among its many features a free-form dialing directory, macro file, and support for visually impaired users. Unique to this telecommunications program is an accompanying special version of the program supporting the HP 95LX palmtop computer.

Easy Comm

Easy Comm is an easy-to-use communication program with the following features:

- User interface has mouse support.
- Pull-down menus.
- Online context-sensitive help facility with a help index.
- Multiple, overlapping, resizable, movable windows.
- Dialog and message boxes.
- Customizable options (colors and mouse parameters).

The built-in, fast and flexible text editor provides:

- Multiple windows and multiple files that can be open at the same time.
- Search and replace functions.
- Text-block selection.
- Copying, moving, and deleting.

Because both file printing and file transfer take place in the background, you can simultaneously print one file, upload or download another file, and continue editing still another file. Communications functions include:

- Dial directory database.
- Selectable communication parameters (e.g., baud rate).
- Auto-login procedure.
- Filter option for filtering out characters in both directions.
- Several file transfer protocols (ASCII).
- Four types of XModem, YModem and YModemG transfer status windows.
- File transfer occurs in the background, so you can edit a text file or do something else during the file transfer.
- Customizable modem control strings.
- Chat mode (separate windows for sent and received characters).
- Autoredial function.

Additional functions include:

- Background file printing.
- ASCII table (you can send characters from the ASCII table to the edit window).
- Calendar.
- Calculator (you can send an expression and a result from the calculator to the edit window).
- Time window.
- Memory information window.
- Change directory function.
- DOS shell function.
- Save and restore desktop functions.

EZX-Talk

EZX-Talk provides an electronic link between your computer and a computer in one of thousands of homes and offices throughout the world. Computer time-sharing services and bulletin board systems are now within the reach of your computer. With EZX-Talk, programs and files of information can be sent back and forth between your home and another computer quickly and easily.

Unlike most other communication systems, EZX-Talk does much of the work for you, so there is virtually no training time to operate the system. EZX-Talk does the dialing and the log-on to the remote computer for you automatically.

EZX-Talk features an easy-to-use menu system that saves you time and frustration over other more complex, but no more useful, systems. You can also use a command mode if you need to implement an option without going through the menu.

GT Power

GT Power's easy setup routine and quick, simple-to-use commands simplify communicating with other computers and BBSs. GT Power is a set of programs and files that provide complete capabilities for your system. It supports a variety of modems, including all Hayes compatibles, US Robotics, and IBM. Here are some of the features of GT Power:

- ANSI standard terminal emulation.
- ANSI.SYS-style color graphics, without the need of ANSI.SYS.
- Multiple telephone directories with 999 entries each.
- Records the password for each BBS, as well as the date and time of last contact. The total number of calls, accumulated time, uploads, and downloads are also recorded.

Numerous protocols are supported, such as: SEAlink, Xmodem/CRC, WXmodem, Ymodem/CRC, Telink/CRC batch file, the CompuServe "B" protocol, and more. Also, ten slots are available for external protocols to be added by the user. GT comes with several of these slots preset for the most popular protocols, i.e., Zmodem, Ymodem-G, Kermit, Jmodem, Super8k, and BiMode. These can be easily changed or added to by modifying the external protocol table in the GT configuration. You can also transmit ASCII files using XON/XOFF protocol and/or line-by-line pacing and/or inter-character delays.

The program provides full control of all communication parameters. There are 40 function keys to store log-on, passwords, etc. It has full printer support using XON/XOFF flow control with the host. You can exit to DOS, then return. GT stays in the background. The interactive script language automates many repetitive tasks. It has an optional LOG file and keeps a complete record of your calls.

MicroLink for Windows

MicroLink for Windows is suitable for both beginners and experts. You can use the program to connect your computer to other computers and to information services such as CompuServe and Genie. At the nucleus of MicroLink is the dialing directory, an easy-to-use database where you can store names, phone numbers, log-on scripts, terminal settings, and port settings.

Select an entry, press the dial button, and MicroLink instructs the modem to dial for you and to redial if the line is busy. Once connected, your computer acts as a terminal, giving you access to the remote computer's resources. Eight built-in transfer protocols provide compatibility with nearly any other system for sending and receiving files.

You can have MicroLink capture incoming information to a disk file with the simple press of a button. With the program's scroll-back feature, you can review or print up to 2,000 lines of text. With its simple but powerful script language, you can prepare responses to a remote computer's prompts before a call. In this way, routine information requests and file transfers are accomplished automatically and completely unattended.

Best of all, you do not need to be a programmer to prepare these scripts. MicroLink can write these scripts for you by monitoring your last call. With MicroLink's configurable button bar, you can send or receive files, print captured text, open your dialing directory, and perform many other functions by simply selecting the appropriate button. You can even define your own "custom" buttons to invoke a script file.

Odyssey Communications

Features that make Odyssey stand out are:

- Instant, painless installation for most computer configurations.
- Support is included for COM1 through COM8.
- Support for modem speeds up to 115,200 baud.
- 43/50 line support on EGA/VGA monitors; MNP levels 2, 4, and 5 built into software.
- Can double the effective speed of non-MNP modems.
- 16550 UART chip support built in.
- Very complete and accurate emulations of ANSI, Data General 200, TTY, DEC VT100, and VT320 terminals. VT100 and VT320 emulations include 132-column capability.
- Screen review (scrollback) buffer with editing and paste, including paste to communications session.
- Edit and paste can also work with any text file.
- A full suite of file transfer protocols: ANSI, CompuServe B+, SuperKermit, Xmodem, Xmodem-1k, Ymodem, Ymodem batch, Ymodem-g batch, Zmodem (including batch). All transfers execute in the background so you can prepare messages or even shell out to DOS while they take place.
- Full Pascal-like script language with learn mode; scripts are fully syntax checked before execution.
- Powerful script language.
- Keyboard remapping facility to customize odd terminal emulation situations.
- Also included are typical telecommunications functions including: full dialing directory with redial and circular dialing queue, keyboard macros, chat mode support, session text logging, online help, and DOS shell.

PC-Dial

PC-Dial lets you send and receive electronic mail, explore the world of electronic bulletin boards, link your computer directly to someone else's computer to exchange data, and use any DOS command (like DIR or COPY) without leaving PC-Dial. You can use a built-in MiniEditor to create or edit log-on files, even write memos or letters, all within PC-Dial.

Ease of use is the keyword for this type of program, and you can access a help screen whenever you need it. This screen also shows the defined smart keys (i.e., macros). Set up smart keys to save keystrokes and commands into a "string" that is permanently saved. As the name implies, smart keys help you work smarter, not harder; you can time how long you are online. This is better known as getting a rough idea of how much your phone bill is.

PC-Dial's on-screen timer shows both time of day and elapsed time; it can be reset at any time. You can communicate at speeds from 300 and 1,200, all the way up to 9,600 baud. The only limitation here is the speed of your serial port and modem. You can exchange (send or receive) text files with other computers. While sending text files, you have the option to send selected lines from a file while skipping past other lines not to be sent.

Pilot

Pilot is a fully functional communications program. With it, you can call other PCs, BBSs, information services (CompuServe), or mainframes to upload and download files. The program also lets your PC receive calls. Pilot has the following features:

- On-screen help.
- Function keys are defined on screen.
- ANSI terminal emulation.
- Uploads and downloads files.
- Supports the major file transfer protocols, including XModem, XModem 1K, YModem, YModem-g, YModem-g Batch, and Kermit.

Any incoming data can be dumped directly to the printer at the same time that the information is being displayed to the screen. You can also dump any incoming data directly to hard or floppy disks at the same time that the information is being displayed on the screen.

The program lets you create different file names for the information you are saving. It supports the major extended modems such as Hayes, US Robotics HST, Telebit, MultiTech, GVC SM-96, Generic, and others. Pilot supports baud rates from 150 to 38.4K and provides the option to operate under a TSR. The TSR capability allows Pilot to pop up on top of most applications on the market. The software program being run is frozen until you exit Pilot. Complete host and remote capabilities include:

- Answer calls.
- Upload and download files.
- Show files with wildcard capabilities in the current directory.
- List subdirectories.
- Change directories.
- Make new directories.

- Chat online with host service.
- Emulate a BBS and receive calls. Your PC can be left on to wait for a call from another PC or a mainframe; supports up to nine serial ports (one at a time).

ProComm

ProComm is a general-purpose program designed to provide easy and convenient access to a broad variety of telecommunications tasks. Most of the program is written in the "C" programming language, with some assembly language routines for optimum performance.

Included in its abilities are the types of features you would expect to find in highly sophisticated telecommunications software:

- Emulation of a number of popular terminals.
- Dialing directory containing 100 entries.
- Automatic redial facilities for connecting with hard-to-reach numbers.
- Several popular file transfer protocols including XMODEM, Kermit, Telink, and more.
- Command files to control automatic log-on and unattended operation.
- DOS gateway that allows you to execute DOS commands or other programs while you are still online.
- Plus many additional features, including keyboard macros and disk and printer logging.

QModem

This is a fast and powerful telecommunications program. Expanded memory usage, windowing, Xmodem protocol, and autoredial are only some of the strengths of this package. The "quick-learn" mode option is perfect for novices. Further training is available with the context-sensitive help function. You can read and write ASCII files with the integrated text editor.

XY and IMODEM protocols are fully supported, as are multitasking environments such as Topview, Desqview, Taskview, and Windows. The dialing directory can be sorted, printed, and searched. Terminal emulations for VT100, ANSI, TTY and Televideo 925 ensure further flexibility. LOG allows you to maintain usage records—who called, time called, time online, and a date/time stamp for each log entry.

Slick

Slick is simple, very powerful, and easy to use, and it has everything you basically need to enjoy telecommunicating. Slick features internal Zmodem protocol with auto-sensing. (You need only instruct the host to begin the Zmodem transfer; Slick does the rest.) A monitor mode allows display of all raw characters without filtering. This is a very useful diagnostic and advanced function for investigating raw data appearing at the port.

Additionally, a literal insert mode facilitates use of control codes, graphic characters, and ANSI. PARITY/DATA/STOPBIT selection can be saved as a permanent run-time default, along with port and baud rate. Baud rates include 4,800, 9,600, and 19,200. Slick is an interrupt-driven (on receive only) terminal that supports the four standard DOS BIOS addresses of 3F8, 2F8, 3E8 and 2E8 respectively for COM1 through 4.

If you use OS/2 or any other system that addresses COM3 through 8 at 3220h through 5228, a special version of Slick/OS2 is available for those circumstances. Today, 99% of all BBSs provide Zmodem file transfers because of the superior performance it has over X and Ymodem. Because many Slick users are running laptops, the program provides X and Ymodem via the external DSZ protocol engine, thus keeping the size of Slick's executable file to a minimum.

Telemate

Telemate is an excellent, full-featured, flexible communication program with an easy-to-learn and easy-to-use integrated environment. With Telemate, you don't need a separate editor, viewer, and mouse driver for smooth communication. All these features are built into Telemate and are accessible through menus and windows.

Multitasking. Telemate has multitasking features built in. While you are dialing or downloading, you can prepare messages, view files, and scroll back to the last connection. Or you can have all of them at the same time; each clearly runs in its own window.

Mouse support. While other communication programs still require a mouse-menu TSR to emulate their function keys, Telemate considers the mouse to be an input device from the very beginning and has an environment suitable for both mouse and keyboard.

Built-in editor and viewer. The editor lets you prepare your messages, edit documents, or type letters, and the viewer displays files as references without influencing the editor.

Back scroll. The very large back scroll buffer works just like an editor. You can save parts of the content to files, find text, and even quote a message by pressing several keys.

Clipboard. Through the clipboard, you can cut and paste text among windows. You can prepare a message in the editor, then copy and paste it to the terminal.

Protocols. Telemate has the most popular protocols built in, including Zmodem, Ymodem-G, SEAlink, and CompuServe QuickB. And there are four slots for external protocols.

Easy-to-learn script language. Writing a Telemate's script program is as simple as writing a BASIC program. If you don't have programming experience, don't worry. The learn script mode generates script files for you.

Enhanced video support. Without additional hardware, you can select a screen height from 25 to 32 lines on your normal CGA adapter and seven choices of screen height on an EGA or VGA adapter.

EMS, XMS, Video Ram support. Telemate uses expanded memory, extended memory, and EGA/VGA video RAM as virtual memory to store data.

Telix. Telix is a full-featured communications program that meets the needs of most users. When Telix was written, a basic idea was kept in mind. It was designed to be a powerful program that would be simple to operate, with no loss of speed in achieving this ease of use.

Telix features include: an extensive range of built-in file transfer protocols, including Zmodem, CompuServe Quick B, Xmodem, Xmodem-1k, Xmodem-1k-g, Ymodem (TRUE), Ymodem-g, Kermit, SEAlink, Telink, Modem7, and ASCII. Telix allows up to four "external" protocols to be defined and called from within Telix. In this way, almost any kind of file transfer is possible from within Telix.

Powerful multiple dialing directories hold up to 1,000 entries each. An entry contains many pieces of information required for communicating with remote services, such as the phone number and communications parameters, automatic queue redialing for hard-to-reach numbers, reliable DEC VT100/102, and VT52 terminal emulation.

Keys can be redefined by the user; extensive SALT script language allows Telix to perform automated log-ons and many more complicated functions. As an example of the power of script files, the Telix host mode is written in the SALT script language. Full access to DOS, including a DOS shell, a DOS Command option, and full path support, are provided. It also has a host mode with file transfers, operator paging, a remote DOS shell, two access levels, a chat mode, keyboard macros, a scroll-back buffer, session capture, usage log, translate table, and a variety of other features. Above all, Telix is fast and friendly.

UniCom for Windows

UniCom performs data communication tasks in the background while other applications are running. UniCom can even transfer Windows-unique data formats between computers, such as the contents of Window's clipboard from one computer directly into the Clipboard of another. UniCom includes the following features:

- X/Y/ZMODEM, Kermit, CompuServe B, Quick B, and ASCII file transfer protocols.
- ANSI-BBS, VT52, and TTY terminal emulation.
- Directory-assisted batch dialing for users with Hayes-compatible modems.
- An online help system.
- Ability to record mouse and keyboard operations for later playback.

- Script file processing for automatic log-on to remote host systems.
- Keyboard macros, file logging, file paste, and print screen.

UniCom can transfer Windows-unique data formats between computers, such as the contents of Window's clipboard from one PC directly into the clipboard of another. Operate your computer in a multiuser mode with UniCom's host mode. A built-in command processor lets a validated remote user examine or transfer files on a designated disk drive while remaining completely transparent to any user who might be at the keyboard operating other Window applications.

Connectivity—LAN/WAN

EasyNet NOS

EasyNet Network Operating Systems is the latest in the EasyNet family of network products. This powerful network operating system allows 2–35 ATs, 386s, and/or compatibles to be connected in an extremely flexible network system. The EasyNet NOS is 100% NetBIOS compatible and works equally well with most NetBIOS-compatible twisted-pair, ArcNet, Ethernet, or Token Ring network communications cards. EasyNet Systems, Inc., also offers its own twisted-pair bus network cards. The EasyNet NOS operating system can operate in a bus, star, or tree configuration, depending on the communications cards selected.

EasyNet NOS allows maximum flexibility in the configuration of the PCs on the system. PCs can be set up as workstations, servers, or in a full peer-to-peer configuration. You can also set up any combination of these to provide maximum flexibility of the network resources.

The EasyNet 116 LAN system includes a 1 Mbps twisted-pair network card and peer-to-peer network software combined in a complete network package for up to 10 PCs. This system can be configured with any combination of workstations, servers, or peer-to-peer users on the bus-type system.

Some of the additional utilities for the Easynet ShareLan package are:

Print capture/spooling (PCS). PCS provides a full print capture system for automatic printer redirection over the network to a remote printer. It allows the local operator to simply print from a program as if a local printer were connected. Transparent to the user or application, PCS reroutes the print job to the specified remote printer. The spool control program allows the user to submit print jobs to the various printers on the network and to monitor and rearrange the print queues as desired.

Security system. The EasyNet Security System controls access to network resources through the use of security filters. Any device on the network can be secured and made unavailable to the other users on the network.

Cabling. An easy-to-use adaptor and cabling kit can be purchased separately for those of you who cannot bear the thought of soldering up another cable assembly.

EasyMail

EasyMail brings the paperless office one step closer to reality. It supplies a means of communication for those on a network by providing a simple electronic mail program. The program interface is easy to use and understand.

This program can be used for the simplest form of "junk" mail or corporate policy documents. Its ability to distinguish between unimportant mail and permanent mail makes it suitable for quick communications and short-term memos as well as important data to be retained.

EasyMail works on most NetBIOS-compatible networks. It has been thoroughly tested on the latest Novell network technology, as well as on the simplest peer-to-peer network (invisible LAN). The program should be installed onto the network server disk. Place the program into its own "EasyMail" subdirectory if possible. To install it, simply copy the contents from the program disk onto your network server. Once copied, it is ready to operate.

U-Net

U-Net is the first complete network available via shareware, and it is also the first network that can be installed by a computer novice. U-Net allows small businesses to install the network and see the advantages of sharing software, data files, and printers between any two PCs without a major investment. The biggest benefit to users is ease of use. Installation takes less than five minutes by running a simple install program on each computer; the program automatically configures the software for the computer ports being used. The cable simply connects between the parallel ports of the two computers. If there is more than one parallel port, U-Net uses LPT2.

No learning is required, as U-Net uses standard DOS commands. The only additional commands are S-Print and W-Print for directing printing. The hard disk drive(s) on the main computer are accessed as if they were physically in the other computer, becoming additional virtual drives and being automatically assigned the next logical letter.

U-Net is very applicable to laptop, notebook, or IBM PS/1 computers with no free slots available for networking adapter cards. Also, it can be used for the transfer of data between computers with incompatible diskette sizes or simply to share a high-cost laser printer between two computers. Small businesses or professional people such as doctors, attorneys, and traveling salespeople who do not need to be constantly using a network will find U-Net extremely useful in being able to link a second computer into the office's main computer and access information from time to time, as required.

Prodigy

ProUtil

Most of the public-domain utilities have also been built into ProUtil. Make sure you check out the ABOUT PRODIGY bulletin board under the topic MEMBER UTILITIES for messages about ProUtil. Jump to ABOUT PRODIGY. Features of ProUtil:

- Extremely memory conscious: uses only 14K of memory in its smallest form, and memory requirements can be changed to fit your system.
- Change the WORKING icon into a clock.
- Print public and private mail wider than 40 columns.
- Print just the text from any screen that uses the 40-column font.
- Restore the Print Screen key. (Now uses PrtScr, not Alt-P).
- Restore the type-ahead buffer.
- Advance the printer one page—form feed.
- Advance the printer one line—line feed.
- Add a signature line to the bottom of a message.
- Print all of the unread private mail with one key (printed in the correct chronological order).
- Wait for a message to arrive in your electronic mailbox.
- Remove the form feed sent after Prodigy prints a message, and remove the extra carriage return sent at the end of each line.
- Upload a private message with the press of a key.
- Upload a public message with the press of a key.
- Up to 50 single key macros that can be recorded while using Prodigy instead of being created offline. There is practically no limit to the size of a macro.
- Cancel the printout in progress. (Ctrl-ESC).
- Removes itself from memory after exiting Prodigy.
- BOOKMARK allows you to mark your place anywhere in Prodigy and jump back at your leisure (similar to Prodigy's LOOK/ZIP; slower, but more powerful).
- Records each keystroke as you use Prodigy. If you are disconnected from Prodigy, simply log back on, press the RESTORE BOOKMARK key (Alt-R), and PRO-UTIL returns to the same location, even if you were in the middle of typing a message. (Please read the warning on this.)
- Press Ctrl-ESC to clear the keyboard buffer, cancel a macro, and cancel almost any other function of ProUtil.

Genie

Ease-Up For Genie

As with most computer software, Ease-Up was born of a need. In this case it was the need to eliminate the aggravation of trying to edit an upload description while online.

Ease-Up is a front end to the script languages of several popular telecommunications programs that support script files for the specific purpose of generating a script that will automatically upload a file for you, once you are online. More importantly, all the online descriptions are entered for you by the script. Ease-Up was originally developed to support Qmodem and Genie and is also available in a CompuServe version. Ease-Up supports the following terminal programs in the Genie version: PROCOMM PLUS, Qmodem, and Crosstalk Mk.4.

The great benefit to using Ease-Up is that you allow yourself the convenience of composing the file description information offline, free of connect charges. You also have at your disposal:

- A sophisticated environment that includes data validation (as far as it makes sense).
- Full editing capabilities for the long description.
- Pop-up choice lists for easy entry of file name extensions, forum names, and forum libraries for the forum you select.

Ease-Up includes a list of most of the forum names used on Genie for you to choose from. (You have to tell Ease-Up what forum to upload the file to.) You might never have to manually enter a forum name, unless you want to. Ease-Up reads an ASCII text file into the long description using a command-line switch. You have the choice of using Genie's menu mode or command mode. You can create as many scripts as you like, reusing the description information or changing it as needed.

Chapter

11

Computer Programming

This chapter covers the art and science of programming, that mystical skill that some have and many wish they had. Programming continues to explode, providing us with the tools to pursue our professions, take care of our finances, manage our homes, and provide some much-needed relaxation and recreation (read "re-creating").

Included in this chapter are several programs that are directed to some of the specific aspects of programming. The coverage ranges from very elementary programming principles to some very in-depth dissertations. Topics covered are: making a software program easy to use, software installation procedures, printing programs, batch files, BASIC, DOS, assembler programs, Fortran, Pascal, Forth, Ada, C programming, and even some advice about how to market the software you write. Even if you don't intend to do any programming, this chapter should help take some of the mystery out of this art and prepare you to better use, even to modify, the software you use.

General

Aortal

Aortal is a graphics editor and CASE tool. Designed using the latest object-oriented techniques, Aortal provides an indispensable tool for the C graphics programmer. With Aortal you can create sophisticated graphic displays that you can use in your own applications.

By allowing you to directly manipulate the graphics on the screen until it looks perfect, Aortal skips the tedious "edit, compile, and run" phase of writing most graphics applications. Then Aortal produces C code that you can embed directly into your own Borland code for the fastest graphics ever. Some graphics attributes can be modified directly on the screen. Other graphics attributes can be changed by clicking on the buttons at the bottom of the screen. To use the Aortal-generated code in your own programs, you must compile and link the ART C code with your programs.

Creating User-Friendly Software

This utility set is a must for everyone writing attractive, user-friendly BASIC or Turbo Pascal programs. Some utilities make your programs more attractive; most make them quicker and more friendly. Source code is available for most, but users should understand ASCII.

Included in this software are several ways to help you write easy-to-use programs to obtain input from users, especially multiple data input. One utility helps set up one-sentence prompts for each input. Another enhances the use of default values. A third makes BASIC and Pascal understand uppercase and lowercase inputs identically. Other utilities help you deal with arrow key inputs, setting up data input screens, using lines and boxes, and displaying a message to buzz off when the computer is busy.

Freeware Hall of Fame

This excellent collection lists programs and databases that are commonly found on public-access bulletin boards. The programs on this list have been tested and found to work as described, and to be worth downloading. This list is limited to freeware. These are programs and databases for which no fee is required for use by individuals.

Names in the circle of great freeware writers for their output and quality are: John McNamara, Raymond Kaya, Vern Buerg, Ward Christensen, Robert Vostreys, Frederick Volking, Michael Mefford, Keith Graham, Keith Ledbetter, and the writer who gave us CED and so much more, Chris Dunford. In all, the Freeware Hall of Fame lists more than 370 entries by 270 authors, indexed 3 ways: by author, program name, and use.

The Freeware Hall of Fame has three sorted ASCII indexes. Each index contains the same information: author, program name, a brief description of the program, and the program category. Index number one is sorted by author, index two by program, and index three by description.

Among the most prominent people in the computer world, past and present, are those programmers who allow us to use their programs at no cost. Writing software and distributing it via bulletin boards (with no request for payment), they enable everyone to share the fruits of their knowledge, the benefits of their discoveries, and the steady advances that have typified the computer industry from the beginning.

This file honors those generous and wonderful people. The file is not finished and never will be. It's a list that will grow every few months as we attempt to catch up with the hundreds of fine people who represent the hobby side of the PC world, some from the era of CP/M (lest we forget), some writing for DOS, and now we're expanding it to include OS/2. Among them are some of the best programmers in the profession, software experts sought after and employed by the most advanced government and industry employers in the world. The freeware they write on their own time (cough) commonly reflects the highest level of concept and execution.

This Freeware Hall of Fame is the only published guide to tested freeware. All of these programs work. All are free. Nearly all are commonly available on public access bulletin boards. I can recommend them as programs that will do the job they were intended for. The file is indexed to make these programs easier to find, but locating them on a bulletin board can require a search.

Freeware refers to programs or databases that an individual can use without payment of money to the author. Commonly the author copyrights his or her work as a way of legally insisting that no one rewrite it prior to obtaining his or her approval. The copyright has nothing to do with a user fee, aside from giving the author a legal right to require one. Freeware authors don't exercise that right, though in some cases a program will be freeware for an individual, but require a site license if used by a business. The Freeware Hall of Fame includes these.

Other freeware is in the public domain; that means the author has donated his/her work to the public and retains no ownership rights. No fee can be required for the use of these programs, though a polite and sometimes humorous request for author support might drift into the grapevine.

Shareware refers to a way of marketing programs on the "try-before-you-buy" plan. Because you must buy it, it isn't freeware. As shareware, many superb programs are made available at reasonable cost and on a much fairer plan than purchasing "commercial" software, where you must buy before knowing if a program will do what you need. Most shareware is a delight to use. Some is not.

Shareware always asks for a fee for continued use of the program and the author says so, usually in a professional way. A few authors aren't professional in the least when discussing payment. They presume you're not going to pay and in their documentation they insult you. Some shareware comes crippled. You get only a limited version of the program until you buy it. Shareware might have annoying delays and payment dunnings built into the program that go away only when you purchase a registration number. Some shareware quits working if you don't buy a code number within a few weeks.

Freeware does none of these things. The author might ask, as one did, that you take him to dinner (he called his method dinnerware). Another suggests you take your spouse out to atone for the time you spend at the computer. Freeware authors often show a great sense of humor. They also frequently provide the source code.

Every program gets a one-line listing. Ward Christensen gets one line for his CP/M communications programs—perhaps the most influential freeware of all time—and you'll get a one-line entry for your program. Nominations are to be made to the compiler of this list, and the compiler needs to know four things: author's name, software name, brief description of what the program does, and your certainty that the program is freeware.

Library of GIF Manipulation Routines

Each utility provided in the "util" directory performs some kind of filtering, generates a new image, or dumps it to a device. A filter prints in increasing order the current input image line number any time it needs to read the input, and it prints the output image line number (in decreasing order) any time it dumps out.

Other utilities that only read or write will always print that half the same way (but in increasing order). This is true only for utilities that decode the data. Utilities like GifPos (which only changes positions) copy the image as a block of compressed data and print nothing. There are two reasons for this: they cannot identify line number, and they are much faster.

Some utilities require memory in the order of the whole screen, while others read one scan line at a time. Each utility DOC file has entry called "memory usage," which will be one of the following: line (memory required is in the order of one scan line), image (order of biggest image in GIF file), or screen (order of GIF file screen). In all cases, a byte is allocated per pixel, so an image of 320 by 200 pixels requires about 64K bytes to hold.

Personal APT

Personal APT offers 2½ axis NC part programming and a powerful set of mathematical functions and geometry descriptions to help solve complex problems often encountered in engineering, design, and drafting functions.

Personal APT uses a subset of the most powerful and most widely used NC programming language; APT Compatibility with APT provides users of Personal APT an upward growth path, as their work increases in complexity, without having to learn a new programming language.

Also, Personal APT is so easy to use that it is ideal for teaching and learning the APT language and programming concepts. Text editing and graphic display of tool paths are built-in features of Personal APT. Personal APT also offers capabilities beyond other programming languages (BASIC, Fortran, C, etc.) for solving complex mathematical and geometric problems.

The Printer

This software eliminates a nightmare that many programmers face: printing formatted text on different printers, with each printer requiring different instructions (i.e., different printer codes). If you want bold italic, underline, etc., you need printer codes to "turn on" these options. And if you distribute your programs, you often need printer codes for the dozens (or hundreds) of different printers your users have—unless you want to limit your audience.

Few programmers have the time and resources to obtain appropriate codes for every printer. Fortunately, you don't have to. The Printer offers you codes for hundreds of printers. And with this software, you can, in many cases, develop simple print routines that you can use, regardless of which printer a user selects. Finally, any programmer writing in any language can use The Printer.

If you can read a fixed-length record, you can use this program. The Printer contains printer codes for more than 720 different printers (more than 160 in user-supported versions). For each printer, the program authors allocated space for about 70 printer codes. Some of these are blank "user" codes so you (or your users) can customize codes for specific needs, or so the authors can have room to expand.

ScreenPaint

ScreenPaint lets you design lightning-fast batch-file-based hard-disk menus, help facilities, pop-up windows, and prototype screens for applications programs. Screens can be quickly "painted" in a full-screen editing environment, complete with pull-down menus, help screens, automatically connecting lines, ten screen pages, and a wide variety of drawing and editing tools. Completed screens can then be displayed from a DOS batch file and "popped" directly to the video display, rather than scrolled using the extremely slow DOS ECHO command.

The ScreenShow program allows screens to be displayed and key codes to be returned to the calling batch file, where they can be tested. Unlike other menu-creation systems, the user has complete control over the style of menu displays, and the menu screens can be totally independent of the user input routines. This allows the user the flexibility to use any batch language, from DOS to batch extenders like Beyond.Bat, Command Plus, and Extended Batch Language.

Snap

This program makes documenting X-Base programs a (surprise?) snap. Snap produces technical documentation for an entire X-Base system. Its reports include:

- Formatted source code listings.
- System summary showing lines of code, file statistics, etc.
- Variable cross-reference report.
- Tree structure of the system. The tree optionally can also show databases, index files, etc., used by each program.
- List of all files in the system, plus a source reprint file.
- Database summary.
- Index file summary (including multiple index files).
- Format file summary.
- Label form summary.
- Report form summary.
- Procedure/function file summary.
- BIN file summary.
- Memory file summary.
- Other files summary.
- Action diagrams.
- Batch files to back up programs, databases, etc.
- Batch file to move Snap output files back to the source subdirectory.

For FoxPro applications:

- Screen file summary.
- Menu file summary.

For Clipper applications:

- Link files (PLINK86, TLINK, MS-LINK).
- Make files.

In addition, on each program file, Snap writes a heading showing: program name; system name; author and copyright notice; which programs, procedures and functions call this program; which programs, procedures, and functions this program calls; databases used; index files used; formats used; report forms used; procedures; BIN files called or loaded; memory files used; other files used (text files, misc. files, and the like); and date and time last modified. If you want, all source code headings can also be echoed to a separate file.

Temescal

If you want to bring that next software development project in on time, and under cost, Temescal can do it for you. With Temescal your project can be on time and on budget.

Temescal is for planning and managing software development projects. Software development projects are notorious for getting bogged down in unforeseen delays and costs. The reason, all to often, is that they are planned by guesswork. Temescal brings you the experience of thousands of development projects.

There is a more orderly approach to planning a software project, and it is based on the "software equation." The software equation was developed after studying more than 800 large development projects performed by many different companies and government agencies. The software equation has been improved and refined to include data on today's popular languages such as ADA, C, Pascal, etc., and 8-, 16-, and 32-bit computers. With Temescal, you can:

- Compare a change in development time for a corresponding change in the number of work months needed for a development project.
- Evaluate the development time required in a software project when different languages are used.
- Calculate the efficiency of each and every programmer and monitor their productivity as the project progresses. If you have a deadline to meet, you can calculate the number and efficiency level of the programmers required to meet it.

The languages supported by Temescal are ADA, Assembly, BASIC, C, COBOL, Forth, Fortran, Modula-2, and PL1. If you plan or manage any software project, you need this program.

The Window BOSS

This is one of the most powerful and cost-effective products available to enhance and accelerate the development of system and applications programs in the "C" language. The BOSS lets you create programs that have the same look and feel as top sellers such as Lotus 1-2-3, Sidekick, dBASE III, and Framework. Pop-up windows, pull-down menus, status lines, and in-context online help functions can be easily implemented. Your applications can drag windows around the screen and automatically sense the video card installed. All of this without snow, flicker, or delay.

The BOSS's assistant, The Data Clerk, is always on call to handle the tasks associated with data entry. Whether they are as simple as fetching a line of text, or as complicated as the coordination of filling out a form, the Data Clerk will be there to assist, and if necessary, validate precious information as it is entered.

The Window BOSS supports PC/MS-DOS for the IBM PC/XT/AT, PS/2, and compatibles. However, you'll need one of the following compilers to take advantage of the state-of-the-art techniques available from The BOSS: TopSpeed C, Zortech C, Borland C++, Turbo C++, Turbo C, Lattice C, Watcom C, Mix Power C, Microsoft C, Quick C, or Quick C for Windows. The BOSS is written in C and assembly language. You'll need the Microsoft Assembler, MASM, to assemble any local changes to the assembler source.

General Programming

At Ease With Computer Programming

Lessons in At Ease flow back and forth between the computer and the workbook. The lessons proceed in the following order: thinking, defining, modeling, formulating, computing, conclusion.

At Ease With Computer Programming teaches three main skills: creative problem solving (in the lessons on thinking, defining, and modeling); how to write solutions that work (in the lesson on formulating); how to translate those solutions into structured computer code, powerful macros, or effective database queries (in the lesson on computing).

During your first viewing of a lesson, you should explore all overviews, examples, and explanations. But wait until your second runthrough before looking at detail items. Details are clearly labeled as such. (The information in the details is important, but it is outside the main flow of the lesson.) The Esc key takes you out of any lesson and back to the main menu.

Programmers Test: Do You Have What It Takes?

Scoring for this test: Count one for each item that you have done, or each question that you can answer correctly. The following gives the score ranges' classifications.

- 0×000 and 0×010—Computer illiterate.
- 0×011 and 0×040—A user.
- 0×041 and 0×080—An operator.
- 0×081 and 0×0C0—A nerd.
- 0×0C1 and 0×100—A hacker.
- 0×101 and 0×180—A guru.
- 0×181 and 0×200—A wizard.

If you don't understand the scoring, stop here. And now for the questions.

- Have you ever used a computer?
- For more than 4 hours continuously?
- More than 8 hours?
- More than 16 hours?
- More than 32 hours?
- Have you ever patched paper tape?
- Have you ever missed a class while programming?
- Missed an examination?
- Missed a wedding?
- Missed your own wedding?
- Have you ever programmed while intoxicated?
- Did it make sense the next day?
- Have you ever written a flight simulator?
- Have you ever voided the warranty on your equipment?
- Ever change the value of 4?
- Unintentionally?
- In a language other than Fortran?
- Do you use DWIM to make life interesting?
- Have you named a computer?
- Do you complain when a "feature" you use gets fixed?
- Do you eat slime-molds?
- Do you know how many days old you are?
- Have you ever wanted to download pizza?
- Have you ever invented a computer joke?
- Did someone not "get" it?
- Can you recite Jabberwocky?
- Backwards?
- Have you seen *Donald Duck in Mathemagic Land*?
- Have you seen *Tron*?
- Have you seen *Wargames*?
- Do you know what ASCII stands for?
- EBCDIC?

And it goes on and on. Every aspiring and practicing programmer should take this test.

Software Installation

The Finishing Touch

The Finishing Touch is a professional software installation program for developers to distribute with their software. It compresses files 50%+, reads across floppies, checks hardware, modifies AUTOEXEC.BAT and CONFIG.SYS, prints or views ASCII files, does error checking, and more. It supports the English, French, German, and Spanish languages.

The purpose of The Finishing Touch is twofold: It reduces the floppy disk space needed by 50% to 75%, allowing distribution of software on fewer disks. Depending on the number of disks in the set, The Finishing Touch should save its purchasers thousands of dollars in software production, mailing, and distribution costs. It gives the end user a simple, friendly, and reliable way to install software. It tests the end user's hardware so installation is as smooth as possible. This pleases the user and reduces support costs. Software developers only have one chance to make a first impression. Why not make it the best one possible?

Install-Pro

Install-Pro is a user-friendly installation program for programmers that safely copies application files from installation diskettes to a hard drive. Far superior to a batch file, it validates and verifies all destinations before starting the file transfer process.

Program Installer

Program Installer has been used by DataWorks, Inc., in some of their own third-party applications as a generic installation utility. This particular version is released into the public domain for use by other developers/programmers. However, Program Installer is not freeware. If you want to use all of the features of the program, register it with The DataWorks, Inc. But don't lose heart. Program Installer is not crippled. It still supports all of the main features of the registered program, except that it is limited to only one program disk. The registered copy allows unlimited program diskettes to be used.

Program Installer is for programmers/third-party developers. Use it to facilitate the installation of a program you have written for either a client or for some vertical market. Program Installer doesn't care about what you have written, or whether or not you want the program(s) you have written installed into a particular directory on the client/customer's hard drive.

Program Installer reads files from either the A: or B: floppy diskette drive and allows the end-user to specify the hard drive letter (C: through Z:) and/or subdirectory into which your programs should be copied. As a programmer, all you need to do is follow the simple setup instructions of Program Installer, then just copy your files (for installation) onto the installation diskette(s) and let Program Installer do the rest for you (and your client/customers).

Q-Install

This is a fast and easy-to-use installation utility. It allows you to create professional-looking, customized installs for your program distribution. Nonprogrammers can use it with ease. It takes just minutes to create a complete, custom-made installation. Like most programs, you need to install them on a customer's hard disk drive.

Batch files are too clumsy to write and do not provide all of the needed functions for user interaction. User input and screen handling are very limited. Most install utilities available on the market today cost between $50 and $250. That cost is too much for most programs. Q-Install is a quick, inexpensive install program that meets 98% of most programmers' needs. It includes:

- A high-quality, professional look.
- Fast, colorful text (colors are easy to change).
- Very small program size (below 24K). This leaves more room for your program files.
- Easy-to-define options and parameters via a menu. (No cryptic programming language to learn!)
- Automatic checking of disk space.
- Automatic detection of source drive.
- Automatic checking of directories and creation.
- User override for target drive and directory destinations.
- Ability to display a text README file from the menu screen.
- Ability to use self-extracting PKZIP or LHA files.
- Allows multiple install-to directories per diskette (up to four).
- Most programs require only one directory per diskette.
- Multiple diskette installations are made simple.
- Royalty-free, unlimited program runtime distribution for those who order/register.
- Plus a free update to the latest version.

Batch Files

Batch File Tutor

Batch file commands are the most powerful and versatile of the DOS commands. DOS has given batch files (called bat for short) special powers to execute any DOS command. Because of this power, bat files are used in almost all programs. This book/disk teaches bat files from the ground up.

Many new techniques have been introduced to simplify the writing of bat files. For the first time, detailed illustrations on color codes are given. The mystery of creating colored screens and colored letters are unmasked by some new formulas. Writing color programs is now an easy step.

Writing batch files is the best way to learn how to use the DOS commands. In these lessons, you write many batch files. You'll have an opportunity to see these batch files come alive with the menus, programs, and color combinations. You'll experience a sense of power that you have over the computer. It is a real accomplishment to learn one of the most powerful and useful parts of DOS.

You'll be one of the few persons who know how to write a batch file, a program, and a menu—and in color. Bat is not a DOS command, but a file that is created with the copy command. It has special powers given to it by DOS. Bat files can instruct the computer (DOS) to carry out commands. Batch file commands can make the computer (DOS):

- Create menu-driven-programs.
- Automate many DOS functions.
- Beep.
- Blink.
- Change colors.
- Show you what's in a file.
- Print a file.
- Perform many other functions.

Extended Batch Language—Plus

Extended Batch Language adds additional features to the batch processing capabilities already supported by DOS. The program guides you in the creation of batch files that prompt for responses and keep track of variables. Use it to set up printers, keep configuration parameters up-to-date, initialize programs, or create other batch files. Optional floating point, enhanced string and arithmetic operations, and file I/O support are provided. It is well-documented and includes example batch programs.

Hacker

If you want to write truly interactive and intelligent batch files, Hacker adds a whole new dimension to batch file programming. Hacker is a multifunction utility for programmers. It communicates with both the user and the system, and it performs many useful tasks. Most Hacker functions are for use within batch files, but command line processing is also supported. Hacker returns information to a batch file through DOS's ERRORLEVEL. Hacker functions include:

- ASK a question from the user.
- BOOT causes a warm or cold reboot.
- CLEAR offers cursor control.
- SOUND gives ability to produce tones or music.
- WHAT provides information about date/time/drive/mode.
- NOTE sends notes to the screen.

- WAIT adds time delays.
- IS gives drive/directory control.

Hacker also provides an extensive online help command that explains the use of each function.

PowerBatch

PowerBatch is a compiler for batch file "type" commands. At the present time, 35 PowerBatch commands work only in a PowerBatch source file. (They will not execute in a standard DOS batch file.) Any valid executable command (.EXE and .COM), as well as any of the DOS intrinsic commands, are valid within a PowerBatch source file and are "compiled." These .EXE, .COM, and DOS intrinsic commands are executed via a shell; the PowerBatch .EXE program continues execution with the next compiled statement following the shelled command.

There are four major reasons for using a compiled PowerBatch program: The resulting program almost certainly executes considerably faster. While it might be possible to construct a compiled PowerBatch program that does not execute faster, it would have to consist entirely of preexisting executable modules. PowerBatch commands provide a great increase in flexibility and power over conventional batch files. PowerBatch programs cannot be changed without making machine-code modifications. This means additional security for distributed batch files. PowerBatch programs can be created quickly for the PowerBatch command set is a very high level language.

ProBat I

If you spend hours making batch files and wish there were a more comprehensive way to construct your own user-friendly batch interface, ProBat has revolutionized the way complex batch files are built. This sophisticated program gives you the power to create batch files with chaining screens, graphic screen layout, keyboard input acceptance, and informational prompting.

It is great for the programmer or system designer who currently uses a mixed bag of tricks to get batch files looking even halfway decent. With ProBat, you work in one comprehensive batch file creator, complete with pull-down menus and extensive editing options. Though ProBat takes some time to learn, the results are the most professional batch files possible.

You can use ProBat to save batch file formats or screen designs to be recalled and edited later. Create your own library of batch formats to be recalled and modified for different purposes. ProBat lets you draw boxes, set screen colors, and create menus. There is even a handsome graphics menu screen that can be integrated into the batch files you create. If you are serious about batch files, ProBat is an involved and in-depth program that allows you to create the best in batch files.

SEBFU

You can create professional, powerful batch files with SEBFU. It improves your batch file menus by giving you the option to control the position of text and colors (or monochrome intensity), create windows and boxes, and scroll text across the screen. SEBFU increases the potential of your batch files by allowing you to make decisions based on your system configuration. Error levels are returned, which indicate configurations such as presence of ANSI.SYS, disk drives, memory usage, mouse driver, printer status, cursor location, keyboard type, and much more. SEBFU is a must for software developers or those who want to greatly improve the efficiency and power of their batch files.

SEBFU also includes the Business Control System (BCS), a simple telephone log database/telecommunications program. BCS dials your number and records who you called, the time and date, and any messages. BCS is easy to learn and is a handy organizational tool. SEBFU includes 51 batch utilities, plus BCS.

Secrets of Batch Files, Color, and Macros

This program contains detailed instructions on how to write simple programs (batch files) for color and one-keystroke speed (macro) keys. It instructs in easy-to-learn language. More than 100 formulas for the computer beginner are included. It includes: how to use the simple formulas (batch files), how to write powerful programs that makes the computer change colors, and how to format a disk or make a duplicate copy of a disk.

The DOS commands have been edited and rewritten in simple English language. No secret formulas. The big treat is when you see programs you wrote light up the screen and display a "menu of one-keystroke program." DOS batch files are the fastest and best way to automate your program menus. Batch files can customize program selections, color your screen, make your computer play a song, or answer you with beeps and/or blinks.

X-Batch

X-Batch contains several utilities that can serve as commands in batch files for the busy, no-frills programmer. While some of the X-Batch commands can be started at the DOS prompt, they are most useful as part of batch commands.

BOX displays boxes with optional borders. CHECKENVIR tests environmental variables. CHKSCREN searches for text, while CLRSCRN clears screens with selected colors. COMPFILE compares data, time, and file size. CREATE creates files, and CURSOR positions the cursor. DISPFILE displays file text in a box, while DISPLAY shows it on the screen. GETCHAR awaits specific keys and reports. SHOW displays date, time, or day of week. SOUND beeps a tone for N seconds, while WAIT waits for time of day, key, or N seconds.

BASIC

ABC

This software is an imperative language originally designed as a replacement for BASIC. The program is interactive, very easy to learn, structured, high-level, and easy to use. ABC has been designed iteratively. It is suitable for general everyday programming, the sort of programming you would use BASIC, Pascal, or AWK for.

It is not a systems-programming language. It is an excellent teaching language, and because it is interactive, it is excellent for prototyping. It is much faster than UNIX "bc" for doing quick calculations.

ABC programs are typically very compact, around a quarter to a fifth the size of the equivalent Pascal or C program. However, this is not at the cost of readability. On the contrary. The program is simple to learn due to the small number of types in the language (five). If you already know Pascal or something similar, you can learn the whole language in an hour or so.

It is easy to use because the data-types are very high-level. The five types are:

- Numbers: unbounded length, with exact arithmetic the rule.
- Texts (strings): also unbounded length.
- Compounds: records without field names.
- Lists: sorted collections of any one type of items (bags or multisets).
- Tables: generalized arrays with any one type of keys, any one type of items (finite mappings).

Add-Power Library for PowerBASIC

The Add-Power Library contains many useful, tested PowerBASIC routines. You can write complete pull-down menu interfaces, rodent-aware, just by modifying the demo program. There are fancy entry routines for string or numeric data, phone numbers, dates, etc. (never use INPUT again). Other features include date arithmetic, get file directories and disk space, open a data-entry window, dialog box, let users customize colors, push and pop screens, and much more—all in PB units. This software has been in daily use by the author and others (posted on CompuServe) for as many as four years.

ASIC

This is a BASIC programming language compiler for IBM PCs and compatibles. It includes an integrated, full-screen editor from which you can edit, compile, and debug your programs. Or at your option, you can use your own text editor and compile your programs from the MS-DOS command line.

ASIC is a subset of BASICA and GW-BASIC, which supports more than 80 BASIC statements, integer and string variables, and integer arrays. It compiles your programs at high speed, and the code it generates is fast and compact. The syntax of each ASIC statement is generally a valid subset of the syntax used by BASICA and GW-BASIC.

The ASIC compiler runs in 305K of memory. The ASIC editor runs in less than 200K of memory. However, to compile and debug within the editor, 405K of available memory is required. If ASIC does not find sufficient memory to call the compiler from within the editor, it ignores the compile request and returns to the editor. ASIC programs look very much like regular "BASIC" programs that you are probably already familiar with. The one very noticeable difference is that ASIC programs have no line numbers.

BasWiz

BasWiz is a library of assembly language and BASIC routines for use with QuickBASIC. Some of you might not have heard of BCD math, or at least not have more than a passing acquaintance with the subject. BCD (short for binary-coded decimal) is a way of encoding numbers. It differs from the normal method of handling numbers in several respects.

On the down side, BCD math is much slower than normal math, and the numbers take up more memory. However, the benefits might far outweigh these disadvantages, depending on your application. BCD math is absolutely precise within your desired specifications, and you can make a BCD number as large as you need. If your applications don't require great range or precision out of numbers, normal BASIC math is probably the best choice.

For scientific applications, accounting, engineering, and other demanding tasks, though, BCD might be just the thing you need. The BCD math routines provided by BasWiz allow numbers of up to 255 digits long (the sign counts as a digit, but the decimal point doesn't). You can set the decimal point to any position you like, as long as there is at least one digit position to the left of the decimal.

The Designer

The Designer is a programming tool for BASIC users working with animated graphics on an IBM PC. The program is of significant value to people writing graphic games and to people interested in experimenting with animation. The Designer is written in IBM advanced BASIC (BASICA). The source code is included for those who want to study and/or modify it. Because The Designer uses dynamic array dimensioning, it cannot be compiled.

Included with The Designer is a demonstration program that illustrates one way to program animation. "Demo" is also provided in source code form, so you can study and/or modify it as you want.

EZ-Windows Pull-Down Menu System

This routine allows a programmer to present the user with options in an organized and esthetic way. The Pull-Down Menu System was indeed modeled after the very pull-down menu used in the QuickBASIC environment.

The Pull-down Menu System is very easy to integrate in a QB program, yet it offers all the options someone might expect in a commercial software package. To use this routine in a program, all that is required is a few arrays containing formatting information for the pull-down menu and an array containing the actual text used. Then a simple CALL statement takes care of the rest. Don't worry. Loading the arrays required for the routine is very straightforward and uncomplicated, and the results are definitely worth it.

TsrBasic

TsrBasic lets you write terminate and stay resident (TSR) programs in BASIC. It supports both pop-up applications that are recalled with a user-defined hot key sequence and background tasks that are recalled after a specified period of time has elapsed. The entire interpreter can be recalled as well, giving you a powerful, programmable calculator at your fingertips.

TsrBasic simplifies BASIC programming by relaxing data type checking restrictions. This means you do not have to use special suffixes like $ and # to specify the type of data a variable will hold. Any variable can hold any type of data, either textual or numeric. Conversions are performed when necessary.

Structured programming is a snap with TsrBasic's IF...THEN...ELSE...END IF statement, which supports multiline THEN and ELSE clauses and nested IFs. The conventional form of the IF...THEN...ELSE statement is also supported.

Windows R-E-Z

This is a collection of QuickBASIC and assembly routines that provide users of QuickBASIC and BASIC (PDS) with a complete window management system. With Windows, R-E-Z users can make, save, restore, and delete up to twenty windows. The memory used to save windowed areas is dynamically allocated and outside of basic's normal data storage area, leaving more room for the basic programs data.

Windows are assigned a number from zero to twenty. Input windows and multifield input routines are also provided. Input windows provide a convenient means to prompt for and receive an input. Multifield input allows users to define up to 10 input screens, each having up to 100 input fields. Numerous options are included for input fields. Windows R-E-Z provides users the ability to incorporate pull-down windows, emulating those used in the QuickBASIC programming environment, in their programs. Directory routines find the default drive and path, disk size and free space, and directory listing for any path. File size, date, time, and attributes can also be found.

Several other routines are included that allow the use of get answer windows and scroll windows. All of the routines require a minimal amount of initialization, and the resulting programs present a professional appearance. Unlike many other basic "add-ons," Windows R-E-Z provides extensive error detection and reporting.

Assembler Programs

A86/D86 Assembler/Debugger

A86 is an excellent assembler that actually performs better and is easier to use than Microsoft's MASM 4.00. It also assembles your programs a lot faster than MASM. It lets you leave out all segmentation directives, if you want. The segmentation default model is compatible with most high-level languages. You can also access subroutines from your own libraries and, if you register, you receive the A86LIB tool for creating your own libraries more easily than with MASM. You can use floating point operands in assembly-time expression arithmetic. This is an A86 exclusive.

The manual (included on the disk) is excellent. You can learn 8086 or 80286 assembly language programming with it. The debugger is great. It has a floating point display window for use with the 8087 or 80287 that will please you.

DPL (Doug's Programming Language)

DPL is a macro language that uses high-level language syntax for its instructions. DPL's language does not limit the programmer to using native 8088 code, but it makes programming easier by automatically declaring segments and the program entry point. All return codes except end-of-file are standard DOS return codes. The program handles 64K of code and data and stack segments. Variables are automatically declared.

DPL supports simple data types and still permits the programmer to build other types. Basic data types supported are 16-bit integers and ASCIIZ strings. The 16-bit integers are considered to be signed integers and therefore will be manipulated appropriately. Files are supported that allow the program to address any disk file or logical device such as the keyboard and screen.

Extensive documentation is available on disk, in both standard ASCII and WordStar format. Also available is the DPL source program written in Turbo C. A large number of application routines and application routine construction files are provided, along with a demonstration of the use of shade and how to manipulate it.

PC Assembler Tutor

The PC Assembler Tutor is an assembler tutorial that covers all 8086 instructions. It starts with the simplest instructions and works its way through the whole instruction set. It also covers details of the hardware involved with the 8086. The PC Assembler Helper (ASMHELP.OBJ) is an object file that does I/O for assembler level routines. It also displays the 8086 registers and flags (if desired) and allows each arithmetic register to be independently formatted.

The intended users are people who already are competent programmers in a high-level language and either are just starting to learn assembler or know fewer than two-thirds of the assembler mnemonics and want to learn more about assembler. The tutor assumes that the user has a general knowledge of the DOS commands, subdirectory structure, and path names. It is helpful if the user has already linked object modules to form an executable file.

Fortran

ACM VOL 1–5

These routines are a collection of Fortran routines/procedures for solving a variety of problems in science, engineering, and numerical analysis. There are routines for the solution of zeros of polynomials, various partial differential equation solvers, linear equation solvers for overdetermined and underdetermined problems (L1, L2 and Chebyshev norms), eigenvalue solvers, function integration, minimization of unconstrained multivariate functions, curve fitting in Chebyshev norm, dependence of solution of nonlinear systems on its parameters, sorting, global error estimation for ordinary differential equations, matrix bandwidth, and profile reduction.

In addition, there are routines for spline interpolation, solution of linear systems for special matrices, cubic curve fitting using local data, computation of condition numbers of a matrix without computing eigenvectors, bessel functions, statistical functions, revised simplex method for constrained resource network scheduling, exact solution of integer systems of linear equations, convex hull algorithm for planar sets, a multiple-precision arithmetic package, adaptive smooth curve fitting, basic linear algebra subprograms (BLAS), optimized mass-storage fit, efficient one-way enciphering algorithm, and more.

F2C—Fortran77 TO C

F2C—Fortran77 TO C converts Fortran 77 source code files to C or C++. A thorough and well-documented piece of software, this program was put together and copyrighted by several programmers and is now available for the serious programmer at no registration fee.

F2C converts Fortran 77 source code in files with names ending in ".f" or ".F" to C (or C++) source files in the current directory, with ".c" substituted for the final ".f" or ".F." If no Fortran files are named, F2C reads Fortran from standard input and writes C on standard output. File names that end with ".p" or ".P" are taken to be prototype files, as produced by option "-P," and these are read first.

GRAFX—Graphic Extension Library for Fortran 77

The GRAFX library consists of more than 50 Fortran 77 callable routines that allow Fortran programs access to the PC's graphic and other capabilities. The graphic routines find application where emphasis is on ease of implementation and rapid presentation of numerical data. Three levels of capability are included. The top level requires no knowledge of the PC or computer graphics and is suitable for pedagogical applications such as numerical methods instruction, quick-look analysis of computational progress, and for the professional who is not primarily a programmer.

The middle level operates in problem space and supports display of points, lines, circles, ellipses, boxes, etc., to allow creation of custom displays such as map projections, multiple graphic windows, and 3-D displays. The lowest level consists of primitives implemented in assembly language that operate in display space to directly access display memory, selected DOS, and BIOS functions.

Graph for Fortran

This collection of Fortran subroutines makes it simple for you to plot high-quality scientific and engineering graphs on dot-matrix printers that have the ability to print graphics. These subroutines can easily be incorporated into any Fortran program so that your own programs can automatically print high-resolution graphs as part of the program output.

Also included is a sample main program to illustrate the use of the subroutines. This sample program works as a stand-alone graph plotter for plotting graphs from a data file. So, as an alternative to incorporating the graph plotting subroutines into your own programs, you could instead get your program to simply write a suitable data file to disk for plotting out later, using the sample main program supplied (GRAPH.for).

Some of the features available are:

- High- or low-resolution options.
- Graph can be plotted vertically or horizontally on the page.
- Linear or logarithmic axes, with labelling of axes.
- Overlaying grid lines on the graph, in various line styles.
- Drawing lines in different styles (solid, dotted etc.).
- Plotting points, with a large range of symbols to choose from.
- Joining points with straight lines or a cubic spline.
- Plotting user-defined functions.
- Passing a list of parameters to the user-defined functions.
- Placing text of different sizes and orientations on the graph.
- Clearing an area of the graph, with optional border.
- Automatic generation of legend tables to describe each curve.

PC Graphics C

PC Graphics is not just for HP BASIC types who need to port code to the IBM PC. These software tools provide an intuitive and very powerful method for graphics programming that makes up for shortcomings in the standard Microsoft C library.

PC Graphics is a subset of the library primitives available in Hewlett-Packard workstation BASIC. Here are procedures for soft-clipping, scaling, line patterns, pen colors, and text rotation that let you write two-dimensional graph plotting programs in Microsoft C. The scaling feature is especially useful; it lets you create your own scale, removing the need to think in pixels. The package can also be used for simple animation.

Personal Communications Library For C (PCL4C)

The Personal Communications Library For C (PCL4C) is an asynchronous communications library designed for experienced software developers programming in Microsoft C or Turbo C. Twenty-one communications functions as well as six support functions are provided. Included are:

- Set Baud Rate for a selected port.
- Monitor CTRL-BREAK key.
- Receiver is interrupt-driven.
- Set/Clear/Read the data terminal ready bit.
- Delay.
- Terminate selected port.

Communications rates from 300 to 115,200 baud are supported. The receive queues are adjustable from 8 bytes to 16KB. All four ports, COM1 through COM4, are supported. (Most libraries only provide support for COM1 and COM2.) Additionally, two ports can be run concurrently.

PCL4C provides complete modem control and status. Twelve different communications errors can be trapped and reported. There is even support for the CTRL-BREAK error exit. The library supports all memory models and was written in optimized assembly language for small size and fast speed.

The source code for a simple terminal emulator program is provided as an example of the use of the library functions. This sample program can be used to call up bulletin board services and mainframe computers, or even to build a specialized communications interface for your application. Four other example programs are provided, as well as 48 pages of documentation.

SBM Fortran

SBM is based on American National Standards Institute Fortran 77 language. It is assumed that the reader has knowledge of an existing Fortran language and is familiar with the MS-DOS or PC-DOS operating system. This manual is not intended to teach Fortran; it is to be used as reference material only.

The original intention of SBM Fortran was the software author's answer to the C language. The cryptic nature of C, while appealing to many programmers, leaves the author cold. While he fully appreciates the power of the C language (as with assembler), he felt that he needed something that would better suit his style. The powerful Fortran systems that are now on the market are designed with the high-level user in mind.

Program size is not important. What is needed is something that generates about the same code as if it were written in assembler. This compiler operates by converting Fortran source code into assembler instructions, which are then converted into machine code using WASM, an assembler written by Eric Tauk, of Merengo, IL. The result is a COM file, which can then be run on any MS-DOS based system. SBM Fortran is a compiler compiler.

Modula-2

M2JFTool

M2JFTool contains library modules that provide new, useful procedures to be incorporated into Modula-2 programs to either produce new effects or make it easier to achieve older effects. Also included are programs to make the programming process easier. This system offers control over:

- Date and time information with user-selectable formats.
- File reading and writing of strings, numbers, and Booleans.
- Printing of strings, numbers, and Booleans.
- Printer control codes.
- Additional string procedures.

Other features include:

- Comprehensive screen control.
- Cursor control.
- Partial or total clearing.
- Boolean prompts.
- Inverse video boxes for data input, including decimal point, data entry, editing, and display of reals.
- Screen editing of strings and integers.

The data functions provide keyboard entry, FileInOut, printing of all types of data items, and entry of logical variables. The Math1 library module provides the trigonometric functions. Those modules that employ real arithmetic are presented in two versions, one for those with a math coprocessor and the other for those without. There is even an improved programming environment that liberates you from constant mouse use and gives you greater editing control. All the tasks that are necessary when developing a Modula-2 program can be executed from one menu.

Modula-2-Compiler

The Modula-2-Compiler (M2C) proves that you don't have to buy a commercial package to get a full-featured compiler. M2C features the compiler combined with an integrated editor and "make" facility, a program linker, an execution profiler, and a make-file generator.

Code generated by the M2C suits either the Intel 8086 "huge" or "large" memory model. More restrictive memory models are not supported. For the "huge" model, each module has its own data and code segment. Each of the segments can be up to 64K in size. For the "large" memory model, static data from all the modules is combined into a single data segment. For either model, pointers are four bytes long, and all remaining memory is available for the "heap." The package includes 37 pages of indexed documentation that is well-written and understandable.

Pascal

DML

Here are 87 Turbo Pascal routines not found in the standard version 4.0 or 5.0 library. Full source code and documentation are included for each routine. The library is divided into four functional modules containing:

- Forty-two general-purpose procedures for such things as system programming extensions, file I/O and protection, text encryption, general-purpose video and video messages, disk and memory sizes, instruction timing, general-purpose file, math, DOS, and version control.
- Eighteen numeric procedures for handling general numeric formatting and conversion, and date and time formatting and conversion.
- Fourteen string procedures for general purpose string functions and conversion.

- Thirteen keyboard procedures for controlling and accessing the keyboard and keyboard buffer.
- The entire library can be accessed by a single statement, or the four unit modules can be accessed individually. A demonstration is included that shows the action of each routine.
- Additional utility included is XREF, a Pascal source code cross-reference and formatting program.

FlashPac Pascal Library

FlashPac Pascal Library contains low-level Turbo Pascal routines to access the video display, keyboard, printer, disk, mouse devices, and a couple of DOS functions. These routines are not intended to replace the standard features of Turbo Pascal, but add to and extend the Turbo Pascal programming environment for the MS-DOS operating system. To simplify the use of these routines, the number of global variables needed for this library has been kept to a minimum. All functions in the library were written in assembly language using the Pascal parameter-passing conventions.

FPDISK supports several DOS function calls that use byte streams when accessing disk files. FPKBD supports BIOS and DOS keyboard function calls. FPMOUSE supports basic mouse functions, including a mouse event handler. FPPRT supports BIOS printer functions. FPVIDEO consists of several video routines that provide direct access to the video display. Routines include saving and restoring the screen, framing windows, setting the absolute position of the cursor, and more.

The Helper

With The Helper, you can create hypertextlike programs and help files with the following features: A help file is created from an ASCII text file, and this help file is then compiled into an interactive hypertext document. The compiled file can be used as an interactive information document where various topics can be selected using the arrow keys and a cursor bar, or it can become a controlling program in its own right, or a combination of both. Each help file can have up to 254 categories of information, with up to 254 pages per category.

Note that you do not need Turbo Pascal to use The Helper. However, if you are a Turbo Pascal programmer, you can incorporate access to your compiled help files directly into your programs. Several sample Turbo Pascal programs are supplied to demonstrate how easy it is to do this. In its simplest application, only two or three lines of source code are required for full access to a help file for your program, including all the features listed above.

Mouse Tools

The use of a mouse as an input device is becoming increasingly common in many types of programs. While this (usually) makes life easier for the user, it certainly does make the programmer's job more difficult.

This set of utilities provides most of the necessary procedures to use a mouse in your own programs, including: display the mouse cursor, read mouse and button positions, display various prompts with push buttons, and select file names.

To use these tools, you need Turbo Pascal 5.5 and you must have the TP5.5 units CRT, DOS, drivers, fonts, and graph available. Two sample programs (mapedit.pas, and cgaexp.pas) are provided, along with the .tpu files for all the mouse tools. This should give you enough to get started writing mouse programs in Turbo Pascal. The source code for the units is not included. New is support for MCGA graphics mode. You'll need the driver that Borland has made available in the BGIKIT to be able to use this mode because it was not included in the standard Turbo Pascal package.

P-Robots

P-Robots is a game that helps teach programming in Pascal. The objective is to design and program a "robot" that triumphs over similar robots designed and programmed by others in a real-time battle of wits and flying missiles.

You control your robot by writing a procedure in Pascal to specify your robot's behavior and strategy in its efforts to vanquish up to three other robots in a battle to the death. A variety of predefined P-Robots Pascal functions allow your robot to track its position on the battlefield, monitor its health or damage condition, and calculate the distance and angle to opponents from its current battlefield position. Each robot is equipped with a cannon to fire missiles, and a motorized drive mechanism to either close in to kill a hapless opponent or flee from a fierce foe.

P-Robots is an excellent way for the novice programmer to sharpen Pascal skills and have fun at the same time. However, P-Robots does assume that the robot designer/programmer already knows the fundamentals of programming in Pascal.

For the experienced programmer, P-Robots offers a chance to see just how well you program in a programming environment where "bad" code can lead to graphic and ignoble defeat and "brilliant" code can bring triumph and glory.

Pascal Translator/Interpreter

This software consists of a Pascal to p-code translator and a p-code interpreter. The translator accepts standard Pascal programs with C masking and shifting extensions and produces text assembly language files that the interpreter program executes directly.

Programs can be compiled from a user's text file or directly one line at a time from the keyboard. (Keyboard input allows only the current line to be edited.) This introductory version allows seven files to be open at a time, limits sets to 64 members, and allows string constants of up to 80 characters.

Pascal Tutor

Pascal Tutor is a text-based tutorial on programming with the Pascal language. The assumption is that you know nothing about Pascal in particular and nothing about programming in general. It begins with the most elementary aspects of programming and takes you through each step from dynamic allocation to the use of linked lists.

The program was specifically written for use with Turbo Pascal versions 3.0, 4.0, and 5.0. Version 5.5 is also supported through two additional chapters covering object-oriented programming. The Pascal source code used for examples throughout the tutorial is included.

PopText

This help-screen generation utility for Turbo Pascal and assembly language programmers allows programmers to easily integrate pop-up style help windows into their own programs. PopText features an easy-to-use editor, a unit to include in Turbo programs, and three bonus units, including some fast writing routines, routines to enhance CRT, and a unit with all of the key definitions.

Each help window is a record consisting of 15 lines and 15 links. The lines are displayed on the screen, and the links are hidden in the background. In addition to being linked by line, windows are linked by the Page Up and Page Down keys. Using the editor, you can create new files and append the files up to 256 records each.

SkyHawk Development Library

This is a collection of units compilable under Borland's Turbo PASCAL 5 and above. Some of the functionalities are to be found in other collections, but they are reimplemented here in an extended form to provide a different approach. Some are unique to the SkyHawk collection. All are provided to remove the necessity of "reinventing the wheel" during the development process.

The collection is dynamic in the sense that it is being constantly augmented by the addition of new units and the improvement of the existing ones. The SkyHawk collection depends on the TurboProfessional 5.0 library from TurboPower Software.

Most of the units were initially developed while working on a series of projects for which conversion to an object-oriented design was not economically feasible. Thus, the dependence on TurboProfessional rather than Object Professional. In most cases, however, you will find that conversion to object-oriented code will be trivial.

TechnoJock's Turbo Toolkit

TechnoJock's Turbo Toolkit (TTT) is a comprehensive set of procedures and functions for Turbo Pascal and Quick Pascal programmers. The Toolkit is designed for professional programmers, as well as for newcomers to the Pascal programming environment.

The main focus of the Toolkit is to give programs a polished and professional appearance. There are routines providing: formatted screen output, windows, many menu routines including pop-up and pull-down, form input procedures, string formatting, a Sidekick Plus style directory lister, and much more. These units are compact and very easy to manipulate and implement. The manual is clear and instructive. All source is included. The abbreviated manual is on disk (more than 100 pages).

Turbo Designer

Turbo Designer is a programmer's toolbox for the Turbo Pascal compiler. It helps you create pull-down menus, pop-up menus, and sophisticated help screens in your applications. Very easy to use, the program asks you to input your data (what you want to appear when the program runs), on a series of "what-you-see-is-what-you-get" full-screen editors.

Once you have set up your menu work environment, which includes windows, icons, and even a trash can, Turbo Designer creates a Pascal source file according to your data. This new file can be compiled using Turbo Pascal. The result is a program with a complete environment and a sophisticated user interface.

Turbo-Lessons

Here are 16 different lessons in Turbo Pascal that teach you more than the elements of the language; you'll also learn programming technique and gain experience through practice. Each Turbo-Lesson is a bite-size tutorial that introduces one or more new concepts. Each lesson is presented in the same format: clear objectives followed by an organized tutorial. Most lessons direct you to work with a sample program that is provided with these lessons.

Programmers who had difficulty learning Pascal from reference books will save time and frustration in future programming. The lessons are designed to encourage you to try things on your own. Lesson titles include: Essential Editing Skills, Program Structure, Declarations, Input, Integer Expressions, Conditional Processing, Repeat Statement, Case Statement, For Statement, While Statement, Introduction to Functions, A Function to Detect Errors, Strings, Introduction to Procedures, Interactive Screen Handling, and Real Numbers.

Turbo Pascal Modulating Printing Filter

Turbo Pascal Modulating Printing Filter (TPMPF) prints Turbo Pascal source code files in an easy-to-read formatted outline. TPMPF first prints the file name and the file's creation date at the top of the page and then proceeds to print the formatted source code.

Procedure and function names are emphasized and underlined, index numbers are printed on the right side of the page, global variables are double-struck, and local variables are italicized. After printing the source code, TPMPF prints a listing of procedures and functions with their index numbers. You can elect to create source code files stripped of one or both comment types allowed in Turbo Pascal.

Vicon

Vicon is a powerful, full-fledged icon editor/designer in a friendly graphical user interface. It also comes with File Manager and Directory Manager. The major features in Vicon are: draw solid or framed lines, rectangles, circles and ellipses; draw rotatable, solid, or framed triangles and diamonds; each shape has its own variable color, stroke size, undo and redo, paint-over and paint-around operations; remapable colors, including changing only one or all color(s) in a user-defined region (circular, rectangular, triangular, etc.) to a specified color within the icon; flood fill and border fill, with undo; kaleidoscopic patterns for all shapes and paintbrushes; four paintbrushes (toggle, graffiti, potpourri, and eraser); speed control for continuous strokes; resizable and scalable icon, up to 64 × 64 pixels in size; a grid over the drawing area with coordinate indicator for accuracy and fine-tuning; snap or free roaming mode to ease cursor movement in grid; protected mode to be used as a safety net so that any drawn portion cannot be overwritten; global undo; edit two icons at once by easy swapping between them; use expanded memory if available; video emulation (VGA, EGA, CGA, and mono) and gray scale with adjustable gradients, plus much more.

Ada

Ada-Tutr

Ada-Tutr, the interactive Ada tutor, teaches you to be an excellent Ada programmer in minimum time. Ada is an ideal programming language, because Ada programs tend to have fewer errors, to be easier to read, and to be much easier to modify later. The U.S. Department of Defense mandated the use of Ada in mission-critical systems (DoD Directive 3405.2, 1987).

Not just a "quiz," Ada-Tutr is a thorough course of interactive instruction that even checks "homework" assignments. Ada-Tutr concentrates on teaching good program design, not just syntax, so that programs take advantage of the features of Ada that make them more reliable and easier to maintain. It was written by John J. Herro, Ph.D., who taught a graduate course in Ada at the State University of New York at Binghamton and taught Ada to employees of General Electric and Grumman Aerospace Corp.

When Ada-Tutr runs on a PC, an Ada compiler is helpful, but not required. A list of Ada compilers available for the PC is included. Because Ada-Tutr comes with Ada source code, it runs on other computers with Ada compilers—even on mainframes.

C Programming

C and C++ Tutor

This program provides two tutorials on the programming languages C and C++. The C Tutorial is an interactive on-screen tutorial designed to help you learn the C programming language; the tutorial includes an introductory course on the syntax of C language and the standard library. The C++ Tutorial, while not interactive, includes lessons, tutorials, and more.

C++Tutorial

Here is a list of the chapter headings:

- Chapter 1. Simple Things.
- Chapter 2. Compound Types.

- Chapter 3. Pointers.
- Chapter 4. Functions.
- Chapter 5. Encapsulation.
- Chapter 6. More Encapsulation.
- Chapter 7. Inheritance.
- Chapter 8. More Inheritance.
- Chapter 9. Multiple Inheritance.
- Chapter 10. Virtual Functions.
- Chapter 11. More Virtual Functions.
- Chapter 12. Flyaway Adventure Game.

C-Wndw Toolkit

C-Wndw Toolkit is for both novice and intermediate C programmers who want full screen, color, cursor control, and windowing abilities, without programming complex escape sequences or DOS-level interrupts. It's also designed for experienced programmers who need a set of fast I/O windowing and disk access functions. A full set of file access functions are given that trap and handle all usual errors. Access is supported to ASCII, dBASE III+, hashed random, and relative files.

C-Wndw Toolkit provides input and output windowing abilities that operate at professional speeds. The output windowing facilities use memory mapping for instant screen display. They take full advantage of color monitors, and they provide automatic editing of entry and output fields. Pull-down and pop-up menu functions, help screens, and multilayered windows provide a sophisticated user interface.

ExtLib

ExtLib is a library of more than 50 C runtime video routines for the advanced MS-DOS C programmer. ExtLib features windows, pop-up and pull-down menus, menu bars, scroll bars, direct video memory I/O, and cursor control routines to be used with Microsoft C or QuickC. In addition to these video routines, there are formatted display input routines (dollars, for instance) and advanced error-handling functions. The small, medium, compact, and large memory models are all supported. ExtLib also comes with the floating point emulator or 80 × 87 packages.

GUI_LIB Library for Borland C++ and Turbo C++

This is one of the finest graphics libraries available for use with Borland C++ and Borland's Turbo C++ compilers. Using the GUI_LIB library enables you to create graphics-based applications to run in DOS just as easily as creating text-based applications. The objects included in the library are all familiar objects that are found in many commercially available software packages.

They include push buttons, icons, text entry fields, and bitmaps, to name just a few. Some of the features in this library are found nowhere else! Now you can harness the power and intuitiveness of the graphical user interface for use in your applications. From paint programs to databases, all of your programs can benefit from the use of graphics.

Marketing Software

Sell More Software

This PC system facilitates the marketing, sales, and administrative activities involved in selling computer software. Created specifically for software developers and value-added resellers, the program helps plan projects, track contacts with prospects and clients, track support, and print custom letters and forms. The program saves you time by simplifying and automating many routine tasks.

For example, when you enter a client's order, the program makes it easy to create a product label, mailing label, and invoice. The revenue information is then passed to the general ledger module. The program's extensive set of reports helps you to review trends in sales and support, plan future projects, control expenses, and manage your time more effectively. Some of the important features are described in the following.

Complete contact histories. The clients and prospects file maintains a record of each contact, thus preventing you from forgetting a client or his or her needs.

Next marketing step. The program keeps track of the next contact date for each client, thus assuring that you always know when to make the next call or send the next letter.

Software support tracking. The support record tracks support for each client. The separate error-tracking record lets you keep up with problems and solutions.

Project planning. The project planning module helps you track hours spent and the status of each project. It also helps to monitor expenses with the cost analysis report.

Mail merge. The mail merge feature automatically merges name and address records with your own letters. The ability to rate your prospects simplifies mass mailings. Laser fonts are supported, thus permitting high-quality output.

Reports. The program's reports help you manage your sales, marketing, and administrative functions. Reports are listed in the functional checklist.

Flexibility. The Sell More Software program can be customized to fit your company's particular needs.

The Software Marketer

Getting your software product into the retail market can be frustrating, especially if you don't have any contacts or formal knowledge of the business. The best route for an independent programmer or small company would be to go to a professional software marketing consultant. A professional has the contacts and the knowledge and experience to get your software onto the store shelves. The only drawback is that the experienced consultants are expensive to deal with. Now there's a solution to this dilemma: The Software Marketer.

This software package is a combination training package and marketing package. It shows you how to get your software onto the store shelves, while at the same time it gives you direction and guidance. Because the average person does not have contacts in the software industry, The Software Marketer contains three databases of contacts for marketing your software. Plus, it allows you to generate form letters to market your software to these three echelons of the software industry.

The Distributor database contains more than 700 distributors' addresses and telephone numbers (Software Plus, Software Resources, etc.). With The Software Marketer, you can merge a form letter with the database and create a custom mailing to attract a distributor to your software product. A distributor is a middle person or a wholesaler that will distribute your software to the dealers.

The Dealer database contains more than 1400 retail software stores and software resellers' addresses and telephone numbers (Egghead Discount Software, ComputerLand, Software City, etc.). You can also merge this database with your custom form letter to market your software to these retail outlets. Most dealers will not carry software unless it is from a distributor, but you should still let them know that your distributor has your software available.

The Magazine database contains more than 1,100 contacts in the computer magazine industry (*PC Magazine*, *PC Computing*, *Byte*, *Info World*, etc.). This software helps you to generate press releases and gain free exposure and marketing for your software. A favorable article in a national magazine pulls in many times more sales than a well-placed advertisement in the same magazine, and at no cost.

For almost two years now, most of the large retail outlets require a UPC Manufacture Identification Number (Bar Code) to be placed on all software packages. The Software Marketer helps you apply for a UPC code and prints UPC labels on an HP or Epson compatible printer. This allows you to begin packaging your product immediately, then just place stickers on the packages that require a code number from the Uniform Code Council.

Supershareware System

SSS is a comprehensive system for shareware authors, enabling them to track customers, prospects, distributors, contacts, products, sales, and expenses. SSS allows you to quickly and easily target mailings to customers, prospects, or distributors using the mailing label or postcard options.

SSS creates 24 reports, some of which include:

- Customer, distributor, and contact listings and directories.
- Geographic summaries for customers and distributors.
- Income over a specified time period with subtotals by day, month, or year.
- Expenses subtotalled by date or expense category.

- Balance sheet.
- Payments due.
- Tickler.

Use the Querying capability of SSS with the postcard, label, and report options for further customization.

SSS allows you to easily enter customer registrations and print invoices with accompanying labels. You can import data in dBASE, Delimited, or SDF formats and export data for use with existing software.

SSS was designed specifically for shareware authors. Its power and flexibility make it a tool that most shareware authors will not want to be without. The intended user is anyone who markets products via shareware. Unique features:

- Tracks distributors, customers, prospects, products, registrations, and expenses.
- Provides querying capability for most files.
- Uses queries to produce targeted mailings to customers, prospects, and distributors.
- Creates and prints postcards using various printer features such as bold, italic, underline, etc.
- Prints addresses on postcards, or print mailing labels.
- Printer drivers provided for more than 430 printers, including laser and inkjet printers. The program's capacity is limited only by available hard drive space.

Software Sales Tracker

SST is a feature-packed program designed for software authors who sell and distribute their own products directly to PC Users. It helps with the sometimes complex tasks of keeping track of the "who, what, when, where, why, and how much" involved in the sale of most products. It also helps to get the most mileage out of your "program distribution dollar." SST has many features that are useful to every program author who distributes software (especially shareware).

Its best and most powerful feature is especially useful to those who spend a lot of time and money promoting their own software products. With many shareware authors, their entire effort at program distribution consists of uploading their latest creation to a few BBSs and Pay Nets. Then they simply sit back and wait for the mail to bring them thousands of orders, each one stuffed with a check. They might as well save themselves the time and trouble and go buy a Lotto ticket instead because it's very rare to obtain the kind of results they hope for without a little more effort than that.

To get anywhere in this crazy market, you have to constantly sell, sell, and resell your product with more dedication and persistence than it might have taken to write it. If you are the kind of author who promotes one or more products, who has spent many sleepless nights on long distance, uploading programs to far-off BBSs, who has spent lots of time and money preparing, packaging, and mailing countless disks to user groups and distributors, then this program is for you.

If you produce updates and revisions on a regular basis to keep your software up-to-date and active, if you try to keep users informed about updates, revisions, and support programs, and, above all, if you're not doing all of this just for fun and you expect to make a buck for all of your efforts, then SST is written especially with you in mind.

Used properly, SST saves you a great deal of time, effort, and money because it helps you organize your sales and program distribution into a tracking system that is much more intelligent, mechanized, and productive than simply using a shotgun approach. Aside from simply keeping track of registered users, you will come to know who is producing sales for you and who is not. SST allows you to organize your sales into separate files for each different product, then switch from file to file with ease through the use of "directory pop-up windows."

When new entries are made, it automatically issues a unique product serial number for each item sold. SST then keeps track of quantities sold, customer addresses, telephone numbers, product serial numbers, product versions or models, prices charged, shipping charges, sales tax, and the dates sold and paid (as well as any purchase order and check numbers). SST lets you automatically patch or modify your sold EXE programs with a customer name, date of sale, and assigned serial number. This feature is primarily designed to work with EXE files that were compiled and linked with Microsoft compilers and linkers.

It prints customer statements for all unpaid records, or it prints a single statement for any record that you select. SST has a variety of other print options as well.

- Customer address book.
- Customer telephone book.
- Address and telephone book.
- Complete information book.
- Index and Rolodex cards.
- Labels (1, 2, 3 across).
- Pick-and-choose records for mailing labels.
- Shipping labels and envelopes.
- Print to a disk file.

In addition, SST provides "gross income" information for each file and gives a variety of totals, subtotals, and averages. Beyond this, SST provides these figures for either the entire file or for any range of dates you choose.

Chapter

12

R & R (Relaxation & Recuperation)

Now that you've spent a lot of time evaluating the software in the earlier chapters, it's time for a little R & R. Relax and recuperate; let those hard-thinking, busy brain cells have some fun, some entertainment, some relaxation! Don't belittle the games and other software in this chapter. Everyone needs a little time away from fancy formulas, sliding schedules, redundant reports, taxing tests, monotonous meetings . . . time to shift your muddled mind onto something else.

One pleasant surprise you'll find by using this software is that your subconscious mind will be constantly working as you play these games or concentrate on your hobbies. In the middle of Jill of the Jungle or Hearts for Windows, your refreshed mind will suddenly explode and unravel that perplexing problem that has puzzled you for a long time.

So read on; try some of this software. Some of it is free, and some of it you should register if you choose to add it to the games directory on your hard drive. You'll be amazed at what a little R & R can do for you.

General

Excuses For Skipping Out Of Work Early

This program provides 99 excuses for skipping out of work early. I've included a few here so you can evaluate them and see if you can use or improve on them. Better yet, buy the disk and get all 99. Here are some samples:

- My kids are locked outside.
- My kids are locked inside.
- My kids are stuck in the door.
- I have to help my grandmother bake cookies.

- I have to help my Aunt Flo in Omaha bake cookies. (She's much better now, and she wants to send thank-you cookies to everyone who came to see her when she thought she was dying.)
- The water company has to read my meter once a year, and this was the only time they would come.
- The gas company has to read my meter once a year, and this was the only time they would come.
- The water meter guy and the gas meter guy were both leaving cards on my door about my not being home, and they got into a fight about whose meter was better, and I have to go home and clean up.
- My daughter is graduating from high school and I'd like to go to the ceremony.
- My daughter is receiving a Nobel Prize and I'd like to go to the ceremony. (Do not use within one month of the previous excuse).
- I have to pick up my car at the shop; if I don't get there in half an hour, my car will be locked up all weekend.
- I have to get my car to the shop; if I don't get it there in half an hour, my car will be locked out all weekend. (Don't use this one if your boss seems wide awake).

Fantastic Freebies

This listing appeals to virtually all people who use computers. Freebies is a valuable source of free promotional items, information, product help, and many other things for computer users. It is the largest centralized source of demo disks in the United States.

Names and descriptions of promotional items offered are included, making this very easy to use. Freebies is not a novelty or gimmick item that is later thrown away or used up. Freebies contains names, addresses, and phone numbers of almost every major computer company in the United States (more than 200 listed in the registered version, all in alphabetical order and catalog format). Names and short descriptions of each company's major products (hundreds in the registered version) are included. Windows-related, DOS, MAC, and most computers, including mainframes, are supported. All companies listed offer some kind of promotional item. Each company can be contacted by phone or letter to request information or free items that they'll gladly send. A description of the advantages, usefulness, and what can be obtained through use of Freebies follows.

- Actual working models of major software programs and demo disks. Many companies give away software as part of an advertising and promotional campaign, especially for new products. As you know, a working model is the very same software sold for a lot of money.
- Best contact telephone numbers and extensions. Most phone numbers are toll-free.
- Access to company experts, free technical support, and the opportunity to make contacts. You'll speak with the department who knows the most about the product. You can ask almost anything, except the most technical questions.

- Video cassettes are also free. They describe the company and/or their products. The tapes can be reused to tape favorite programs and will fit into any VCR recorder (VHS or Beta).
- Books, literature, discount catalogs, posters, personal organizers, etc. Many companies listed send thick, informative booklets and literature describing many topics in the world of computers. To educate their users, or as part of advertising campaigns, they send basic "how-to" booklets. For example, some titles are, "Getting the Most from Your PC, Network, Mainframe, Connectivity, Telephone," etc. (titles depend on the company). The discount catalogs listed here often have every available software program (and most hardware). You'll find many discount software stores.

No purchase is required at any time from the company. Take as much time as necessary to review the demo disks and other material.

Satellite

To use this program, you need to know the geographical coordinates for the location of your satellite dish. The better the accuracy, the more accurate the output data. If you don't know the coordinates, you can find them on any topographical map. The coordinates of the nearest airport can also be used if it's within a few miles.

This program has been tested throughout the northwestern hemisphere and has been found to be quite accurate. Although it calculates dish aiming for the other hemispheres, no guarantee can be made of its accuracy. Also, the Canadian-U.S. satellites are aimed at the northern hemisphere and might not be viewable elsewhere.

When you start the program, you are asked if you want hard copy. The program displays the aiming coordinates for all the geostationary satellites currently in the North American section of the Clark belt. If hard copy is requested, the printout also contains areas, so that technical values of each system can be recorded.

VGAWheel

This wheel-of-fortune game for Windows is a hangman-type word game. It features:

- A colorful spinning wheel.
- One to three players.
- Five rounds of play.
- A used letter board to separate the vowels from the consonants.
- A sound on/off toggle option.
- Automatic win after all the letters have been bought and guessed.
- A high score table for the top ten wheelers.
- An editor to add/delete/modify puzzles.

Instructions are included in the game. An editor facilitates adding new clues to your game. When adding new puzzles, the database is checked to prevent duplication. The ability to edit previously added clues is included in case of misspellings. However, be careful not to change a puzzle into a duplicate because duplicate checking is not done at this point. More than 1,000 puzzles are provided.

Travel

Airline 800 numbers

This software simplifies finding 800 numbers for airlines. It contains a large database of airline 800 numbers. To locate the airline you're looking for, simply type AIR800 and wait for the search prompt. At this point, enter what airline you're looking for; a part of a name is okay. (Uppercase and lowercase are both acceptable.) The screen shows all airlines found with that name or phrase. Enter "Japan," for example, and Japan Airlines comes up.

Hotel800

Hotel800 simplifies finding 800 numbers for hotels. It contains a large database of airline 800 numbers. To find the airline you're looking for, simply type Hotel800 and wait for the search prompt. At this point, enter what hotel you're looking for; a part of a name is okay. (Uppercase and lowercase are both acceptable.) The screen shows all hotels found with that name or phrase.

Interstate Traveler

This software plans motor trips across the contiguous United States using the interstate highway system and roadways you add to the screen maps. The final products of the program are an itinerary and an estimate of expenses for each day of your trip. The program presents a map of the United States and allows you to zoom in for a closer look at the features of a region. A trip is planned by specifying the route you want to take in a point-to-point manner.

For each day of the trip, you identify an origin city, each intermediate city and highway junction along the way, and a destination. The view of the map can be changed as the trip progresses. You can plan stops for meals, gas, rest, and visits en route to a location or at that location. In addition, you can plan overnight stops and full-day stays at a location.

Once an overnight stop is scheduled, or the end of the trip planned, an itinerary of the trip for that day can be completed and printed, if desired. Likewise, expenses for the day can be estimated and a printed report produced. At the end of the trip, an estimate of the total expenses for the trip can be made and printed. Because your home or destinations might not be on or near the interstate highway system, you are provided with the means to add cities, highways, and junctions to the maps on the screen. You can use these new features in the current run of the program only or make them a permanent part of the program data files.

Your hometown, and any locations you expect to visit frequently, would be good permanent additions. The program also provides maps of selected individual U.S.

cities. These maps show the locations of major highways, features such as rivers and lakes, a list of points of interest and their locations, plus a brief description. In addition, information about population and weather is displayed.

With these maps you can find routes around the larger cities, or you can locate places you might want to visit. The maps displayed on the screen are intended for planning purposes only. To keep them readable, the amount of information shown on them must be limited. On an actual trip, you should take along more detailed maps.

Travel Points of Interest

With this software, you can find a variety of points of interest in the United States. Travel Points of Interest searches for the areas and types you want. It provides names, addresses, and descriptions. Travel Points of Interest searches with geographic areas, and with types, to find places you want. This software has about 1,000 points of interest in the United States. You get names, addresses, types, days open, telephones, and descriptions.

This is a great way to find interesting and fun travel points of interest. With this software, you can focus on what you want and ignore the rest. You can view the descriptions on your screen or print them out to take along on your trip.

Card Games

Calculation

This program (also known as Broken Intervals) is a well-done solitaire card game that requires the utmost in skill and logic. The software offers challenging but rewarding play and special on-screen assistance to make it easier to play. You can learn the essentials of the game quickly, but it takes a while to become a good player. It rewards skill, experience, concentration, and luck.

Calculation uses a standard 52-card deck (no jokers). To begin the game, the deck is shuffled, then any ace, two, three, and four are pulled from the deck and form the start of four foundations. These foundations are built up according to the following sequence, regardless of suit or color.

Cards are turned up from the deck, one at a time, and can be played on the foundations or on any of four waste piles located just below the foundations. The top of the waste piles are always available for play to a foundation, but once a card is played to a waste pile, it cannot be moved again, except to a foundation.

The game is won by building all four foundations up to King. In addition to just keeping track of what card goes next on each foundation, the skill involved in playing Calculation is in avoiding covering up a card you'll need soon, with a card that cannot be used for some time, if at all. Kings are especially troublesome because they cannot be moved until at least one foundation has been fully built up. The general strategy is to build on the waste piles in descending order, but the cards are seldom so kind as to allow this to any great degree.

Darn It Solitaire

You shuffle the deck and, one at a time, take cards from it and lay them out in a 4 × 4 grid. Nonface cards can go in any position. Face cards have to go into fixed positions. After filling the grid, you can then discard all the tens and all the pairs of nonface cards whose values add up to ten (aces count as ones). Then you fill the grid again, and discard, and fill until the face cards are all out in the above pattern.

But you lose if you can't place a face card, and you lose if you fill the grid and it has no tens or pairs that add up to ten. It's rough; it's tough. But the thrill of victory makes all the agony of defeat worthwhile (usually). Variations: Some people wimp out and play by the rules that let you not only eliminate tens and pairs that add to ten, but also to eliminate any groups that add to ten—say a six, a three, and an ace. This is an unbelievably cowardly thing to do, but the program offers it as an option. You never know with some people.

Hearts for Windows

Hearts for Windows (my favorite game) is a four-player card game in which the goal is to get as few points as possible. Cards of the heart suit are worth 1 point, and the queen of spades is worth 13 points. You want to avoid these 14 cards.

The highest card of the lead suit takes the four cards in the trick, and the player who does this gets to lead for the next trick. You must play the lead suit if you have it, or else you can play any card. After 13 tricks are played, the points are added up and cards are dealt again. The game is over when a player gets 100 points. (He or she is the loser.)

Klondike

This game is also known under a variety of local names such as Fascination and China-Man. It is probably the best-known solitaire card game. (Actually, many people are not aware of the fact that Klondike is just one of dozens of solitaire games, and they use the name "Solitaire" to refer to it.) Several implementations of Klondike are available for MS-DOS machines, but most have very crude graphics (when they do have graphics at all), and they have cumbersome user interfaces. This program has excellent color, graphics, and music and offers Las Vegas scoring.

To play, one regular pack of 52 cards is used. You deal 28 cards into 7 piles. The first pile at the left has one card, the second two, and so on, up to seven in the last pile. The top card of each pile is face up; all the rest are face down. On the piles, you can build in descending sequence, red on black or black on red. For example, on the eight of spades you can place either the seven of hearts or the seven of diamonds.

All the face-up cards on a pile can be moved as a unit and matched according to the bottom card. When you uncover a face-down card on a pile, you turn it face up. If one pile is entirely removed, you can put a King in the space. Each time you free an ace, you move it to the foundation, a row located above the piles. On the foundation, you build up in suit and numerical sequence.

Klondike, as played by its original rules, is very hard to win. It has been estimated that the chances of winning are one in every thirty games, and thus it can be a bit frustrating. Two popular variations greatly increase the odds of winning. Under the first variation, cards are moved from the pack to the waste pile in groups of three, and you can go through the pack as many times as you want. For example, once the pack is exhausted, you move all the cards from the waste pile back to the pack. Under the second variation, cards are turned up from the pack one by one, as in the original rules, but you can go through the pack an unlimited number of times, as in the first variation.

Klondike uses Vegas-style scoring. You pay $50 for the deck, and get $5 back for each card you can place on the foundation piles. Under variations one and two, you can go through the pack multiple times, but you lose $3 for each extra pass on variation one, and $10 for each extra pass on variation two. Note that you cannot lose more than the initial $50; if your score is already minus $50, you won't lose anything for extra passes. The maximum possible score is $210 (52 cards in the foundation piles at $5 each, minus $50 for the deck). If you win the game in a single pass through the deck, you receive a $90 bonus, for a winnings of $300.

Arcade Games

A huge variety of arcade games are available; here are a couple of the best of the genre.

CD-MAN

This ultimate PacMan-type game is fantastic. The object of the game is to clear the screen of dots while avoiding the nasties. However, by eating one of the pulsating energy pills, you get stronger, and the roles are reversed for a short time. Then you can gain points by catching the fleeing monsters. When the screen is cleared, you advance to the next level. To gain bonus points, collect goodies that appear every now and then.

If you use the Esc key to exit from game play, you can always continue your game by choosing CONTINUE. The high-score table shows the ten best scores, the player's name, and the time and date the game was played. Each screen has three levels; they differ in difficulty and bonus points. The aggressiveness of the monsters increases for each higher-level screen.

Jill of the Jungle

Jill of the Jungle is a Nintendo-style arcade adventure game with 256-color VGA graphics, great sound effects, and a Sound Blaster-compatible musical sound track. This immense game world, Volume I in the Jill series, challenges both your mind and your arcade reflexes; it's fun for the whole family.

Music

CDPlay

CDPlay, a full-featured program, allows your CD-ROM player to play a CD audio disc. CDPlay (herein referred to as CDP) needs to have the following:

- An IBM-compatible computer.
- A CD-ROM drive capable of audio output.
- MSCDEX v2.1 or later.

If you have the items just listed, you are all set to start enjoying CDP. Note: Your CD-ROM device driver must support audio playback. A few device drivers do not support audio. If CDPlay does not run, check with the drive manufacturer for an updated device driver.

Optional equipment. A Microsoft/Logitech-compatible mouse and a hard drive are optional. (A floppy will work.)

The Guitar Chord Hypertext Dictionary

This program is a hypertext reference of the most commonly used guitar chords. Five chords are available in each of the 12 keys. These are the guitarist's "workhorses:" the major, seventh, minor, minor seventh, and major seventh. Each chord is shown in three different positions on the fingerboard.

A total of 60 chords are included (12 keys, 5 chords, 3 positions). GCHD uses hypertext for fast, easy movement between chords. Each screen is fully menued and user intuitive. The help system provides explanations of all aspects of GCHD. Chords can be marked and later played back in sequence with GCHD's "forward" and "back" commands. It can be used with any monitor (monochrome to VGA), and screen mode is automatically detected. A mouse is supported but not required.

Guitar Teacher

Guitar Teacher includes two main features: a guitar tuner that is accurate to $1/10000$ of a cycle per second, and a large database of guitar chord diagrams. Each diagram can be quickly selected and instantly displayed on the screen. The manual covers the following sections: getting started, using Guitar Teacher, tuning, chord structure, and technical support.

The Music Transcription System

The Music Transcription System is an innovative product designed for entertainment, education, and beginning-level music composition and vocal instruction. Developed especially for music with lyrics, it covers the full vocal range from bass through soprano and displays scores on any of five different musical clefs.

The program features extremely fast graphics that allow you to follow along in a song as it plays. A unique "bouncing ball" cursor points to each word in the song as the corresponding note plays, and the paging-score display lets you read the music as easily as you can from a book or piece of sheet music.

The Music Transcription System includes a powerful yet easy-to-use music editor. With it you can enter your favorite songs or create new songs of your own by playing on your computer keyboard like a piano. The program supports songs up to 550 notes long with lyrics, guitar chords, and basic harmony in any of 15 different key signatures, 12 different time signatures, and 5 different clefs. You can have your computer play songs for you, store them on disk, or print them out as high-quality sheet music.

Using the advanced sound input features of these computers, The Music Transcription System becomes the perfect tool for learning to sing or for improving on your current skills. You can sing along as your computer plays a song, and the program will measure the pitch of your voice at a rate of almost 20 times a second. Unique, real-time "pitch-graph" displays allow you to see what you're singing, while you're singing it. Beginning-level singers and self-professed "nonsingers" especially benefit from this instant feedback that shows when notes are sharp, flat, or on-key.

If you're a songwriter, or aspire to be one, The Music Transcription System helps turn your ideas into printed musical scores—even if you can't read music.

PC-Guitar

This program helps the user learn basic chords and notes on the guitar. A guitar consists of three basic parts:

- The body, over which strings are plucked or strummed. In an acoustic guitar, the body amplifies the sound. Electric guitars usually have a solid body and use a pickup with an amplifier.
- The fingerboard ("neck"), where the fingers are used to select notes.
- The tuning head, where tuning pegs are used to tension the strings to change the tone of the strings.

Guitars come in both 6-string and 12-string varieties. However, a 12-string guitar has 6 pairs of strings, and each pair is tuned to the same note. This software assumes a six-string guitar, but the principles are the same. The fingerboard has 6 strings and either 12 or 14 frets on the fingerboard itself. The guitar body also has several frets that are made accessible with cutaways on some guitars.

Chords. Some notes in a key sound good when played at the same time (this is called *harmony*). When three or more such notes are played simultaneously, they are called a *chord*.

Chords are given letter names in the same manner as notes. A chord is named after the dominant, or *root*, note in the chord. Thus, a "C" chord contains the note "C" plus at least two other notes that harmonize with it.

Pianoman

Pianoman lets you play the keyboard of an IBM-PC (or close compatible) like an electronic piano. The keys sound a tone as long as you hold them down, and they stop sounding when you release them. You can record and edit music, insert and delete notes, adjust pitch and length, make global changes, and save/retrieve files. The maximum number of notes Pianoman keeps in memory is 63,488 (64 kilonotes minus 2,048 notes). This maximum is available for more than 450K of RAM.

Playing a tune with Pianoman is simple. You see a picture of the keyboard on your monitor, with the notes written on the key tops. When you press a key, it lights up in the picture. There are three octaves available at a time, and you can shift those octaves up or down using the up and down arrows.

Hobbies

Computer Gardening Data Disk

Computer Gardening Data Disk is a DIF (data interchange format) or mail merge (delimitted by quotation marks ("), and commas (,)) listing of more than 400 flowers. Computer Gardening Data Disk can be imported into your own database system. The mail-merge format extension is (ws), and the DIF extension is (dif).

You have instant access to a disk full of cultural information on many flowers. The information can be sorted and printed any way you specify, and it can be sent to screen or printer. Via the add or edit features of your database, you can add information on your own flowers and their special cultural needs. Information includes common name, botanical name, other names, class, germination, sowing guide, transplant, height, colors, light needs, soil, propagation techniques, and other comments. Importing the data into your own database eliminates the need to learn to use another database system. You already know your own.

Crafty Li'l Helper

Crafty Li'l Helper is a home business system that lets you concentrate on your craft, not your paperwork. The lure of running a business, making some extra money, or helping with a charity has overcome all of us at one time or another. So what went wrong? Crafty Li'l Helper wasn't available to do all the hard work. The only thing that's "little" about it is its registration fee. Among its features is an easy-to-use menu system that guides you painlessly through inventory and invoicing, sales tax, and UPS shipping rates.

Crafty Li'l Helper keeps track of your inventory size and cost; tracks your retail, wholesale, and consignment customers, and follows all outstanding orders and billing. Although it can run on just one floppy drive, its number of records is limited only by your disk storage size. With Crafty Li'l Helper, all of the really difficult parts of running your own small business are taken care of, and you're left with all the fun. Crafty Li'l Helper is a big aid for small businesses with items to sell. It shows exactly where money is spent and how much has been made. And it's very easy to use, with clear instructions on moving around in the program. It gives you encouraging words when you make a sale. If you are going into a craft business, take along Crafty Li'l Helper.

Data Manager

This 19-in-one program help you organize much of the day-to-day information you need at your fingertips. Menu options for Data Manager include: automobile, calendar, directory, financial, inventory, library (audio/video), messages, to-do lists, outlines, shopping, and word processing (with mail merge to directory and business contacts). Additional menu choices specifically for the office are business contacts and training classes. Specifically for home are: TV, recipes, foreign language vocabulary testing, weight, and exercise.

This program is extremely easy to use, yet very powerful. It is an outstanding learning tool for novices, and it's appealing and functional for the pros. Most menu options work the same way. Once you learn one, you have learned them all. Other features include: security levels, the ability to add notes of information (up to 20 pages per note, depending on disk space) to most data entries, print report options, online help, scrolling windows, and the ability to easily change operating options from within the program.

Gardener's Assistant

This is a system for planning a garden. Describe where you live, the size of your garden, and choose what vegetables and fruits you want to grow. The Gardener's Assistant helps you to:

- Calculate how many of each type of plant you need to grow in order to yield a certain size of the crop.
- Design your garden layout while considering space availability, planting density, and expected yield.
- Calculate planting dates and expected harvest times.
- Add to/modify planting information and vegetable chart.

The Gardener's Assistant is easy to use and can be a valuable tool for designing and managing your garden.

Garden Guide

The Garden Guide is an informational database program that contains vital information for the home gardener. The program helps you determine the location and size of your garden, as well as how to improve the soil quality and use the correct tools for the job. It tells you the proper methods for planting and tilling and how to prepare for winter. Other topics covered are: composting, herbs, flower bulbs, and cut flower gardens. This program also has a worksheet to plan your garden on paper and includes some various garden recipes.

Home Control

Welcome to the world of the computerized home. Someday, all homes will be intelligent. You can get on board now with Home Control, an excellent introduction to home automation. Light/appliance control is accomplished by using the X-10 Powerhouse Computer Interface. Home Control consists of three independent programs.

SENDX10 turns lights and appliances on/off from the DOS command line or from batch files. This program requires the X-10 Powerhouse Computer Interface.

SYNCHX10 synchronizes the clock inside the X-10 Powerhouse Computer Interface with that of your PC. This program also requires the X-10 Powerhouse Computer Interface. GAMEPORT explores the analog and digital input capabilities of the PC game port. The Game Port Monitor lets you hook up sensors to your PC's game port so that you can begin to experiment with environment monitoring. You can measure light intensity in a room, monitor windows and doors, measure temperature, and observe human movement through open spaces.

Home Helper

Home Helper computerizes routine household tasks. The program includes an electronic phone book with mailing labels, a recipe filer, an audio and videotape library database, and a record system for the valuable appliances in your home and more.

In addition, your commitments and personal appointments can be recorded with the personal scheduler. Home Helper also stores credit card information that will prove invaluable if your cards are ever lost or stolen. Finally, with Home Helper, emergency telephone numbers are but a keystroke away. The program is completely menu-driven with context-sensitive, pop-up help screens.

Home Inventory

Home Inventory lets you keep a database of your important possessions. The number of entries is limited only by disk space. Home Inventory is completely menu-driven and very easy to operate. To begin, you create a new inventory file or load an existing one. Then choose "add a record" to add records to the database. It's just that simple to build your own home inventory database.

You can change or delete a record in the database with the "find a specific record" option. There are also three search options: search on item description field, search on a specific field, and search on all fields. Print options provide a listing of your entire database, or just a single record in the database. This program provides an excellent way to gather and maintain an inventory of your valuable possessions.

Knitware Sweater Design

The Knitware Sweater Design program has been written by a knitter for knitters. The author knits both by hand and by using a Bond Classic knitting frame. The author was frustrated by the continual need to convert patterns for different yarns, for different sizes, for changing a neckline, or for lengthening a sleeve. The Knitware Sweater Design program was the eventual outcome of this frustration.

Using the data that you give it and tables of standard body measurements and needle sizes, this program prints a knitting pattern for you to use in knitting a customized sweater. The program uses your measurements, your style selections, and your gauge and yarn information to create a unique pattern. This pattern also gives you a permanent record of what you have knitted, and when.

Pattern instructions are written for either hand or machine knitters. Machine knitting instructions are very simplified, assuming you are using a basic knitting frame without a ribber. If you have a ribber (lucky you), feel free to use it. There are no specific instructions for punch-card or electronic machines.

Knitware Sweater Design handles most combinations of the following:

- Knit body of sweater in stocking stitch by machine or hand.
- Knit ribbing by machine or hand.
- Pullover or cardigan style.
- Drop shoulder, raglan, or set-in (fitted) sleeve style.
- Crew (round) or V neckline, or rolled cowl collar.
- Body length: waist, regular, hip, or coat.
- Children's sizes 20–32, women's sizes 30–48, men's sizes 32–52.
- Variable amount of ease, from very tight to oversized fit.
- All default finished measurements (e.g., length of body, ribbing, sleeve, depth or width of neck, etc.) can be changed.
- Measurements in imperial (inches) or metric (cm).
- Calculate amount of yarn required.

Board Games

Ed's Chess

Ed's Chess (or EdChess) is a computer program that plays the game of chess. The main menu offers the following selections: The CONFIGURATION MENU lets the user adjust the skill of the computer to that of the users. EDIT BOARD allows board positions to be set up to help solve chess problems, or to see what the computer would have done in a particular situation. FILES allows for saving and restoring games. It also prints a list of moves made during a game. OPPONENT determines whether you want to watch the computer play itself, or if you want to play the computer or another opponent.

PLAY MENU provides the game playing functions. It lets you enter a move or instruct the computer to take the next move. QUIT lets you exit the game and return to DOS. It does not automatically save the game currently in progress, but it prompts you to make sure you don't quit accidentally.

GO Tutor

This tutor is for playing the ancient Oriental game of GO. A complete tutorial, along with sample games, are provided with this disk. Here are the basic elements of GO:

- Liberties.
- Connections.
- Capture.
- Life and death.

- To escape.
- Two eyes.
- End/score.
- Rule of KO.

GO is sometimes called a war game. The players battle to control "territory" on the game board. "Territory" is unoccupied points fully enclosed by a player's own stones at the game's end. The winner is the player who ends up with the most territory, i.e., surrounded empty intersection points. The game starts with an empty board. At each turn, the player places a marker (called a "stone") of his or her color wherever he or she chooses on the board. Once placed, a stone is never moved.

Each player attempts to build fences that surround more empty points than the opponent's do. The objective is to surround more territory than your opponent. Stones can be captured and held as "prisoners" during the game and, at game's end, a count of prisoners held is added to that player's score. A full size GO board grid has 19 lines each way, forming a total of 361 points where the grid lines touch one another.

Monopoly

This Monopoly adaptation plays just like the Parker Brothers game. It is designed to be played intuitively and really should not require any explanation or rules. Even if you are not familiar with the original board game, it is simple to catch on from the program. The computer won't let you break any rules, and it keeps track of all your properties, houses, hotels, rents, and money.

Graphics are phenomenal, and player pieces are beautifully animated; this game is full of wonderful surprises. Two to four players can participate.

PC Backgammon

This software provides an easy-to-use menu system. The space bar is used to roll the dice. When it's your turn to move, hit the space bar to roll the dice. You can't roll the dice more than once per move. The dice are rolled at the beginning of each game to see who goes first (yours is the left die). The player who wins that roll must use the roll as shown.

Additionally, if the first roll is a double, the doubling cube is incremented to two (provided auto-doubles are enabled), and you roll again to see who goes first. Mouse users can roll the dice by clicking the left mouse button anywhere on the playing board, or in the dice area when it is their turn to roll the dice.

For keyboard users, to move a man, type an M. A form appears near the bottom of the screen. Enter the location you want to move from, and the value of the dice you want to use, separated by a comma (e.g., entering a "12,5" would move your man from the 12 slot 5 spaces). The bar is designated as slot 0. Press Enter to complete the move or the Esc key to cancel a move.

For mouse users, to move a man, simply click the mouse cursor on the man you want to move; a dot appears in the center of the man to indicate it has been selected.

Then click the mouse cursor on the appropriate slot you want to move to. To deselect a man (i.e., you changed your mind and don't want to move the man you selected), simply click the mouse cursor on the selected man a second time. (The dot disappears.)

Shogi Japanese Chess

This is the most popular strategy game in Japan. Shogi has many similarities to English chess. Both are descended from third-century, chesslike games that developed in China and India. Shogi is the most highly developed game of the chess family and is much more dynamic and exciting than its western English counterpart. The main reason for this is the drop move. When a player captures an enemy piece, this piece becomes his or hers and can be dropped into play anytime and nearly anywhere later in the game. This ingenious rule greatly increases the number of possible moves players have to choose from.

The object is the same as in western chess. Capture of the king is the whole game. Just as in chess, the king must be "checkmated" or cornered. When a king cannot avoid being captured on the next move, he is "checkmated." Computer Shogi has one distinct advantage to an actual shogi tabletop board; a computer can keep track of which pieces belong to which player and change the colors accordingly.

With a real Shogi tabletop board, the pieces must be all the same color so they can be "pointed" in either direction as the ownership changes with the drop rule. Pieces move very simply. They are generally less powerful than the western chess pieces are overall.

The pieces move as follows: The king can move one square in any direction (horizontally, vertically or diagonally). The rook can move any number of squares horizontally or vertically without jumping over any pieces. The bishop can move any number of squares diagonally without jumping over other pieces. The gold general can move one square horizontally or vertically or one square diagonally forward. The silver general can move one square diagonally or one square vertically forward. The lance can move any number of squares vertically forward without jumping over other pieces. The knight can move by direct jump, even if other pieces are in its path. It moves to either of the two squares that can be reached by making an L-shaped move that is two squares forward and one square horizontally. The pawn can only move one square vertically forward. The promoted pawn, lance, knight, and silver all move exactly as a gold would move. When captured, of course, they revert to their original rank. The Promoted rook can move one space diagonally in any direction. The promoted bishop can move one space, horizontally or vertically.

Sports

1st & Goal, fantasy football simulation

1st & Goal is fantasy football in a nutshell, a contest to see who can draft and manage the best team made up of real-life football players. Each competitor in the league becomes an "owner," and each owner selects the players that will become his or her "team" on paper (or on disk in the case of 1st & Goal). After teams are drafted, statistics are kept for each player every week in predetermined categories, and at the end of the season, the owner whose team produces the best composite statistics wins.

The program also has an option to pay cash prizes based on weekly standings, as well as for the entire season. The challenge comes in selecting the right players and choosing when and where to use them. 1st & Goal allows each owner to select up to 30 players. Depending on the way you configure the program, only about 12 of these players are "active" for any one game.

Cunning Football

Cunning Football is a dynamic, interactive, menu-driven football game. It tests both skill and strategy, requiring a user to guide players and call plays. CFB lets the user select either the keypad, a mouse, or a joystick as the controlling input device. CFB simulates a real football game as closely as possible. It includes interceptions, delay of game penalties, and other subtleties found in football.

However, CFB is easy to use, so some realism was sacrificed for the sake of user-friendliness. Cunning Football is considered by many to be the best interactive football game available, including any commercial games.

CFB is a fun and exciting game oriented toward those football fans who have a relatively good understanding of the game of football and really enjoy it. The user is the visiting team, and the program is the home team.

Defense

This is a game of strategy, skill, and quick reflexes. Just as in real football, the offense and defense try to outguess each other. The biggest gains are made by calling plays that fool the defense, and vice versa. When the play starts, you (not the computer) play the game.

Tackles can be made or missed depending on your skill in moving defenders and ball carriers on the screen and how fast the player on defense can hit the tackle button. And the player on offense can block, so that the reaction times of both players become key factors.

You make passes complete or incomplete. You cause sacks, interceptions, and fumbles. You can block kicks. Infractions of the rules result in penalties. Even the pace at which the game is played is about the same as that of a real game. It is not a hard game to play, but to play it well takes practice and knowing the subtle tricks of the game.

Mah Jongg

Mah Jongg is based on an ancient Chinese game that has been around for centuries. The object of the game is to remove as many matching tiles from the playing board as possible. Sounds easy, huh? Not quite.

The board has five levels of tiles, and the tile you might need might be at the bottom level, out of your reach. There are 65,535 different boards that can be played, which will keep you glued to your computer for many hours. The game supports Hercules graphics as well as EGA. An undo facility, mouse support, and game statistics make it more interesting than ever.

FastBreak Fantasy Basketball

FastBreak Fantasy Basketball brings the exciting world of Fantasy Basketball to your computer. No longer will you need to arm yourself with reams of paper, several pencils, and a five-pound box of erasers and try to get a handle on your league's statistics by poring over the sports section for hours. FastBreak organizes your league and reduces the time spent entering statistical information to about an hour a week.

FastBreak conducts your player draft, keeps track of team rosters, starting lineups, player contracts, league standings, fees and fines, and it prints out your league newsletter for you. The program allows rosters of up to 18 players per team and can be played for big bucks or just for fun. If the program is set up as an auction league (the big-bucks version), it can redefine the office pool. Players no longer need to sit passively and watch the teams they drew in the pool; they can build their own team, wheeling and dealing their way to weekly prizes as well as the season-end pot.

Field General

This program is unequalled in pro-football simulation. The plays, results, and statistics mirror those of actual pro-football games. The program's realistic play includes nearly everything that can happen on a football field. By letting you call the plays, Field General creates an atmosphere where the players can experience the pressure and excitement of the great game of football.

However, you have no control over the player while the play is taking place. Do you play conservative and try to protect your lead, or do you go all out until the final gun? You must not only battle the wits of your opponent, but you must also race against the clock.

Personal Pro

Most existing golf instruction shows you how to build a swing by taking you through an explanation of the perfect grip, address, backswing, downswing, and follow-through. Personal Pro uses a different approach. It recognizes that you have problems, problems that might change from round to round, and that you need immediate help in diagnosing and correcting those changing problems.

The expert golf knowledge contained in this program represents years of research to gain an understanding of the proper golf swing. The ongoing research includes computer analysis, observing student response to various teaching methods, reading books, watching videotapes, taking lessons, experimenting on the practice range, playing, and more playing. The research includes thinking about, discovering, and above all, organizing the cause-and-effect relationships present in the golf swing.

Personal Pro uses artificial intelligence techniques to represent the expert knowledge of several of golf's best resources, organized in such a way that their collective advice can be applied instantly to your particular golf swing problem of the moment. This software gathers symptoms of your problem, performs a diagnosis, and recommends corrective changes in your swing. Recommendations are presented first in checklist form, then reinforced through the display of animated diagrams. The recommendations can also be printed for use on the practice range.

But please remember that there is no replacement for help from your golf professional. See your golf professional, then consult Personal Pro as the serious golfer's form of the "19th hole," where you can review the mistakes of each round and get instant advice on how to correct them before your next outing.

PC ProGolf

PC ProGolf graphically depicts in four colors an actual golf course with water hazards, sand bunkers, wooded areas, buildings, roads, etc. With each stroke, the player watches the course of the ball and sees where it lands. When your ball comes within a certain distance from the hole, the program will display a close-up view of the hole so that you can putt the ball in.

Penalty strokes are added under the correct conditions, and an updated scorecard is available for review at any time, as are the help screen and club-values list. An unfinished game can be saved for resumption at a later time.

Casino Games

Casino Slots

Here are three slot machine simulations for the casino lover. Besides a basic slot machine that matches dollar amounts, three across for your payoff, there are also slot versions of keno and low-ball poker. These games would be ideal for people planning their own casino night. While the ASCII graphics don't compare to the video slots you find in the casinos, the action is very much the same. And if you play long enough, you might find these slots are a little loose.

Craps

The casino game of craps is played in all of the major gambling establishments throughout the world. This is a fast-paced game where only the knowledgeable gambler has a chance of winning. This software provides an excellent pictorial representation of a casino craps table.

The program has several unique features such as showing the dice actually flying across the table, instead of just popping the numbers up on the screen. The program thoroughly explains all of the different kinds of bets, odds, payoffs, etc. If you are the only player at the table, you must throw the dice. With more players at the table, you need not throw the dice to play craps.

Poker Slot

This game is widely played throughout the gambling casinos in Nevada and New Jersey. The Poker Slot layout resembles a video poker slot machine. It has high-quality graphics and is easy to use. You should have a knowledge of poker before playing.

The object of the game is to get a poker hand of jacks or better before you can win any money.

You will be paid off depending on your hand. The better the hand, the more money you'll win. Just like the casino poker slot machine, you must first insert money. You'll be playing a dollar poker slot, so each coin is worth $1. You can insert any amount from $1 to $5 into the poker slot machine. Once you've placed your coins into the machine, you are ready to deal a hand by pressing the Deal button.

Once the hand has been dealt, you select which of the five cards you want to hold. Press the Deal button again, and the cards not being held are discarded. They are then replaced with new cards from the deck. You can play poker slot with either a mouse or your keyboard.

UltraSlot

This is a slot-machine gambling game that allows the user to select from a large variety of slot machines. You can play on a three, four, or five-reel slot, and you can select from one to five pay lines. The best part of UltraSlot is that you can construct your own slot machine with your favorite symbols and payoffs. You can add or remove any symbol from any of the slot machines, or you can even change the payoffs. You can do almost anything with UltraSlot, except cash in your winnings.

Your screen displays a three-reel slot machine with one payoff line for your first game. This screen varies depending on which of the slot machines was played last. To select an option, move your mouse cursor over the option of your choice and press either the left or right button. When you select Insert Coin, one coin is inserted into the coin slot. If you have more than one pay line on the slot machine, you can continue inserting coins until you reach the maximum allowable coins.

The maximum allowable coins are determined by the amount of pay lines on the slot machine. When you select Maximum Coins, the maximum allowable coins are inserted into the coin slot. When you select Spin, the reels of the slot machine start turning. You can either select the Spin option or pull the slot machine handle or arm. To pull the handle, move your mouse cursor to the blank knob of the handle and press a mouse button. Pulling the handle animates the movement of the handle; selecting the Spin option does not. You can't spin or pull the handle unless you have inserted at least one coin in the coin slot.

VGA Roulette

VGA Roulette simulates the popular betting game, just like Las Vegas. VGA Roulette provides a very graphic and colorful screen that resembles a real roulette table (except for the wheel).

The screen is divided into six areas: the roulette table, the command lines, the clock, the wheel, the scoreboard, and the message area. The clock displays the current time and is for information only.

All bets and commands are entered by using the mouse. There are no complicated meanings to the mouse buttons as in some programs. Any or all buttons simply mean SELECT. Bets or commands can be selected by pointing to the bet location or command and pressing any or all of the mouse buttons. Placing bets involves three areas of the screen: the roulette table, the command area, and the scoreboard. Bets can be placed on the table in three denominations: $100, $250, and $500. These are represented by betting chips and are selected by pointing at the desired denomination on the second command line and selecting it by pressing a mouse button.

Adventure, Fantasy, and Role-Playing Games

Adventure Game Toolkit

Want to create your own text adventure games? Now with the Adventure Game Toolkit (AGT) you can, and you don't need to know programming. In text adventures, players assume the roles of characters in a story that they read. But unlike a book story, the reader interacts with the plot, goes in different directions, asks questions, and does things that might lead them to treasure or doom.

Using AGT, game enthusiasts can transform their wildest fantasies into high-quality adventure games. And, with a little imagination, teachers and parents now have a wonderful tool for making child-riveting games that can teach basic subjects such as history, math, foreign languages, and vocabulary development in a fun and engaging way.

AGT can create complex worlds with up to 200 locations, 100 inanimate objects (swords, trees, cars, lasers, etc.), and 100 animate objects (hobbits, orcs, famous persons, etc.). An AGT game can understand a 400-word vocabulary that can include complex (multiple-action) and function-key macros for common commands. This gives a polished result that conforms to the Infocom standard in adventure games, so anyone can learn to play quickly.

This AGT version of the original game was written by Crowther and Woods and was the first adventure game ever written. It is a classic. The AGT files for this game can be played and enjoyed simply as great entertainment, or as a platform for the AGT game designer to build on and/or learn from.

Caves of Thor

On an expedition far within the icy mountains of Norway, you discover a crevice leading deep into the jagged rock. You find yourself within a vast underground empire, an enormous network of caves, passages, and chambers. Local legends reveal that a famous mythological deity once ruled these peaks. Could you have stumbled across the mystical ruler's kingdom? You decide to find out . . .

Caves of Thor is a brilliant 16-color arcade/graphics game with a four-way scrolling play field. You'll need to explore seven huge levels for the three symbols of virtue. These three virtues are needed to finish Caves of Thor.

As you explore the empire, you must overcome the many creatures that patrol it. Use your gun (unlimited bullets) to shoot creatures and creature generators. Other hazards include underground flowing streams (with currents that can drag you away!), bubbling lava pits and lakes, and the very dangerous Thor beast, which creates creatures with devastating speed.

This program features some of the best music yet heard in a shareware game. Other notable features include a save/restore option, easy keyboard controls, a demonstration mode, permanent high score, and an easy-playing cheat mode. For those players looking for a game similar to the Kroz games, this is it.

Commander Keen

Commander Keen pushes graphics to the limit. It's truly a "commercial-quality" game that Nintendo and arcade fans will appreciate. It features ultra high-speed, smooth-scrolling EGA graphics and superb sound effects. The animation and scrolling is updated at 60 frames per second. (Motion-picture-quality cartoons are filmed at only 24 frames per second.)

The story of Commander Keen is epic in the style of the memorable matinee serials like Flash Gordon. You play the role of Billy Blaze, an eight-year-old child genius who builds an interstellar ship when not working at home on his fast-track college degree. Among other household objects, Billy uses his Nintendo joystick for flight control and his mom's vacuum cleaner (heavily modified) for his ship's ion propulsion system (with pile height adjustment).

At the hint of galactic trouble, Billy dons his brother's football helmet and becomes "Commander Keen," defender of earth! Ever on the side of justice, fairness, and high-calorie junk food, Keen dispenses justice with an iron hand.

"Marooned on Mars" is volume one of the Commander Keen trilogy. (Only part one is available via shareware distribution; the other two volumes can be purchased only from Apogee Software Productions.) Each volume has its own unique appearance, with different goals, graphics, game characters, animation, and locations.

Duke Nukem

After the tremendous response to Commander Keen, Apogee's follow-up game had to be really spectacular. And Duke Nukem is just that—a superb showstopper with more than one MB of animation and graphics, four times more than Commander Keen had. This arcade/adventure game compares to the Sega Genesis home systems. (It's far superior to games typically seen on the Nintendo home system.)

Duke Nukem also has four-way, dual-scrolling play fields, and does it better than any other IBM game. Features:

- Large, animated characters.
- Reflections.
- Arcade sound effects.
- Built-in hint mode.
- Joystick support.
- Cinematic sequences.
- Self-running demo mode.
- Save/restore.
- High scores.

Rarely is a game released with such a well-animated main character. Duke Nukem does realistic somersaults and uses his grappling hooks to hand-walk along ceilings. Duke is a hero everyone will like. Duke Nukem is rendered in vivid EGA graphics, with many special screen effects. As usual, Apogee offers two more episodes if players want to continue their adventures with Duke. This is truly a game that demonstrates the graphical abilities of your computer if you have EGA or VGA. Get ready for a fun time as you rescue the world from the latest madman, Dr. Proton.

Genealogy

Accept

Accept is a genealogical research program. Features include:

- Soundexed query on last, first, and middle names, plus name reversal.
- Removable Soundex fonts to enable you to design your own.
- Plus or minus 50-year search parameter on five dates.
- Four area search parameters on five places.
- Automatic family linking during entry, plus multiple marriages.
- A set of user-definable fields with the ability to switch sets.
- Multiple parallel note files that can be switched in or out.
- Multiple marker files that allow you to extract names, etc.
- Transmission of an extracted/linked/stand-alone database via a modem.
- Files can be appended to the receiving database.
- Files can be converted to GEDCOM format.
- Files can be transferred to other database management programs.
- Multiuser capability with automatic relationship generation.
- Full hardware configuration for each routine.
- Full Non-LDS/LDS compatibility on entry and form generation.
- Several unique C&SC forms.
- Soundexing on last, first and middle names, with a reverse test on the first and middle names, just in case all three names are surnames. This works in conjunction with an alternate name table for names that are the same but do not have the same spelling or soundex values. (For example, Sanctoclaro and Sinclair.) The program also searches on initials.

This program was not designed with small-time users in mind, but it works for them also. You will notice that all the vowels except "e" were removed. If not, every four-letter word in the dictionary would appear as an ID#. This is the first use of this numbering system we know of. A set of user-definable fields can be changed any time you want. They can be used by the LDS users to enter temple data. You can also have multiple sets in case you want to use your file for other studies.

Brother's Keeper

Brother's Keeper is a set of programs that helps you organize your family tree information and prints the information in several ways. You can print descendant charts that show how anyone in the file is related to everyone else. You can also print ancestor charts, family group sheets, alphabetical name lists, descendant trees, birthday lists, ahnentafel charts, and customized reports.

For each person entered, you can include a date of birth, a date of death, and three other dates; you can also store a place of birth, a place of death, and a place for three other events. Also, you can enter two additional fields that you might define. Each person can have up to seven message lines of additional data, or a text file of unlimited size containing additional data. Each person can have up to eight marriages stored. Each marriage record can hold a marriage date, a place of marriage, one additional date and place, and information on up to 24 children.

Brother's Keeper—View PCX Files. Brother's Keeper has a feature that lets you display graphics. It displays PCX files on a computer with a VGA monitor. You can use your scanner to scan pictures, documents, maps, etc. Save the graphics file as a PCX file.

If your scanner does not save in the PCX format, read your graphics file with a program such as PC Paintbrush, then convert it to a PCX file. Even if your scanner does save files as PCX files, you will probably have to read the file with a program such as PC Paintbrush. Save the file. This places the correct information in the file as to your screen size.

Cumberland Tree

This is an application program for doing genealogy work. It allows entry of individual names, birth, christening, marriage, death, and burial dates and places. The program ties all this information together automatically as a family and extended family structure. It prints out pedigree charts, descendant charts, family group sheets, individual and marriage listings, birthdays, anniversaries, ordinance checklists, and many other reports.

Visually attractive pop-up menus let you always know where you are and where you came from in the program. Simple pop-up screens make this complicated program appear to be much simpler than it really is. The program automatically keeps track of relationships of all individuals. This is accomplished because an individual is added only by entering him/her into the existing family structure.

There is no lengthy sequence of menus to track through to identify someone in the family lineage. Repeat typing of surnames is greatly reduced by automatically adding the father's surname to a new individual, or adding an individual's surname when a father is added. Flexible, worldwide individual name entry is accomplished by allowing a surname to appear anywhere in the name.

This program also imports and exports GEDCOM (Genealogical Data Communication) files. GEDCOM is the standard format for interchanging data among various genealogy programs. Most major genealogy programs support importing and exporting to GEDCOM files. This allows the transference of data between programs. Limitations: 20 children per marriage and 8 spouses per individual.

Cumberland Story

Write your own or your ancestor's life story. Cumberland Story is a program to help you write and print a structured book about your life or the lives of your ancestors. It contains many ideas and suggestions and a chapter and topic outliner containing memory-jogging prompts. Its features include:

- Friendly pop-up menus and screens with extensive online help and a word processor with cut-and-paste features.
- A spell-check dictionary (available to registered users).
- Suggested or user-defined chapters and topics with a printable table of contents.
- Imports text from Cumberland Diary for inclusion into your book.
- ASCII text can be exported or imported, allowing swapping of data with other word processors.
- A multientry text search allows you to find embedded phrases within the diary entries.
- Allows creation of multiple books.
- A printable manual is also included.

EZ-Tree

With EZ-Tree, you can keep track of up to 16 marriages per person and up to 32 children per marriage. There is no logical limit to the number of generations you can have for each family tree. Each record contains first name, middle name, last name, maiden name, sex, birth date and information, death date and information, personal comments, marriage date and information, divorce date and information, and marriage and divorce notes.

From the main menu you can retrieve an existing family tree, create a new family tree, work on the current family tree, delete the current family tree, save the current family tree, or modify the program's configuration, such as colors, disk directory, and screen-blanking timeout. You can easily add parents, spouse, and children to the family tree.

For convenience, function keys are used to select family tree options such as edit a record, find a specific record (person), zoom in on a particular person, etc. EZ-Tree produces three different reports that can be directed to the screen or the printer. A report containing detailed information about a person can be produced; a report on all people in a family tree can be produced, or a list of all descendants of a person can be produced.

Family History

Ever wonder who's lurking in the upper branches of your family tree? The Family History system, a modular genealogy program, helps you organize and refine your search. The system stores information about individuals, including all family relationships. You are not limited in the number of relationships that can exist for each individual, or in the number of generations of ancestors or descendants. Information for up to 3200 individuals can be stored on a diskette, and information on up to 9999 people can be stored on a hard disk.

You can produce ancestor, descendant, relative, and family group reports and ancestor (tree) charts. You can print out blank worksheets and charts to record information for entry into the system. Sample files are included for the descendants of Adam (as recorded in Genesis) and the family of Prince Charles. A GEDCOM export/import program permits sharing information with other genealogy software. Research your roots and know where you came from. A great gift for the next generation.

Family Ties

The Family Ties genealogy program is for the organized compilation of personal genealogical information. All of the names entered in this program are automatically linked to each other. Starting with one individual (usually yourself), use the edit mode to enter your information. Then add a person linked to this first individual (e.g., parents, spouse, other spouse, or child) and enter that person's information. You can start with any male person, and then ancestors or descendants can be added in any order. You can jump around and add as you please. This program also does a nice job printing out a family tree.

The program also has an optional mode that conforms to the preferences of the Church of Jesus Christ of Latter Day Saints (LDS), and the program maintains records according to sealings, baptisms, endowments, and temple dates. Other histories of relatives can be entered in the record for every person covered.

Family Tree Journal

This software allows you to create family history books that document the lives of more than 32,000 family members. For each individual, you can include up to nine pages of biographical information and five general-purpose information fields to record such things as cause of death, occupation, church affiliation, or any other items of your choosing, as well as standard genealogical data. The program then does the tedious jobs of sorting, merging, and compiling the information into books.

Family Tree Journal targets all levels of expertise in both genealogy and computer literacy. The ease of data entry allows the novice to quickly grasp the concepts of operation, and the versatility of the program satisfies the needs of the most serious user. There are many genealogical database programs on the market, but Family Tree Journal is the only one that presents the information formatted into book form.

The books can be the object of conversation at family reunions, holidays, weddings, or even funerals. Copies can be used as unique gifts and can become active and dynamic catalysts for correspondence among relatives. An annual update of the book based on new information accumulated throughout the year could become a family tradition. Major features are: consolidation of input functions, additional data-entry hot keys, and internal speed enhancements.

Genealogy Data Indexer

Genealogy Data Indexer (GDI) is a special index for indexing large collections of genealogical information. It allows rapid data entry in a flexible and functional format. Information can be sorted and output in a variety of useful forms.

Despite its power and versatility, GDI is surprisingly simple to use. It can be used to create a simple index to a book or periodical, but it can do far more than that. GDI allows you to include the date and location for an event, besides the usual name and reference citation. It is ideal for indexing original records such as census, parish registers, land records, etc. It can also be used to index a variety of sources such as all the local history books for a specific locality.

To keep GDI as flexible as possible, so that it will run on the largest variety of machines with as few limitations as possible, GDI uses DOS functions and capabilities for many of its operations. Graphics have been kept simple. For example, no special EGA adapters are required, even though most current computers have EGA adapters.

Puzzles

Compuzzle

This challenging crossword puzzle game combines the favorite pastime of crossword puzzle playing with the color graphics and fun of a computer game. It's easy to use. Challenging, top-quality puzzles coupled with a unique numberless design elevate this ancient 2000+-year-old game to a new level of enjoyment.

Twenty puzzles are included with this program. The appropriate across and down word clues, which are determined by the cursor position on the puzzle, are automatically displayed. This eliminates the tedious correlation of clue numbers to puzzle position numbers. The program optionally verifies your answers as you enter each character, or at the entire puzzle level. Other features include:

- Two puzzle sizes: 15 by 15 and 21 by 21.
- Working puzzles can be saved and restored for play at a later time.
- Choice of two fonts on the VGA and EGA video adapters.
- Switching to across or down mode allows the Backspace, Home End, Insert, and Arrow keys to function horizontally when entering an across word, and vertically when in down mode.

CrossDown

Create crossword puzzles like the newspaper with this software. It presents real crossword puzzles on your computer. A brand new PC software package, CrossDown brings all the challenge and enjoyment of solving crosswords exactly like the ones found in your daily or Sunday newspapers.

CrossDown employs a sophisticated graphical interface that is fun and intuitive for anyone to use who has ever picked up a pencil or pen and filled in a ten-letter solution. And now, fans who love to decipher clues will be happy to know that Cross-

Down includes cryptic crosswords in its repertoire. You navigate this desktop with the mouse. You simply click on a space, choose an across or down block, and type in your answer. A help mode that reveals one letter at a time is available when you get stuck.

There are probably as many ways to construct a crossword puzzle as there are crossword constructors; the author describes the technique he uses. American crossword puzzles conform to a set of established standards.

Crossword Power

This program creates crossword puzzles from your words and clues. It combines entertainment and educational value and was originally developed for teachers to help students learn vocabulary. This software can be used by anyone who enjoys making and solving crossword puzzles.

Puzzles can be created for parties, games, and study aids. You could even set up your own part-time business making crosswords for business organizations and parties. Businesses and consultants have used this program for training sessions and advertising promotions. Businesses can also use this program for their internal newsletters.

Crossword Power is specifically designed to be simple to understand and easy to use. It requires very little knowledge of computers; no commands need to be memorized. The menus allow easy movement through the program, so you can begin producing and printing puzzles in just a few minutes. Help is always available.

Words can also be manually entered. If your monitor supports EGA screens, puzzles can also be displayed and captured as a PCX graphic. These graphics can be imported into desktop publishing (and some word processing) programs. A variety of puzzles can be created and printed from each list of words and clues. This option allows teachers to stress the same vocabulary but give different puzzles to their students. This has been proven to be an effective method for learning vocabulary words.

Professional Crossword Creator

This is the easiest to use, most powerful crossword puzzle editor available. Crossword Creator is a crossword puzzle "editor" that provides the tools needed to lay out a crossword puzzle on your computer's screen, define a short clue for each word in the puzzle, and save the puzzle to disk. If you are an educator, an instructional materials professional, or a newsletter publisher, you will find that Crossword Creator has been tailored to your unique needs, especially your "need for speed." CWC's "robot" puzzle builder can generate a crossword from a list of words in a matter of seconds.

And if you are simply an enthusiast, Crossword Creator provides you with everything you need to design and print puzzles for yourself, family, and friends. This "cookbook" is a collection of short tutorials. Each section is devoted to a common Crossword Creator task and contains step-by-step instructions to help guide you through. The lessons are "technique-oriented" and fairly general in nature.

If you want more detailed information, you can always refer to Crossword Creator's online help. Because CWC prints TEXT, not graphics, the best results are obtained from printers that support the IBM PC character set or a proprietary line-drawing character set. Printers that support only the ASCII character set are suitable for draft output. Features:

- Generate free-form puzzles from a word list or ASCII file.
- Pull-down menus.
- Mouse support.
- Comprehensive online help.
- National language support for French and Spanish.
- Supports all dot-matrix printers, plus HP inkjet/laser printers.
- 36 × 23 puzzle matrix.
- Up to four lines of text per clue.
- Compatible with all popular networks.

Food

CompuChef

CompuChef is a feature-laden program. Here's a brief list of its many features.

- Add recipes with the following information available: title, three categories, serving size, time to prepare, fat, cholesterol, calories, ingredients, directions, and source of recipe.
- You can edit existing recipes from the main menu by simply highlighting the recipe and pressing the Enter key or the letter E. Once you pull the recipe up on the screen, you can edit ingredients or directions and perform many other functions.
- You can delete any recipe from the main menu by pressing the letter D. Before the recipe is deleted, you will see the entire recipe for you to confirm the deletion.
- You can mark recipes in many different ways. You can mark your favorites or mark recipes for a shopping list, export, or printout.
- You can export any recipe by highlighting it and pressing the letter X from the main menu. The program will ask you for a file name to which it can be exported. This feature is handy to export a recipe for transferring to a bulletin board system.
- Any recipe can be printed from the main screen by just highlighting the one you want to print and pressing the letter P. Recipes can be printed on paper or almost any size index card.

Computer Baker

The Computer Baker is a computer recipe book containing 99 recipes for homemade (made from scratch) baked goods that are very quick and easy to prepare. The Computer Baker is completely menu-driven and provides the ability to select recipes from six categories: snacks (brownies and pastries), fudge and candy, muffins and biscuits, cookies, cakes and frostings, and pies.

Edna's Cookbook

Edna's Cookbook stores and organizes all your recipes and even provides password protection to guard your family-recipe secrets from prying eyes. Now you no longer have to deal with disorganized slips of paper and clippings thrown together in a drawer or box, only to be frustrated when you can't find a particular recipe. Edna's Cookbook is a program for collecting, filing, sorting, finding, and printing recipes. Essentially, it is the equivalent of the cookbook in which you now keep all your recipes, except that it is electronic. It can also be used to prepare cookbooks for publication.

Edna's Cookbook automatically sorts and arranges your recipes for you alphabetically by category, then by subcategory, and then by recipe name, making it easy to find things. So that certain recipes can be kept private, you can also encrypt any recipe with a password.

Master Meal Manager

One of the toughest parts of being in charge of family meals is deciding, week after week, what you're going to cook. Another tough part is trying to remember to put all the important ingredients down on one shopping list so you can actually cook the meals. If you're rich, you avoid the whole problem and get a housekeeper. If you're smart, you get Master Meal Manager.

Master Meal Manager is the program that home users dreamt about when the computer industry said, " and you can keep recipes in it, too!" Now you really can. Master Meal Manager stores up to 500 recipes, automatically schedules up to 14 days of meals, and creates and prints a shopping list of ingredients in all the meals on your schedule.

This program is easy to get along with, too. You have total control over the recipes, schedules, and lists. You can print any or all of your recipes and change the number of servings they're written for. You can easily edit your shopping list before you print it, so it only shows the items you need to buy rather than everything you're going to cook.

Even if normal meal planning is a breeze for you, Master Meal Manager can prove invaluable for those special occasions when you find yourself catering a family reunion or having to plan two weeks worth of meals for a nursery school program. Whatever size family you're feeding, Master Meal Manager makes the process almost painless.

Meal Master

Meal Master is a menu-driven database system tailored to manage recipes. It helps store, update, and print your recipes in a variety of formats. But the real power of Meal Master lies in its search and utility functions. Find a recipe by ingredient, adjust a recipe by servings desired, mark recipes to create a shopping list, or exchange recipes with other Meal Master users without a lot of work.

Appendix A

Shareware Sources

As mentioned earlier, PC-SIG has some excellent software collections on CD-ROM. Here is their writeup about their collection. I couldn't say it any better.

> The PC-SIG Library on CD-ROM. This might well be the greatest collection of software ever compiled. The PC-SIG Library on CD-ROM is a collection of the entire PC-SIG library. Every shareware program we have is on one disc; more than 700 megabytes of software you can have immediate access to.
>
> It's easy to use; all you have to do is insert the disc and type GO at the DOS prompt. We've included all the tools you need for fast, easy access to the programs and their descriptions. The WordCruncher text retrieval software allows you to quickly access the description for any program you want. All the programs are indexed by title, file name, PC-SIG disk number, and every word within the program description. There couldn't be an easier way to find just the right program.
>
> The Narc utility, newly instituted with the tenth edition, allows you to look directly at all the compressed program files on the CD without having to download the program to your hard disk. So if you want to look at the programmer's notes, or online documentation before using valuable hard disk space, you can.
>
> All of the programs are logically divided into one-hundred twenty categories, making it easier to find just the right software. Whether you're interested in Accounting or Windows, there's a shareware program for you on the PC-SIG Library on CD-ROM. Winner of the Optical Publishing Association's Best Consumer Product Award. Shareware for every imaginable need: accounting, chemistry, educational, engineering, games, hypertext, networks, programming, spreadsheets, utilities, word processing.
>
> To order in the United States, call 800-245-6717 and ask for Customer Service. Outside the United States, call (408) 730-9291 for the name of the dealer in your country.

Important Addresses

United States

ASSOCIATION OF SHAREWARE PROFESSIONALS (ASP)
545 Grover Road
Muskegon, MI 49442
FAX 616-788-2765
CompuServe PPN: 70007,3536

Discount Micro
10881-B Portal
Los Alamitos, CA 90720
1-800-364-2725
An excellent and huge selection of software

PC-SIG
1030-D East Duane Avenue
Sunnyvale, CA 94086
Orders 800-245-6717 Ask for Operator 2228
Information 408-730-9291
FAX orders 408-730-2107

PSL (PUBLIC SOFTWARE LAB)
P.O. Box 35705
Houston, TX 77235-5705
Orders: 800-2424-PSL
Info/Help 713-524-6394
FAX Orders 713-524-6398

PUBLIC BRAND SOFTWARE
P.O. Box 51315
Indianapolis, IN 46251
Orders 800-426-3475
Information 317-856-7571
FAX orders 317-856-2086
An excellent selection of quality programs.

QUANTUM TECHNOLOGIES
10985 Pinehigh Drive
Alpharetta, GA 30201
(404) 664-1097
They bundle sets of archived shareware, the result is shareware for about a dollar a program.

REASONABLE SOLUTIONS
1221 Disk Drive
Medford, OR 97501-6639
Orders 800-876-3475
Information 503-776-5777
FAX Orders 503-773-7803
A very good, but limited selection.

THE SOFTWARE LABS
3767 Overland Avenue, Suite 112
Los Angeles, CA 90034
Orders 800-359-9998

Information 310-559-5456
FAX orders 310-559-3405
A choice selection of software

Canada

Advance Computing
P.O. Box 2, Stn. M
Toronto, Ontario
M6S 4T2

Alternative Personal Software
269 Springside Drive Suite C-2
Hamilton, Ontario
L9B 1P8

Desktop Magic
931 Pensacola Way S.E.
Calgary, Alberta
T2A 2G8

ESCO Electronics
Box 2258
Pincher Creek, Alberta
T0K 1W0

Interface Software & Systems
P.O. Box 329
Cookston, Ontario
L0L 1L0

West Hill Sales
P.O. Box 304
West Hill, Ontario
M1E 4R8

United Kingdom

College Shareware UK
Dept PCW12
The College Business Centre
Uttoxeter New Road
Derby DE3 3WZ

JD Computers Ltd Shareware
Cromer House
1 Caxton Way
Stevenage, Herts SG1 2DF

NEEDFORWARD LTD
52 Pennsylvania Avenue
Hesters Way
Cheltenham GL51 7JP

Premium Software
24 Station Road
Barton Halsall
Nr. Ormskirk
Lancs L39 7JN

SMS Shareware
19 Carshalton Road
Camberley
Surrey GU154AQ

Germany

Computer Solutions GmbH
Postfact 22 12 53
8000 München 22
Germany

Computer-Electronik-Erkins
Eythstr. 16-18
500 Köln 91
Germany

Kirschbaum Software GmbH
Kronau 15
8091 Emmering
Germany

Redysoft Software GmbH
Postfach 12 61
8150 Holzkirchen
Germany

Soft-Mail AG
Postfach 30
7701 Büsingen
Germany

Appendix

B

Telelibraries

America Online
8619 Westwood Center Drive
Vienna, VA 22182-2285
(800) 227-6364
(703) 448-8700

BRS (Bibliographic Retrieval Services) Information Technologies
Division of Maxwell Online
8000 Westpark Drive
McLean, VA 22102
(800) 955-0906
(703) 442-0900

CompuServe Information Services
5000 Arlington Centre Blvd.
P.O. Box 20212
Columbus, OH 43220
(800) 848-8199
(614) 457-8600
Electronic Mail: 70003,2154

DATA-STAR
485 Devon Park Drive #110
Wayne, PA 19087
(800) 221-7754
(215) 687-6777

DATATIMES
14000 Quail Springs Pkwy Suite 450
Oklahoma City, OK 73134
(800) 642-2525
(405) 751-6400

Delphi
1030 Massachusetts Avenue
Cambridge, MA 02138-5302
(800) 544-4005
(617) 491-3342
(617) 491-6642 (fax)

DIALOG Information Services, Inc.
3460 Hillview Avenue
Palo Alto, CA 94304
(800) 3-DIALOG or 334-2564
(415) 858-2700
(415) 858-7069 (fax)

DOW JONES NEWS/RETRIEVAL
Dow Jones Information Services
P.O. Box 300
Princeton, NJ 08543-0300
(800) 522-3567
(609) 520-4000
(609) 520-4660 (fax)

Dun's Marketing Services
Three Sylvan Way
Parsippany, NJ 07054-3896
(800) 223-1026
(201) 605-6000 Ext. 3809

EASYNET
(see Telebase Systems)

EPIC
Online Computer Library Center
6565 Frantz Road
Dublin, OH 43017-3395
(800) 848-5800
(614) 764-6000

Gale Research, Inc.
835 Pennobscot Bldg.
Detroit, MI 48277-0748
(800) 877-GALE
(313) 961-2242

GEnie
General Electric Information Services
401 N. Washington Street
Rockville, MD 20850
(800) 638-9636
(301) 340-4000

KNOWLEDGE INDEX
Dialog Information Services, Inc.
3460 Hillview Avenue
Palo Alto, CA 94304
(800) 334-2564
(415) 858-3785

MCI INTERNATIONAL, INC.
Two International Drive
Rye Brook, NY 10573-1098
(914) 937-3444
Telex: 666101

MEAD DATA CENTRAL, INC.
9393 Springboro Pike
P.O. Box 933
Dayton, OH 45401
(800) 227-4908
(513) 865-6800

NEWSNET
945 Haverford Road
Bryn Mawr, PA 19010
(800) 345-1301
(215) 527-8030

Pergamon ORBIT Infoline, Inc.
8000 Westpark Drive
McLean, VA 22102
(800) 955-0906
(703) 442-0900

PORTAL COMMUNICATIONS CO.
20863 Stevens Creek Blvd. Suite 260
Cupertino, CA 95014
(408) 973-9111 (voice)
(408) 725-0561 (data)
(408) 725-1580 (fax)

Prodigy
445 Hamilton Avenue
White Plains, NY 10601
1-800-PRODIGY
(914) 993-8000

Reuter: File Ltd.
2 First Canadian Place Suite 1900
Toronto, Ontario M5X 1E3 Canada
(800) 387-1588
(416) 364-5361

R.R. BOWKER COMPANY
205 E. 42nd Street
New York, NY 10017
(212) 916-1600

TELEBASE SYSTEMS, INC.
435 Devon Park Drive, Suite 600
Wayne, PA 19087
(215) 293-4700
(215) 341-9660 (fax)

Thomson Financial Networks
11 Farnsworth Street
Boston, MA 02210
(716) 3435-2000

VU/TEXT Information Services, Inc.
325 Chestnut Street, Suite 1300
Philadelphia, PA 19106
(800) 323-2940
(215) 574-4400

WILSONLINE
The H.W. Wilson Co.
950 University Avenue
Bronx, NY 10452
(800) 462-6060
(800) 367-6770 (in New York state)
(212) 588-8400

Appendix

C

Notes to Software Entrepreneurs

This appendix covers some of the more common grammatical errors made in the documentation accompanying software programs. I include some suggestions on how you can improve your text, your sales, and your customer satisfaction.

One of the most common grammatical errors is the use of its and it's. The word "its" is possessive and is used as follows: The software knows its limitations. It's is a contraction for "it is" (the apostrophe indicates that one or more letters is missing from the word), and is used as follows: The software gets exhausted and knows when it's (it is) time to quit.

A common punctuation error with respect to quotes is to include punctuation marks outside of the quotes. For example, all punctuation marks, except the colon, are always included inside the quotes, such as: It is "over," said the fox to the grapes. When I say "Jump!" the boss said, "I want you to ask, 'How high?' " Also, when a parenthetical expression (that's one enclosed in a pair of parentheses) is used, the period used to end the sentence is placed outside the parentheses, as follows: Utilize your 640K RAM efficiently for best results (or you might have to add a few more megs).

Here's another important one: Please, please limit the use of exclamation marks!!!!! I know you get very excited when you write about your innovative creations, but try to convey this unbridled excitement with words and deeds, facts that tell what your software can do and how it does it. You're limited to only one exclamation mark per description.

Text Content on Disk

After reviewing hundreds of programs and their accompanying writeups, here are my recommendations for files that should be on every software disk, in addition to your main program:

MANUAL.TXT
VENDOR.TXT
REGISTER.TXT
INSTALL.BAT
PRINTMAN.BAT
README.TXT

The files with the TXT extension should be ASCII files, which everyone can read and print out without requiring a special word processor.

MANUAL.TXT

Every disk should have a manual that gives a fairly detailed explanation of what the program is and what it does. Don't annoy your potential customers by including the manual in a file that has to be unzipped or installed on their disks to be read. Their disks already have more than they can handle, and they don't want to unnecessarily have to load it up with another manual and fragment their disk more. The manual is your best sales document and should include a well-thought-out and complete introduction to your program. The introduction should summarize what the software does and list its main features. It should also list what the PC requirements are (e.g., how much RAM is needed, if a hard or floppy disk is required, EGA/VGA/Super VGA, color or mono, etc.).

A list of the files on the disks and a detailed table of contents should also be included, no matter what its size. If your manual is big enough (more than about 30 pages), also include an index. There are plenty of index-generating programs available to automate much of this task.

VENDOR.TXT

Be sure to include a VENDOR.TXT. This is an important document because this is the text your shareware vendor will probably use to market your software, so this is the text that your potential customers read first. Put a little extra effort into this vital sales pitch. VENDOR.TXT should contain a one-line description of what the program is and what it accomplishes. It should also contain a short description (a paragraph or two is adequate for this).

Finally, this document should also contain a long description consisting of several paragraphs that cover in much more detail what the software is and what it does. About a page is adequate, and this text should not be archived. So everyone can read this information without the need for conversion, the text should be in ASCII format rather than in some special word processor format. This information is needed both by the shareware distributors who are going to include your descriptions in their catalogs and ads, and also for the users so they can check what the program is going to do before they try it out.

REGISTER.TXT

Use a separate file for your registration form. Title it REGIS.TXT so your customers can easily print it out.

INSTALL.BAT

This program should make it easy for the user to install your software from any disk he or she chooses. And it should be able to be installed from either drive A or drive B, whichever one the user chooses. There are several good install programs in chapter 11. Check them out.

PRINTMAN.BAT

This batch file should print your ASCII manual to any standard printer.

README.TXT

README.TXT files, of course, are for OBTWs (oh, by the way), afterthoughts that you didn't get an opportunity to incorporate into your other files. They can be designated read1st.txt, etc., for subsequent OBTWs.

SHAREWRE.TXT

All of your shareware legal information and disclaimers should be not only in your manual, but also in a separate file designated SHAREWRE.TXT.

How to Write in One Easy Lesson

When you describe how your software functions, be positive. Use the active tense. Instead of, "This software will catalog the disk," be positive and say "This software catalogs the disk." Because you're confident that your software accomplishes what you claim it does, get rid of all the weak, effete, and unsure "wills." And don't say, "This program was designed to speed up . . .." Instead, be positive about its performance: Say, "This program speeds up . . ." Get rid of the words "is designed to."

And don't forget to paragraph your text often. There are few things more depressing and formidable than being confronted with a huge amount of text in which a paragraph occupies an entire page. A reader usually falls asleep after trying to read a long paragraph. A paragraph is a rest stop so readers can rest their eyes and give their minds an opportunity to absorb, contemplate, and understand what they just read.

And, above all, keep your sentences short; an average of about 17 words per sentence makes for easier reading. With a long sentence meandering all over the page or screen, by the time your reader gets to the end, he or she has probably forgotten what the first part of the sentence was about.

Run your text through a spell checker. (There are several described in chapter 5.) Your customers might be a little uncertain of the effectiveness of your software if they discover several misspelled words in your documents.

Save Words

Table C.1 gives some specific suggestions that not only save your words, but they also make your text more readable and easy to understand.

TABLE C.1 Save Your Words

Instead of:	Use:
has been designed to search...	searches...
will be able to print...	prints...
has the ability to calculate...	calculates...
due to the fact...	since...
in order to locate...	to locate...
in order to use...	to use...
is useful for locating...	locates...
assists you in finding...	finds...
can be varied...	varies...
to do editing tasks...	to edit...

Glossary

abort The procedure for terminating a program when a mistake, malfunction, or error occurs.

AKA Also known as. Another name for it.

aliasing Assigning a longer, detailed command to a short set of keystrokes, similar to a macro. Assigning a different letter (i.e., D:) to create a partition on a physically existing disk (C:) so D: acts like a separate disk.

alphanumeric characters A character set containing letters, numbers, and other special symbols, i.e., punctuation marks.

AND A Boolean operator. A AND B provides an output only if both A and B are present.

application program Program written to enable the computer to perform useful work. For example, a word processing program is an application program.

archived files The process of using an algorithm to encode the pattern of frequently repeated letters, words, or pixels to compress data to occupy less space.

ASCII American Standard Code for Information Interchange. A standard used for all communication to represent letters, numbers, and symbols as bit patterns. ASCII files can be read by practically all software.

aspect ratio The ratio of the height-to-width on a display screen or other device.

asynchronous Not synchronous. A method of transmitting signals in which data is sent one bit at a time, with an indeterminate amount of time between characters. The beginning and end of a character are indicated by special start and stop bits.

autodial A modem feature that lets the computer automatically dial a prerecorded phone number for connection to a host computer.

AUTOEXEC.BAT The configuration file that DOS executes every time it is booted.

backing up Making copies of files for use in case original data is lost or damaged.

batch file A file that stores and executes often-used DOS executable commands or programs without user intervention.

baud rate A technical term applied to the rate at which data is transmitted across telephone lines. In most cases, baud is almost the same as bits per second.

bit rate Speed at which bits are transmitted over a communication link.

block In word processing, a specific set of text that is to be manipulated (i.e., copied, moved, or erased).

Boolean algebra Math for expressing logical relationships between truth values using AND, OR, and NOT logical operators.

boot The process by which a computer turns itself on, literally pulling itself by its own bootstraps.

bps Bits per second. The speed at which data is transmitted.

bulletin board A system consisting of a computer and modem with appropriate software that makes available messages over public phone lines at no charge.

byte Smallest addressable unit of storage, usually eight bits long.

CGA Color graphics adapter. A low/medium-resolution color graphics system for PCs.

click The process of pressing a mouse button once.

cluster A group of continuous sectors on a disk. A sector is the smallest unit of allocation for files on a disk.

COM port A communications port. An interface on the computer that is used for transferring information to another piece of equipment such as a printer or modem.

command A short program in DOS designed to perform a specific task.

CONFIG.SYS The file that DOS executes every time to set up certain characteristics of your computer or to control additional devices connected to your computer.

crippleware A type of shareware in which the author offers a limited or "crippled" version of the software.

database A collection of interrelated data that can be accessed by computer.

database vendor A company or organization that markets online databases.

data bit One of the seven (or eight) bits that makes up the code for an alphanumeric or non-printing character.

data compression A process that uses an algorithm to encode the pattern of frequently repeated letters, words, or pixels to compress data to occupy less space.

de-archiving The reverse process of archiving, i.e., returning the archived data to its original form.

debugging In computing, the search for and correction of a mistake in a computer program.

default A preset value used by a computer until it is specifically changed by the user.

device A piece of hardware attached to a computer (usually a peripheral such as a printer or a mouse).

device driver A special program that controls a hardware device or operation. Device drivers are usually loaded with a DEVICE-directive via the CONFIG.SYS file.

DIP switches DIP is an acronym for digital inline package. DIP switches are small switches in a printed circuit board that allow users to set certain hardware options.

directory A place on a disk where file names are stored so DOS can find the files when needed.

disk density Density indicates how closely magnetic information can be stored on a disk. The most common disk types are double- and high-density disks. Double-density disks store 360K (5¼" disk) or 720K (3½" disk) of data, while high-density disks store up to 1.2 MB (5¼") or 1.44 MB (3½") of information.

DOS Disk operating system. A specialized, disk-oriented program that provides an easy-to-use link between the user and the computer's disk drive.

dot-matrix printer A printer in which each character is represented by a pattern of dots and imprinted on paper.

double click The process of rapidly pressing a mouse button twice.

downloading The process of retrieving information from a distant computer and loading and storing it into your own. Opposite of uploading.

drag The process of pressing down a mouse button and, while still holding down the mouse button, moving the mouse to highlight text or move objects.

DR DOS A competitive operating system to MS-DOS created and marketed by Digital Research.

DSSD Double-sided single-density disk.

DSDD Double-sided double-density disks.

echo When a program's or user's characters are displayed on the screen.

duplex A communications system that can transmit in both directions.

EGA Enhanced color graphics adapter. A medium resolution color system for PCs.

electronic mail Also known as E-mail. A service that lets you send messages to other users on the same online system.

ellipsis In printing, a symbol (. . .) used to indicate that something has been left out of a phrase or sentence.

executable Programs that can be run by a computer.

external command Commands that are used less often than other commands and therefore are not automatically loaded in memory.

FAT File allocation table. FAT is a list maintained in DOS of all clusters on a disk. It includes their locations and tells whether they are available, bad, or in use by a file. It also records the order in which clusters link to make a particular file.

fax machine Also facsimile machine. A device that scans documents, transmits them over phone lines, and recreates them at the other end.

file A collection of related records treated as a basic unit of storage.

file name and file name extension The name assigned to a file; it can be one to eight characters with an optional file extension of one to three letters. A period (.) separates the filename from the extension.

floppy disk A thin, flexible, magnetic disk for storing digital data. Available in three sizes: 8, 5.25, and 3.5 inches.

font A complete alphabet of a particular typeface in a given point size— uppercase, lowercase, numerals, punctuation marks, etc.

footer Text automatically printed at the bottom of several pages.

formatting The process of preparing a disk so it can store data to a specific format.

freeware A special class of shareware in which the software is free; however, the author has retained the copyright on the program.

full duplex Simultaneous, independent transmission of data in both directions on the same line.

hard disk A rigid storage disk capable of storing much more data than floppies and which can be accessed in much less time.

hardware The physical components or equipment that make up a computer system.

header Text automatically printed at the top of several pages.

hertz Abbreviated Hz. Frequency in cycles per second.

half duplex The transmission of data in only one direction at a time on the same line.

hard copy A printed copy of writing.

Hayes compatible Modems that use commands originated by Hayes Microcomputer Products. Hayes compatibility is not an absolute requirement, but is a *de facto* industry standard.

header or heading Title or caption at the head of a chapter, section, column, list, table, or illustration.

icon A graphic (picture) used to represent an object or idea.

ink jet printer An output device that prints by spraying a thin stream of ink onto the paper.

interface The point at which two systems interact. It can be a physical interface involving a connector or a logical interface involving software.

internal command A DOS command that is frequently used, so it usually is stored in RAM.

I/O Abbreviation for input/output.

K Abbreviation for kilo or 1000 in decimal notation. 100K means 100,000.

LAN Local area network. A communication system that connects several computers so they may communicate with each other and share data and resources (such as a laser printer and a hard disk).

laser printer A high-speed, nonimpact, quality printer in which a laser "writes" on a drum and prints alphanumerics, graphics, and special fonts.

macro A special preprogrammed series of instructions initiated by activating specific key combinations.

megabyte (MB) A measure of computer memory, disk space, etc. Equal to 1024K or about 175,000 words.

menu A list of available functions for selection by an operator.

modem A contraction of modulator/demodulator. The interface box or card that converts a signal into two different frequencies so that it can be transmitted over telephone lines. It also performs the reverse function, converting the signal on the phone lines to ones that can be handled by a computer. The modem converts the computer's 1s and 0s into two musical tones that whistle down the telephone line and are turned back into 1s and 0s at the receiving end.

monochrome monitor A monitor that displays information in only one color. Monochrome monitors generally have black backgrounds with white, amber, or green letters.

MS-DOS A DOS program created by Microsoft Corporation. Includes both DOS 4.1 and DOS 5.

null modem cable A serial communications cable used to connect two computers via their communication ports so they can send and receive.

online database A computer system that a user can dial into through the public telephone network with or without first paying a fee.

OR A Boolean logical operator that indicates that the presence of either one or both terms provides an output. A OR B means that either or both A or B will provide an output. Also called "inclusive OR." However, an "exclusive OR" gives an output only if A OR B, but not both, is present.

parity A setting for an error-checking bit during transmission of data. Can be odd or even.

password A special word or code that must be presented to a computer system to gain access to its resources.

PC Abbreviation for personal computer.

PD (public domain) Software in which the author has renounced all copyright rights.

peripheral An external device connected to a computer. For example, printers and modems are peripherals.

pixel The smallest dot on a video screen. A contraction of "picture element." On a color monitor screen, each pixel is made up of one or more triads.

PC-DOS Also known as DOS 3.3, an operating system used on IBM computers.

prompt A symbol that appears on the video screen asking the user to enter information. Typically > or] is used.

protocol A set of rules for transmitting data between computers to detect and correct errors that occur during transmission.

port A jack where you can plug in cables to connect other computer equipment.

RAM Random access memory. Storage in which data can be input and read out independent of its location in memory. That is, it can be randomly accessed. RAM is normally "volatile." It loses memory when power is turned off.

read-only An attribute that can be assigned to a file or directory so it can only be read from, and not written to.

root directory The main directory that DOS creates on a disk when the disk is formatted.

ROM Read only memory. Static memory that cannot be changed by conventional procedures. Retains memory contents with power on or off.

RS-232C The standard established by the Electronic Industries Association for serial transmission of data for telecommunications.

serial transmission A method of transmission in which the bits carrying the information are sent one at a time on a channel such as a telephone line.

shareware A method of marketing software where customers purchase a copy of the program to test. Then they are requested to send in a registration fee if they decide to keep using the program.

soft copy A copy of written material that is stored in digital form on a diskette or other storage media. A printer converts soft copy to hard copy.

software Nonhardware component of computers, the "music" that makes computers play.

SSSD Single-sided single-density.

SSDD Single-sided double-density.

string A sequence of characters such as “method” or “1Nov1999.”

SVGA Super VGA, a very high resolution color display system.

synchronous A system of transmission where data is sent in blocks and for which the receiver and sender must be in synchronism.

system prompt The display on the screen that shows you that the operating system is ready to receive a command.

telecomm Telecommunications. Communications over a distance.

terminal A device that can send and receive information from a computer.

terminal emulation When a terminal is online to another system, the computer is in a terminal mode and emulates a specific terminal operation.

TSR Terminate-and-stay-resident. A program that leaves some of its code in the computer’s memory when you exit the program and return to the DOS command prompt. This makes it easy to reactivate the program when you are using a different program.

variable In batch files, a variable is a character substituted for a file name when the batch file is run.

VDU Visual display unit. Monitor or TV display.

VGA Video graphics adapter. A high-resolution color system for PCs.

virus A program whose sole purpose is to surreptitiously invade computers and modify the data stored on them, usually destructively.

Windows An icon-based graphical user interface.

Bibliography

Bjelland, Harley L. and Dan Gookin. 1993. *Outrageous DR DOS batch file programming*. Blue Ridge Summit, PA: TAB/McGraw Hill.

Bjelland, Harley L. 1992. *Using online scientific & engineering databases*. Blue Ridge Summit, PA: TAB/McGraw-Hill.

Glossbrenner, Alfred. 1989. *Free software for IBMs and compatible computers*. New York: Saint Martin's Press.

The PC-SIG encyclopedia of shareware, 3rd edition. 1990. Sunnyvale, CA: PC-SIG.

Schlentner, Klaus, *Finding Free Software*. 1990. Grand Rapids, MI: Abacus.

Source book of free and low-cost software, 4th edition. 1992. Houston, TX: Nelson Ford, Public (software) Library.

Index

About the Author

Harley Bjelland has more than 25 years of experience as a scientist and engineer. He has worked on some major technical projects, including atomic submarines and the Stealth Bomber. He is a free-lance technical writer who has written several books for TAB/McGraw-Hill and other publishers.

DISK WARRANTY

This software is protected by both United States copyright law and international copyright treaty provision. You must treat this software just like a book, except that you may copy it into a computer in order to be used and you may make archival copies of the software for the sole purpose of backing up our software and protecting your investment from loss.

By saying "just like a book," McGraw-Hill means, for example, that this software may be used by any number of people and may be freely moved from one computer location to another, so long as there is no possibility of its being used at one location or on one computer while it also is being used at another. Just as a book cannot be read by two different people in two different places at the same time, neither can the software be used by two different people in two different places at the same time (unless, of course, McGraw-Hill's copyright is being violated).

LIMITED WARRANTY

Windcrest/McGraw-Hill takes great care to provide you with top-quality software, thoroughly checked to prevent virus infections. McGraw-Hill warrants the physical diskette(s) contained herein to be free of defects in materials and workmanship for a period of sixty days from the purchase date. If McGraw-Hill receives written notification within the warranty period of defects in materials or workmanship, and such notification is determined by McGraw-Hill to be correct, McGraw-Hill will replace the defective diskette(s). Send requests to:

Customer Service
Windcrest/McGraw-Hill
13311 Monterey Lane
Blue Ridge Summit, PA 17294-0850

The entire and exclusive liability and remedy for breach of this Limited Warranty shall be limited to replacement of defective diskette(s) and shall not include or extend to any claim for or right to cover any other damages, including but not limited to, loss of profit, data, or use of the software, or special, incidental, or consequential damages or other similar claims, even if McGraw-Hill has been specifically advised of the possibility of such damages. In no event will McGraw-Hill's liability for any damages to you or any other person ever exceed the lower of suggested list price or actual price paid for the license to use the software, regardless of any form of the claim.

McGRAW-HILL, INC. SPECIFICALLY DISCLAIMS ALL OTHER WARRANTIES, EXPRESS OR IMPLIED, INCLUDING, BUT NOT LIMITED TO, ANY IMPLIED WARRANTY OF MERCHANTABILITY OR FITNESS FOR A PARTICULAR PURPOSE.

Specifically, McGraw-Hill makes no representation or warranty that the software is fit for any particular purpose and any implied warranty of merchantability is limited to the sixty-day duration of the Limited Warranty covering the physical diskette(s) only (and not the software) and is otherwise expressly and specifically disclaimed.

This limited warranty gives you specific legal rights; you may have others which may vary from state to state. Some states do not allow the exclusion of incidental or consequential damages, or the limitation on how long an implied warranty lasts, so some of the above may not apply to you.

WHAT'S ON THE DISK

- AsEasyAs, a spreadsheet program that's compatible with Lotus 1-2-3.
- Wyndfields, an excellent database program.
- RGB TechWriter, a scientific word-processing program.

IMPORTANT

Read the Disk Warranty terms on the previous page before opening the disk envelope. Opening the envelope constitutes acceptance of these terms and renders this entire book-disk package nonreturnable except for replacement in kind due to material defect.